Lecture Notes in Mathematics

Edited by A. Dold and B. Eckmann

Subseries: Mathematica Gottingensis

T0184774

1361

T. tom Dieck (Ed.)

Algebraic Topology and Transformation Groups

Proceedings of a Conference held in Göttingen, FRG, August 23–29, 1987

Springer-Verlag

Berlin Heidelberg New York London Paris Tokyo

Editor

Tammo tom Dieck
Mathematisches Institut, Universität Göttingen
Bunsenstr. 3–5, 3400 Göttingen, Federal Republic of Germany

Mathematics Subject Classification (1980): 57SXX, 55-XX

ISBN 3-540-50528-8 Springer-Verlag Berlin Heidelberg New York
ISBN 0-387-50528-8 Springer-Verlag New York Berlin Heidelberg

Printing and binding: Druckhaus Beltz, Hemsbach/Bergstr.
2146/3140-543210

TABLE OF CONTENTS

S. Bauer: The homotopy type of a 4-manifold with finite fundamental group. 1

C.-F. Bödigheimer and F.R. Cohen: Rational cohomology of configuration spaces of surfaces. 7

G. Dylawerski: An S^1-degree and S^1-maps between representation spheres. 14

R. Lee and S.H. Weintraub: On certain Siegel modular varieties of genus two and levels above two. 29

L.G. Lewis, Jr.: The RO(G)-graded equivariant ordinary cohomology of complex projective spaces with linear $\mathbb{Z}/p$ actions. 53

W. Lück: The equivariant degree. 123

W. Lück and A. Ranicki: Surgery transfer. 167

R.J. Milgram: Some remarks on the Kirby - Siebenmann class. 247

D. Notbohm: The fixed-point conjecture for p-toral groups. 253

V. Puppe: Simply connected manifolds without S^1-symmetry. 261

P. Vogel: 2 × 2 - matrices and application to link theory. 269

List of Participants

ANDERSON, Douglas R.
Dept. of Mathematics
Syracuse University
Syracuse, N.Y. 13210
USA

BAK, Anthony
Fakultät für Mathematik
Universität Bielefeld
Universitätsstr. 1
4800 Bielefeld 1
W-Germany

BAUER, Stefan
Mathematisches Institut
SFB 170
Universität Göttingen
Bunsenstr. 3-5
3400 Göttingen
W-Germany

BÖDIGHEIMER, Carl-Friedrich
Mathematisches Institut
SFB 170
Universität Göttingen
Bunsenstr. 3-5
3400 Göttingen
W-Germany

COHEN, Frederick
Dept. of Mathematics
Universitiy of Kentucky
Lexington, Ky 40506
USA

CONNOLLY, Frank
Dept. of Mathematics
University of Notre Dame
P.O. Box 398
Notre Dame, Ind. 46556
USA

DAVIS, James
Dept. of Mathematics
Indiana University
Bloomington, Ind. 47405
USA

tom DIECK, Tammo
Mathematisches Institut
SFB 170
Universität Göttingen
Bunsenstr. 3-5
3400 Göttingen
W-Germany

DOVERMANN, Karl-Heinz
Dept. of Mathematics
University of Hawaii
Honolulu, HI 96822
USA

DYLAWERSKI, Grzegorz
Inst. Math.
Uniwersytet Gedanski
ul. Wita Stwosza 57
P-80-952 Gdansk
Polen

EWING, John
Dept. of Mathematics
Swain Hall East
Indiana University
Bloomington, Ind. 47405
USA

FERRY, Steven
Dept. of Mathematics
University of Kentucky
Lexington, KY 40506
USA

FRANJOU, Vincent
Institut de Mathematique
Universite de Nantes
2, rue de la Houssiniere
F-44072 Nantes cedex
France

HUEBSCHMANN, Johannes
Mathematisches Institut
Im Neuenheimer Feld 288
6900 Heidelberg
W.-Germany

IGODT, Paul
Mathematik
Katholieke Universitet
Leuven
Fakulteit Wetenschappen
Campus Kortrijk
B- 8500 Kortrijk
Belgium

JACKOWSKI, Stefan
Wydz. Mat. i Mech.
Instytut Matematyki
Uniwersytet Warszawski
P-00-901 Warszawa
Polen

JODEL, Jerzy
Inst. Math.
Uniwersytet Gdanski
ul. Wita Stwosya 57
P-80-952 Gdansk
Polen

KOSCHORKE, U.
Lehrst. f. Mathematik V
Universität Siegen
Hölderlinstr. 3
5900 Siegen
W-Germany

LAITINEN, Erkki
Mathematik
University of Helsinki
Helsinki
Finland

LANNES, Jean
Ecole Polytechnique
- Mathematique -
F-91129 Palaiseau
France

LEE, Ronnie
Dept. of Mathematics
Yale University
Box 2155, Yale Station
New Haven, Conn 06520
USA

LEWIS, Gaunce
Dept. of Mathematics
Syracuse University
Syracuse, N.Y. 13210
USA

LÖFFLER, Peter
Mathematisches Institut
SFB 170
Universität Göttingen
Bunsenstr. 3-5
3400 Göttingen
W-Germany

LÜCK, Wolfgang
Mathematisches Institut
SFB 170
Universität Göttingen
Bunsenstr. 3-5
3400 Göttingen
W-Germany

LUSTIG, Martin
Fakultät für Mathematik
Universitätsstr. 150, Gebäude NA
4630 Bochum 1
W.-Germany

McCLURE, James E.
Dept. of Mathematics
University of Kentucky
Lexington, KY 40506
USA

MAYER, K.H.
Institut f. Mathematik
Universität Dortmund
Postfach 500 500
4600 Dortmund 50
W-Germany

MILGRAM, R.J.
Dept. of Mathematics
Bldg. 380
Stanford University
Stanford, Cal. 94305
USA

MUNKHOLM, Hans J.
Matematisk Institut
Odense Universitet
Dk-5230 Odense M
Denmark

NOTBOHM, Dietrich
Mathematisches Institut
Bunsenstr. 3-5
3400 Göttingen
W.-Germany

ODA, Nobuyuki
Dept. of Appl. Mathematics
Jonan-ku
Fukuoka, 814-01
Japan

OLIVER, Robert
Matematisk Institut
Aarhus Universitet
Dk-8000 Aarhus C
Denmark

PEDERSEN, Erik
Matematisk Institut
Odense Universitet
Dk-5230 Odense M
Denmark

PESCHKE, Georg
Dept. of Mathematics
University of Alberta
Edmonton, Alberta
Canada, T 6 G 261

PETRIE, Ted
Dept. of Mathematics
Rutgers University
New Brunswick, N.J. 08903
USA

PUPPE, Volker
Fakultät für Mathematik
Universität Konstanz
Postfach 5560
7750 Konstanz
W-Germany

RANICKI, Andrew
Math. Dept.
The University
Mayfield Rd.
Edinburgh EH9 3JZ
Scotland

RAUSSEN, Martin
Inst. f. Elektr. Systemer
Aalborg Universitetscenter
Strandvejen 19
Dk-9000 Aalborg
Denmark

ROTHENBERG, Mel
Dept. of Mathematics
University of Chicago
5734 University Avenue
Chicago, Ill. 60637
USA

SCHAFER, James A.
Dept. of Mathematics
College Park Campus
Mathematics Bldg. 084
College Park
Maryland 20742
USA

SCHNEIDER, Albert
Mathematisches Institut
Universität Göttingen
Bunsenstr. 3-5
3400 Göttingen
W-Germany

SCHWARTZ, Lionel
Dept. de Mathematique
Univ. de Paris/Sud, Bat. 425
F-91405 Orsay cedex
France

SMITH, Lawrence
Mathematisches Institut
SFB 170
Universität Göttingen
Bunsenstr. 3-5
3400 Göttingen
W-Germany

SWITZER, Robert
Mathematisches Institut
SFB 170
Universität Göttingen
Bunsenstr. 3-5
3400 Göttingen
W-Germany

TWISSELMANN, Ute
Mathematisches Institut
SFB 170
Universität Göttingen
Bunsenstr. 3-5
3400 Göttingen
W-Germany

VALLEJO, Ernesto
Mathematisches Institut
Im Neuenheimer Feld 288
6900 Mannheim
W-Germany

VOGEL, Pierre
Dept. de Mathématiques
Université de Nantes
2, rue de La Houssinière
F - 44072 Nantes
France

WEINTRAUB, Steven H.
Dept. of Mathematics
Louisiana State University
Baton Rouge LA 70803
USA

WEISS, Michael
Mathematisches Institut
SFB 170
Universität Göttingen
Bunsenstr. 3-5
3400 Göttingen
W-Germany

ZARATI, Said
Dept. de Mathematique
Universite de Tunis
1060 Tunis
Tunesia

The Homotopy Type of a 4-Manifold with finite Fundamental Group

by Stefan Bauer*

ABSTRACT: ... is determined by its quadratic 2-type, if the 2-Sylow subgroup has 4-periodic cohomology.

The homotopy type of simply connected 4-manifolds is determined by the intersection form. This is a well-known result of J.H.C. Whitehead and J. Milnor. In the non-simply connected case the homotopy groups π_1 and π_2 and the first k-invariant $k \in H^3(\pi_1, \pi_2)$ give other homotopy invariants. The **quadratic 2-type** of an oriented closed 4-manifold is the isometry class of the quadruple $[\pi_1(M), \pi_2(M), k(M), \gamma(\tilde{M})]$, where $\gamma(\tilde{M})$ denotes the intersection form on $\pi_2(M) \cong H_2(\tilde{M})$. An isometry of two such quadruples is an isomorphism of π_1 and π_2 which induces an isometry on γ and respects the k-invariant.

Recently $[H-K]$ I. Hambleton and M. Kreck, studying the homeomorphism types of 4-manifolds, showed that for groups with periodic cohomology of period 4 the quadratic 2-type determines the homotopy type.

This result can be improved away from the prime 2.

Theorem: Suppose the 2-Sylow subgroup of G has 4-periodic cohomology. Then the homotopy type of an oriented 4-dimensional Poincaré complex with fundamental group G is determined by its quadratic 2-type.

I am indebted to Richard Swan for showing me proposition 6. Furthermore I am grateful to the department of mathematics at the University of Chicago for its hospitality during the last year.

* **Supported by the DFG**

Let X be an oriented 4-dimensional Poincaré complex with finite fundamental group, $f : X \to B$ its 2-stage Postnikov approximation, determined by π_1, π_2, and k, and let $\gamma(X)$ denote the intersection form on $H_2(\tilde{X})$. Then $\mathbf{S}_4^{PD}(B, \gamma(X))$ denotes the set of homotopy types of 4-dimensional Poincaré complexes Y, together with 3-equivalences $g : Y \to B$, such that f and g induce an isometry of the quadratic 2-types.
The universal cover $\tilde{B}$ is an Eilenberg-MacLane space and hence, by [MacL], $H_4(\tilde{B}) \cong \Gamma(\pi_2(B))$, the $\mathbf{Z}\pi_1(B)$-module $\Gamma(\pi_2(B))$ being the module of symmetric 2-tensors, i.e. the kernel of the map $(1-\tau)\colon \pi_2(B) \otimes \pi_2(B) \to \pi_2(B) \otimes \pi_2(B)$, $(1-\tau)(a \otimes b) = a \otimes b - b \otimes a$. The intersection form on $\tilde{X}$ corresponds to $\tilde{f}_*[\tilde{X}]$ of the fundamental class $[\tilde{X}] \in H_4(\tilde{X}; \mathbf{Z})$. Let $\hat{H}_*$ denote Tate homology.

Proposition 1: If X is a Poincaré space with finite fundamental group G, then there is a bijection $\hat{H}_0(G; \pi_3(X)) \longleftrightarrow \mathbf{S}_4^{PD}(B, \gamma(X))$.

The proof uses a lemma of [H-K]:

Lemma 2: Let (X, f) and (Y, g) be elements in $\mathbf{S}_4^{PD}(B, \gamma(X))$. Then the only obstruction for the existence of a homotopy equivalence $h : X \to Y$ over B is the vanishing of$g_*[Y] - f_*[X] \in H_4(B)$.

Lemma 3: Given a diagram

$$\begin{array}{ccc} \mathbf{Z} & \overset{\alpha}{\hookrightarrow} & M \\ \cdot n \downarrow & & \\ \mathbf{Z} & & \end{array} ,$$

such that the torsion in the cokernel of α is annihilated by n, then the torsion subgroup in the pushout K is isomorphic to the torsion subgroup of $coker(\alpha)$.

Proof of 3: Since the torsion subgroup of M maps injectively into K as well as into $coker(\alpha)$, we may assume it trivial. Then M is isomorphic to $N \oplus < x >$ with $\alpha(1) = mx$ for an integer m dividing n. The pushout then is isomorphic to $(N \oplus \mathbf{Z} \oplus \mathbf{Z}) / < (0, m, n) > \cong M \oplus \mathbf{Z}/m$. ♣

Proof of proposition 1: Let (X, f) and (Y, g) be elements in $S_4^{PD}(B)$ such that f and g induce an isometry of the quadratic 2-types. Let $\gamma(X) = \gamma(Y) = \gamma$ denote the intersection form on $H_2(\tilde{X})$ and $H_2(\tilde{Y})$. By [W] one has $\pi_3(X) \cong \Gamma(\pi_2(X))/\langle\gamma\rangle \cong H_4(\tilde{B}, \tilde{X})$

and $\pi_3(X) \otimes_{\mathbf{Z}G} \mathbf{Z} \cong H_4(B,X)$. In the pushout diagramm:

$$\begin{array}{ccccccccc}
 & & 0 & & 0 & & 0 & & \\
 & & \downarrow & & \downarrow & & \downarrow & & \\
0 & \longrightarrow & H_4(\tilde{X}) \otimes_{\mathbf{Z}G} \mathbf{Z} & \longrightarrow & H_4(\tilde{B}) \otimes_{\mathbf{Z}G} \mathbf{Z} & \longrightarrow & H_4(\tilde{B},\tilde{X}) \otimes_{\mathbf{Z}G} \mathbf{Z} & \longrightarrow & 0 \\
 & & \phi\downarrow & & \downarrow & & \downarrow\cong & & \\
0 & \longrightarrow & H_4(X) & \longrightarrow & H_4(B) & \longrightarrow & H_4(B,X) & \longrightarrow & 0 \\
 & & \downarrow & & \downarrow & & \downarrow & & \\
0 & \longrightarrow & H_4(X,\tilde{X}) & \xrightarrow{\cong} & H_4(B,\tilde{B}) & \longrightarrow & 0 & & \\
 & & \downarrow & & \downarrow & & & & \\
 & & 0 & & 0 & & & &
\end{array}$$

the torsion subgroup of $H_4(B,X)$ is isomorphic to the torsion subgroup of $H_4(B)$ by lemma 3: The module $H_4(\tilde{B},\tilde{X})$ is torsion free. Hence the torsion subgroup of $H_4(\tilde{B},\tilde{X}) \otimes_{\mathbf{Z}G} \mathbf{Z}$ is annihilated by the order n of the group G. Note that ϕ is just multiplication by n. In particular one has

$$Torsion(H_4(B)) \cong Torsion(H_4(B,X)) \cong \hat{H}_0(G;\pi_3(X))$$

Since X and Y have the same quadratic 2-type, $\tilde{f}_*[\tilde{X}] = \tilde{g}_*[\tilde{Y}]$, hence we have $f_*[X] - g_*[Y] \in Torsion(H_4 B)$. This gives an injection

$$S_4^{PD}(B,\gamma) \hookrightarrow \hat{H}_0(G;\pi_3(X)).$$

What about surjectivity? Let $K \subset \tilde{X}$ denote a subspace, where one single orbit is deleted. Let $\alpha \in \pi_3(K)$ map via the surjection $\pi_3(K) \to \pi_3(X) \to \pi_3(X) \otimes_{\mathbf{Z}G} \mathbf{Z}$ to a given element $\hat{\alpha} \in \hat{H}_0(G;\pi_3(X))$. Let β be the image of $1 \in \mathbf{Z}G \cong H_4(\tilde{X},K) \cong \pi_4(\tilde{X},K) \hookrightarrow \pi_3(K)$. Now let $k : S^3 \to K$ represent $\alpha + \beta$ and define $X_\alpha := (K \cup_k (G \times D^4))/G$. One has to show that X_α is an orientable Poincaré space. Orientability is clear, since $H_4(X_\alpha) \cong H_4(X_\alpha, K) \cong \mathbf{Z}$. Let $f : X_\alpha \to B$ extend $f|_{K/G}$. The intersection form on $\tilde{X}_\alpha$ is determined by

$$\tilde{f}_{\alpha*}[\tilde{X}_\alpha] = trf(f_{\alpha*}[X_\alpha]) \in H_4(\tilde{X}).$$

But we have $f_{\alpha*}[X_\alpha] = f_*[X] + \alpha$: In the following diagram $1 \in \mathbf{Z} \cong \pi_4(X,B)$ is mapped to $f_*[X] \in H_4(B)$.

$$\begin{array}{ccccccccc}
H_4(X) & \cong & H_4(X,K/G) & \longleftarrow & H_4(\tilde{X},K) & \cong & \pi_4(\tilde{X},K) & \longrightarrow & \pi_3(K) \\
f\downarrow & & \downarrow & & \downarrow & & \downarrow & & \downarrow= \\
H_4(B) & \cong & H_4(B,K/G) & \longleftarrow & H_4(\tilde{B},K) & \cong & \pi_4(B,K) & \cong & \pi_3(K)
\end{array}$$

If the upper row is replaced by the corresponding row for X_α and the vertical maps by the ones induced by f_α, then $1 \in \mathbf{Z}G$ is mapped (counterclockwise) to $f_{\alpha*}[X_\alpha]$ on the one hand, on the other hand (clockwise) to $f_*[X] + \alpha$.
Since the torsion element α lies in the kernel of the transfer, one immediately gets $\tilde{f}_{\alpha*}[\tilde{X}_\alpha] = \tilde{f}_*[\tilde{X}]$. ♣

In the sequel all $\mathbf{Z}G$-modules have underlying a free abelian group.

The short exact sequence

$$0 \to \mathbf{Z} \xrightarrow{\gamma} \Gamma(\pi_2 X) \longrightarrow \pi_3(X) \to 0$$

gives rise to an exact sequence in Tate homology:

$$\hat{H}_0(G;\mathbf{Z}) \longrightarrow \hat{H}_0(G;\Gamma(\pi_2 X)) \longrightarrow \hat{H}_0(G;\pi_3(X)) \longrightarrow \hat{H}_{-1}(G;\mathbf{Z}) \xrightarrow{\gamma} \hat{H}_{-1}(G;\Gamma(\pi_2 X))$$

Here $\hat{H}_0(G;\mathbf{Z}) = 0$ and $\hat{H}_{-1}(G;\mathbf{Z}) \cong \mathbf{Z}/|G|$.The sequence above gives the connection to [H-K], theorem(1.1).

In order to analyze this sequence, I recall some facts from [H-K],§§2 and 3.

Facts:

1) $\Gamma(\mathbf{Z}G) = \oplus_i \mathbf{Z}[G/H_i] \oplus F$, where the summation is over all subgroups H_i of order 2 and F is a free $\mathbf{Z}G$-module.

2) $\Gamma(\mathbf{Z}G) \cong \Gamma(I) \oplus \mathbf{Z}G \cong \Gamma(I^*) \oplus \mathbf{Z}G$.Here I denotes the augmentation ideal, I^* its dual.

3) The modules $\Omega^3\mathbf{Z}$ and $S^3\mathbf{Z}$ are (stably!) defined by exact sequences

$$0 \to \Omega^3\mathbf{Z} \to F_2 \to F_1 \to F_0 \to \mathbf{Z} \to 0$$

and

$$0 \to \mathbf{Z} \to F_1 \to F_2 \to F_3 \to S^3\mathbf{Z} \to 0$$

with free modules F_i.

There is an exact sequence

$$0 \to \Omega^3\mathbf{Z} \longrightarrow \pi_2(X) \oplus r\mathbf{Z}G \longrightarrow S^3\mathbf{Z} \to 0$$

Lemma 4: If $0 \to A \to B \to C \to 0$ is a short exact sequence of $\mathbf{Z}G$-modules, which are free over $\mathbf{Z}$, then there are short exact sequences

$$0 \to \Gamma(A) \longrightarrow \Gamma(B) \longrightarrow D \to 0$$

and

$$0 \to A \otimes_{\mathbf{Z}} C \longrightarrow D \longrightarrow \Gamma(C) \to 0.$$

Proof: Given $\mathbf{Z}$-bases $\{a_i\}$, $\{c_j\}$ and $\{a_i, \tilde{c}_j\}$ of A, C and B, the map $h : a_i \otimes c_j \to a_i \otimes \tilde{c}_j + \tilde{c}_j \otimes a_i$ is well-defined and equivariant modulo $\Gamma(A)$. ♣

To prove the theorem, it suffices to show that $\hat{H}_0(G;\pi_3(X)) = 0$. This in turn can be done separately for each p-Sylow subgroup G_p of G.

Proposition 5: The map $\gamma_* : \hat{H}_{-1}(G_p;\mathbf{Z}) \longrightarrow \hat{H}_{-1}(G_p;\Gamma(\pi_2(X)))$ is injective, if either p is odd or $res^G_{G_p}\pi_2(X) \cong A \oplus B$ splits such that the rank of B over $\mathbf{Z}$ is odd. In general the kernel is at most of order 2.

Proof: For the sake of brevity, let π denote $\pi_2(X)$ and also let Γ denote the module $\Gamma(\pi)$. Now look at the following sequence of maps:

$$\psi : \mathbf{Z} \xrightarrow{\gamma} \Gamma \hookrightarrow \pi \otimes \pi \cong Hom(\pi^*, \pi) \xleftarrow{\alpha^* \cong} Hom(\pi, \pi) \xrightarrow{trace} \mathbf{Z}.$$

A generator of $\mathbf{Z}$ is mapped in $Hom(\pi^*, \pi)$ to the Poincaré map $\alpha : \pi^* \cong H^2(\tilde{X}) \xrightarrow{\cong} H_2(\tilde{X}) \cong \pi$, and then to the element $id \in Hom(\pi, \pi)$. So we have $\psi(1) = rank_{\mathbf{Z}}(\pi)$.

Fact 3) gives $rank_{\mathbf{Z}}(\pi) \equiv -2 \ mod \mid G \mid$, hence the induced selfmap ψ_* of $\mathbf{Z}/|G_p| \cong \hat{H}_{-1}(G_p; \mathbf{Z})$ is multiplication by -2. This proves, that the kernel is at most of order 2. In particular it is trivial, if p is odd.

In case p $= 2$ and $res^G_{G_p}\pi \cong A \oplus B$, such that the rank of the underlying group of B is odd, one can replace the map $Hom(\pi, \pi) \xrightarrow{trace} \mathbf{Z}$ by the map $Hom(\pi, \pi) \xrightarrow{p_* oi^*} Hom(B, B) \xrightarrow{trace} \mathbf{Z}$ in the defining sequence for ψ. A similar argument as above for p odd gives the claim. ♣

Remark: The module $res^G_{G_2}\pi_2(X)$ always splits, if $H_4(G; \mathbf{Z}) \cong Ext^1_{\mathbf{Z}G}(S^3\mathbf{Z}, \Omega^3\mathbf{Z})$ has no 2-torsion, in particular if G_2 has 4-periodic cohomology.

Proposition 6: Let A denote either $\Omega^n\mathbf{Z}$ or $S^n\mathbf{Z}$ and let τ be the selfmap of $A \otimes A$ which permutes the factors. Then $(-1)^n\tau$ induces the identity on $\hat{H}_0(G; A \otimes A)$.

Proof: Let $F. \to \mathbf{Z}$ be a free resolution of $\mathbf{Z}$ and let $\tilde{F}.$ be the truncated complex with $\tilde{F}_i = F_i$ for $i \leq n-1$, $\tilde{F}_n = \Omega^n$ and $\tilde{F} = 0$ else. There is an obvious projection $f : F. \to \tilde{F}.$, such that $f_n = \partial_n$. The tensor product $F. \otimes F. = F.^2$ again is a free resolution of $\mathbf{Z}$ and $\tilde{F}.^2$ is a truncatedfree resolution of $\mathbf{Z}$ with $\tilde{F}^2_{2n} = \Omega\mathbf{Z} \otimes \Omega\mathbf{Z}$. The chain map $f \otimes f$ induces an isomorphism of $H_*(F.^2 \otimes_{\mathbf{Z}G} \mathbf{Z})$ and $H_*(\tilde{F}.^2 \otimes_{\mathbf{Z}G} \mathbf{Z})$ in the dimensions $* \leq 2n$. The selfmap t of $F^2.$, as usual defined by $t(x \otimes y) = (-1)^{deg(x)deg(y)} x \otimes y$, is a chain automorphism, inducing the identity on the augmentation, hence on all derived functors, in particular on $H_*(F.^2 \otimes_{\mathbf{Z}G} \mathbf{Z}) = H_*(G; \mathbf{Z})$. In the same way an involution t can be defined on $\tilde{F}.^2$ and $f \otimes f$ commutes with t. Obviously $t_{2n} = (-1)^n\tau$. Hence $(-1)^n\tau$ induces the identity on $H_{2n}(\tilde{F}^2_{2n} \otimes_{\mathbf{Z}G} \mathbf{Z}) = \hat{H}_0(G; \mathbf{Z})$.
The proof for $S^n\mathbf{Z}$ is dual. ♣

Proof of the theorem: By proposition 1, it suffices to show that $\hat{H}_0(G; \pi_3(X))$ vanishes. By proposition 4 and the remark following it, this group is isomorphic to $\hat{H}_0(G; \Gamma(\pi_2(X)))$. In order to show that this group vanishes it suffices, by lemma 3, to show that $\hat{H}_0(G; A)$ vanishes for $A \in \{\Gamma(\Omega^3\mathbf{Z}), \Gamma(S^3\mathbf{Z}), \Omega^3\mathbf{Z} \otimes S^3\mathbf{Z}\}$ But $\hat{H}_0(G; \Omega^3\mathbf{Z} \otimes S^3\mathbf{Z}) \cong \hat{H}_0(G; \mathbf{Z}) = 0$. Given a module B (with underlying free abelian group), there is a short exact sequence

$$0 \to \Gamma(B) \longrightarrow B \otimes B \longrightarrow \Lambda^2(B) \to 0.$$

The map τ, which flips the both factors, induces, if applied to $B \in \{\Omega^3\mathbf{Z}, S^3\mathbf{Z}\}$ the following diagram:

$$\begin{array}{ccccccc}
\to & \hat{H}_1(G;\Lambda(B)) & \longrightarrow & \hat{H}_0(G;\Gamma(B)) & \longrightarrow & \hat{H}_0(G;B\otimes B) & \to \\
 & \downarrow(-id) & & \downarrow id & & \downarrow(-id) & \\
\to & \hat{H}_1(G;\Lambda(B)) & \longrightarrow & \hat{H}_0(G;\Gamma(B)) & \longrightarrow & \hat{H}_0(G;B\otimes B) & \to
\end{array}$$

The right vertical map is $(-id)$ by proposition 5. This diagram shows that any element in $\hat{H}_0(G;\Gamma(B))$ is annihilated by 4.In particular this group vanishes, if G is a p-group for an odd prime p. That $\hat{H}_0(G_2;\Gamma(B))$ vanishes, if G_2 has 4-periodic cohomology, follows at once from the facts 1 - 3, since in this case $\Omega^3\mathbf{Z} = I^* \oplus n\mathbf{Z}G$ and $S^3\mathbf{Z} = I \oplus n\mathbf{Z}G$ ♣

Final Remark: An elementary but lengthy computation shows $\Gamma(S^3\mathbf{Z}) \cong \mathbf{Z}/2\oplus\mathbf{Z}/2$ and $\Gamma(\Omega^3\mathbf{Z}) = 0$ for $G = \mathbf{Z}/2\oplus\mathbf{Z}/2$. In particular the group $\hat{H}_0(\mathbf{Z}/2\oplus\mathbf{Z}/2;\Gamma(\Omega^3\mathbf{Z}\otimes S^3\mathbf{Z}))$ is nontrivial. Hence the argument above won't work in general.

REFERENCES

[B 1] K.S. Brown: *Cohomology of groups.* GTM 87, Springer-Verlag, N.Y. 1982

[B 2] R. Brown: *Elements of Modern Topology.* McGraw - Hill, London, 1968

[H-K] I. Hambleton and M. Kreck: On the Classification of Topological 4-Manifolds with finite Fundamental Group. Preprint, 1986

[MacL] S. MacLane: Cohomology theory of abelian groups. Proc. Int. Math. Congress, vol. 2 (1950), pp 8 - 14

[W] J.H.C. Whitehead: On simply connected 4-dimensional polyhedra. Comment. Math. Helv., 22 (1949), pp 48 - 92.

Sonderforschungsbereich 170
Geometrie und Analysis
Mathematisches Institut
Bunsenstr. 3 - 5
D-3400 Göttingen, FRG

Rational Cohomology of Configuration Spaces of Surfaces

C.-F. Bödigheimer and F.R. Cohen

1. Introduction. The k-th configuration space $C^k(M)$ of a manifold M is the space of all unordered k-tuples of distinct points in M. In previous work [BCT] we have determined the rank of $H_*(C^k(M);\mathbb{F})$ for various fields $\mathbb{F}$. However, for even dimensional M the method worked for $\mathbb{F}=\mathbb{F}_2$ only. The following is a report on calculations of $H^*(C^k(M);\mathbb{Q})$ for M a deleted, orientable surface. This case is of considerable interest because of its applications to mapping class groups, see [BCP]. Similar results for (m-1)-connected, deleted 2m-manifolds will appear in [BCM].

2. Statement of results. The symmetric group Σ_k acts freely on the space $\tilde{C}^k(M)$ of all ordered k-tuples $(z_1,\dots,z_k)$, $z_i \in M$, such that $z_i \neq z_j$ for $i \neq j$. The orbit space is $C^k(M)$. As in [BCT] we will determine the rational vector space $H^*(C^k(M);\mathbb{Q})$ as part of the cohomology of a much larger space. Namely, if X is any space with basepoint x_o, we consider the space

$$(1) \qquad C(M;X) = \left(\coprod_{k\geq 1} \tilde{C}^k(M) \times_{\Sigma_k} X^k \right) \Big/ \approx$$

where $(z_1,\dots,z_k;x_1,\dots,x_k) \approx (z_1,\dots,z_{n-1};x_1,\dots,x_{k-1})$ if $x_k = x_o$. The space C is filtered by subspaces

$$(2) \qquad F_kC(M;X) = \left(\coprod_{j=1}^{k} \tilde{C}^j(M) \times_{\Sigma_j} X^j \right) \Big/ \approx$$

and the quotients $F_kC/F_{k-1}C$ are denoted by $D_k(M;X)$.

Let $\bar{M}_g$ denote a closed, orientable surface of genus g, and M_g is $\bar{M}_g$ minus a point. We study $C(M_g;S^{2n})$ for $n\geq 1$. H^* will always stand for rational cohomology, and P[ ] resp. E[ ] for polynomial resp. exterior algebras over $\mathbb{Q}$.

Theorem A. *There is an isomorphism of vector spaces*

(3) $$H^*C(M_g;S^{2n}) \cong P[v,u_1,\ldots,u_{2g}]\otimes H_*(E[w,z_1,\ldots,z_{2g}],d)$$

with $|v|=2n$, $|u_i|=4n+2$, $|w|=4n+1$, $|z_i|=2n+1$, *and the differential* d *is given by* $d(w) = 2(z_1z_2 + \ldots + z_{2g-1}z_{2g})$.

Giving the generators weights, $\mathrm{wght}(v) = \mathrm{wght}(z_i) = 1$ and $\mathrm{wght}(u_i) = \mathrm{wght}(w) = 2$, makes H^*C into a filtered vector space. We denote this weight filtration by F_kH^*C. The length filtration F_kC of C defines a second filtration H^*F_kC of H^*C.

Theorem B. *As vector spaces*

(4) $$H^*F_kC(M_g;S^{2n}) = F_kH^*C(M_g;S^{2n}).$$

It follows that $H^*D_k(M_g;S^{2n})$ is isomorphic to the vector subspace of $H^*(g,n) = P[v,u_i]\otimes H_*(E[w,z_i],d)$ spanned by all monomials of weight exactly k. To obtain the cohomology of $C^k(M_g)$ itself, we consider the vector bundle

(5) $$\eta^k: \tilde{C}^k(M_g)\times_{\Sigma_k}\mathbb{R}^k \to C^k(M_g)_+$$

which has the following properties. First, the Thom space of m times η^k is homomorphic to $D_k(M_g;S^m)$. Secondly, it has finite even order, see [CCKN]. Hence

(6) $$D_k(M_g;S^{2nk}) = \Sigma^{2n_k\cdot k}C^k(M_g)_+$$

for $2n_k = \mathrm{ord}(\eta^k)$. Thus we have

Theorem C. As a vector space, $H^*C^k(M_g)$ *is isomorphic to the vector subspace generated by all monomials of weight* k *in* $H^*(g,n_k)$, *desuspended* $2n_k k$ *times.*

Regarding the homology of $E = E[w,z_1,\ldots,z_{ig}]$ we have

Theorem D. The homology $H_*(E,d)$ *is as follows:*

(7) $\operatorname{rank} H_{i(2n+1)} = \binom{2g}{i} - \binom{2g}{i-2}$ for $i = 0,1,\ldots g$, *and all (non-zero) elements have weight* i;

(8) $\operatorname{rank} H_{i(2n+1)+4n+1} = \binom{2g}{i} - \binom{2g}{i+2}$ *for* $i = g,\ldots,2g$, *and all (non-zero) elements have weight* $i+2$;

(9) $\operatorname{rank} H_j = 0$ *in all other degrees* j.

Note the apparent duality $\operatorname{rank} H_j = \operatorname{rank} H_{N-j}$ for $N = 2g(2n+1)+4n+1$.

We will give the proof of Theorem A in the next section. The proof of Theorem B is the same as for [BCT, Thm.B]. By what we said above Theorem C folows from Theorem B. And Theorem D will be derived in the last section.

3. Mapping spaces and fibrations.

Let D denote an embedded disc in M_g. There is a commutative diagram

(10)
$$\begin{array}{ccc} C(D;S^{2n}) & \longrightarrow & \Omega^2 S^{2n+2} \\ \downarrow & & \downarrow \\ C(M_g;S^{2n}) & \longrightarrow & \operatorname{map}_o(\bar{M}_g;S^{2n+2}) \\ \downarrow & & \downarrow \\ C(M_g,D;S^{2n}) & \longrightarrow & (\Omega S^{2n+2})^{2g} \end{array}$$

where map_o stands for based maps. The right column is induced by restricting to the 1-section, and is a fibration. The left column is a quasifibration. Since S^{2n} is connected, all three horizontal maps are equivalences, see [M], [B] for details.

The E_2-term of the Serre spectral sequence of these (quasi)fibrations is as follows. From the base we have 2g-fold tensor product of

$$(11) \qquad H^*\Omega S^{2n+2} = H^*(S^{2n+1} \times \Omega S^{4n+3}) = E[z_i] \otimes P[u_i] \quad (i = 1, \ldots 2g),$$

where $|z_i| = 2n+1$ and $|u_i| = 4n+2$. From the fibre we have

$$(12) \qquad H^*\Omega^2 S^{2n+2} = H^*(\Omega S^{2n+1} \times \Omega^2 S^{4n+3}) = H^*(\Omega S^{2n+1} \times S^{4n+1})$$

$$= P[v] \otimes E[w] \;,$$

where $|v| = 2n$ and $|w| = 4n+1$. The following determines all differentials in this spectral sequence.

__Lemma. The differentials are as follows:__

$$(13) \qquad d_{2n+1}(v) = 0$$

$$(14) \qquad d_{4n+2}(w) = 2z_1 z_2 + 2z_2 z_3 + \ldots + 2z_{2g-1} z_{2g}$$

Proof: Assertion (13) follows from the stable splitting of $C(M_g; S^{2n})$, on [B]. (14) results from symmetries of M_g and of the fibrations (10) which leave d invariant. ■

The lemma implies $E_{4n+3} = E_\infty = H^*C(M_g; S^{2n})$. Furthermore, E_{4n+3} is a tensor product of the polynomial algebra $P[v, u_1, \ldots u_{2g}]$ and the homology module $H_*(E, d)$ of the exterior algebra $E = E[w, z_1, \ldots, z_{2g}]$ with differential d. This proves Theorem A.

__4. Homology of E.__ Let us write $x_i = z_{2i-1}$ and $y_i = z_{2i}$ for $i = 1, \ldots g$. The form $d(w) = 2z_1 z_2 + 2z_2 z_3 + \ldots + 2z_{2g'} z_{2g}$ is equivalent to the standard symplectic form $x_1 y_1 + x_2 y_2 + \ldots + x_g y_g$. The vector space

$E[g] = L[g] \oplus wL[g]$ with $L[g] = E[x_1y_1, \ldots, x_gy_g]$. The differential is zero on the first summand, and sends the second to the first. Hence we regard d as an endomorphism of $L[g]$, given by multiplication with $d(w) = x_1y_1 + \ldots + x_gy_g$.

Let $L_k[g]$ denote the vector subspace spanned by all k-fold products

(15) $\quad z_{i_1}z_{i_2} \ldots z_{i_n}$ with $1 \le i_1 < i_2 < \ldots < i_k \le 2_g$.

Since d(w) is homogeneous of weight 2, we have

(16) $\quad d = d[g] = \bigoplus_{k=0}^{2g} d_k[g], \qquad d_k[g]: L_k[g] \longrightarrow L_{k+2}[g]$.

The (co)kernels of $d_k[g]$ is determined by the (co)kernel of $d_l[g-1]$ and $d_l[g-1]^2$ for $l = k, k-1, k-2$. The (co)kernel of $d_l[g-1]^2$ in turn is determined by the (co)kernels of $d_m[g-2]^2$ and $d_m[g-2]^3$ for $m = l, l-1, l-2$. Therefore we will study all powers $d_k[g]^r$ and prove the following (Lefschetz) lemma by simultaneous induction on g, k and r.

Lemma. For $0 \le k \le g$ the differential

$$d_k[g]^r : L_k[g] \longrightarrow L_{k+2r}[g] \text{ is}$$

(17) a monomorphism for $0 \le k < g-r$,

(18) an isomorphism for $k = g-r$

(19) an epimorphism for $g-r < k \le 2g$

Proof: For $\lambda_g = \sum_{i=1}^{g} x_iy_i$ we have $\lambda_g = \lambda_{g-1} + x_gy_g$ and $\lambda_g^r = \lambda_{g-1}^r + r\lambda_{g-1}^{r-1}x_gy_g$, in particular $\lambda_g^g = g!\omega_g$ where $\omega_g = x_1y_1x_2y_2 \ldots x_gy_g$ is the volume element. To facilitate the induction, we decompose $L_k[g]$ further by partitioning the canonical basis elements (15) into four types.

(20) $\quad i_k \le 2g-2$

(21) $\quad i_{k-1} \le 2g-2$ and $i_k = 2g-1$,

(22) $\quad i_{k-1} \le 2g-2$ and $i_k = 2g$,

(23) $\quad i_{k-1} = 2g-1$ and $i_k = 2g$.

Hence $L_k[g] = L_k[g-1] \oplus L_{k-1}[g-1]x_g \oplus L_{k-1}[g-1]y_g \oplus L_{k-2}[g-1]x_g y_g$. With respect to this decomposition $d_k[g]^r$ has the following matrix form

(24)
$$d_k[g]^r = \begin{bmatrix} d_k[g-1]^r & 0 & 0 & rd_k[g-1]^{r-1} \\ 0 & d_{k-1}[g-1]^r & 0 & 0 \\ 0 & 0 & d_{k-1}[g-1]^r & 0 \\ 0 & 0 & 0 & d_{k-2}[g-1]^r \end{bmatrix} = \begin{bmatrix} A & 0 & 0 & A' \\ 0 & B & 0 & 0 \\ 0 & 0 & B & 0 \\ 0 & 0 & 0 & C \end{bmatrix}$$

To start the induction consider the case $g = 1$. The only non-zero differential $d_0[1] : L_0[1] \to L_2[1]$ is an isomorphism. For $g \ge 2$ and $k = 0$, $d_0[g]^r$ sends the generator of $L_0[g]$ to λ_g^r, and thus is monic. Assume the lemma holds for $g-1$. We distinguish three cases.

<u>Case $k < g-r$</u>: Then A, A', B as well as C in (24) are all monomorphisms by hypothesis. Hence, from $0 = d_k[g]^r(a,b_1,b_2,c) = (A(a), B(b_1), B(b_2), A'(a)+C(c))$ we conclude $a = b_1 = b_2 = 0$, and so $c = 0$ as well. Thus $d_k[g]^r$ is a monomorphism.

<u>Case $k = g-r$</u>: Here A is an epimorphism, A' and B are isomorphisms, and C is a monomorphism. Assume $0 = d_k[g]^r(a,b_1,b_2,c) = (A(a), B(b_1), A'(a)+C(c))$. First, $b_1 = b_2 = 0$. We now have $A(a) = d_k[g-1]^r a = 0$ and $d_{k-2}[g-1]^r c = -rd_k[g-1]^{r-1}a$; writing this as $d_k[g-1]^r(-ra) = A(-ra) = 0$. Thus, since $d_{k-2}[g-1]^{r+1}$ is an isomorphism, $c = 0$. Therefore, $-rd_{g-r}[g-1]^{r-1}a = 0$, and $a = 0$ since $d_{g-r}[g-1]^{r-1}$ is an isomorphism. We see that $d_k[g]^r$ is a monomorphism between vector spaces of equal dimensions, hence an isomorphism.

Case $k > g-r$: This time A, A', B, C are epimorphisms. Given $(\bar{a},\bar{b}_1,\bar{b}_2,\bar{c}) \in L_{k+2r}[g]$ we can first find a, b_1, b_2 satisfying $A(a) = \bar{a}$, $B(b_1) = \bar{b}_1$ and $B(b_2) = \bar{b}_2$. Then we choose c such that $C(c) = \bar{c} - A'(a)$. Hence $d_k[g]^r$ is epimorphic. ∎

The lemma completely determines $H_*(E,d)$ as a vector space over $\mathbb{Q}$. Theorem D now follows.

References

[B] C.-F. Bödigheimer: Stable splittings of mapping spaces. Proc. Seattle (1985), Springer LNM 1286, p. 174-187.

[BCM] C.-F. Bödigheimer, F.R. Cohen, R.J. Milgram: On deleted symmetric products. In preparation.

[BCP] C.-F. Bödigheimer, F.R. Cohen, M. Peim: Mapping spaces and the hyperelliptic mapping class group. In preparation.

[BCT] C.-F. Bödigheimer, F.R. Cohen, L. Taylor: Homology of configuration spaces. To appear in Topology.

[CCKN] F.R. Cohen, R. Cohen, N. Kuhn, J. Neisendorfer: Bundles over configuration spaces. Pac. J. Math. 104 (1983), p. 47-54.

[M] D. McDuff: Configuration spaces of positive and negative particles. Topology 14 (1975), p. 91-107.

C.-F. Bödigheimer
Mathematisches Institut
Bunsenstraße 3-5
D-3400 Göttingen
West Germany

F.R. Cohen
Department of Mathematics
University of Kentucky
Lexington, KY 40506
USA

An S^1-Degree and S^1-Maps Between Representation Spheres

by

Grzegorz Dylawerski

Abstract. Let V be an orthogonal representation of $G=S^1$ and let $S(V)$, $S(V\oplus R)$ be the unit spheres in V , $V\oplus R$ respectively. In this paper we classify S^1-equivariant maps $S(V\oplus R) \longrightarrow S(V)$. More precisely we construct an isomorphism $[S(V\oplus R),S(V)]_G \longrightarrow A(V)$ where $A(V) = [S(V\oplus R)^G,S(V)^G] \oplus (\bigoplus_H Z)$, $H\subset S^1$ runs over all isotropy subgroups of V different from S^1 .

Introduction. Let V be an orthogonal finite-dimensional representation of $G=S^1$, $\Omega\subset(V\oplus R)$ an open bounded invariant subset and $f:(\Omega,\partial\Omega) \longrightarrow (V,V\setminus\{0\})$ an equivariant map. For the above f an S^1-degree , denoted $Deg(f,\Omega)$, was defined in work [3]. This is an element of the group $Z_2 \oplus (\bigoplus_H Z)$, where $H\subset S^1$ runs over all the isotropy subgroups of V different from S^1 . It is natural to ask whether this degree classifies the homotopy classes of equivariant maps $(B,\partial B) \longrightarrow (V,V\setminus\{0\})$, where B denotes the unit ball in $V + R$. This is not true ingeneral. Anyway slightly modifying the first coordinate of $Deg(f,\Omega)$ we get a new invariant which classifies these G-homotopy classes . Since $[(B,\partial B),(V,V\setminus\{0\})]_G \approx [S(V\oplus R),S(V)]_G$, the new S^1-degree classifies the Ghomotopy classes of G-maps between spheres.

The problem of classification of G-maps between representation spheres was studied by G.B. Segal [5] , T. tom Dieck [2] , R.Rubinsztein [4]. We would like to mention that we came to the method used here in a result of studying [2] .

In Section 0. we introduce notations , compile some basic facts concerning group actions and the obstruction theory . In Section 1. we

recall the properties of $Deg(f,\Omega)$, sketch the definition of $Deg(f,\Omega)$ in the special case and define a new S^1-degree $D(f,B)$, ($Deg(f,B)$ and $D(f,B)$ are distinct on the first coordinate only) .
In Section 2. we relate $Deg(f,\Omega)$ to the obstruction theory . In Section 3. we describe the group structure on $[S(V \oplus R), S(V)]_G$ and define a homomorphism $D : [S(V + R), S(V)]_G \longrightarrow A(V)$. Section 4. is devoted to the proof of the main theorem of this paper . This theorem says that homomorphism D is an isomorphism .

0. Preliminaries. We begin by recalling some terminology and facts concerning gropp action and obstruction theory .

Let G be a compact Lie group , and X a left G-space . We shall denote by G_x the isotropy subgroup of $x \in X$ and by Gx the orbit of x . For each subgroup H of G let X^H denote the fixed point set of H i.e. $X^H = \{x \in X , H \subset G_x\}$. The set of points of X for which G_x is precisely H will be denoted by X_H .

Let V be an orthogonal representation of G and $x \in V$. We denote by N_x the normal space to Gx at x and $B_N(x,r) = \{y \in N_x , |y-x| < r\}$. If $X \subset V$ is an invariant subset such that $G(X_H) = X$, then the projection $\pi : X \longrightarrow X/G$ is an $N(H)/H$ - bundle and the homeomorphism

$$\psi_\alpha : G \times_H U_\alpha \approx (G/H) \times U \longrightarrow G(U_\alpha) = \pi^{-1}(U_\alpha)$$

(where $U_\alpha = X \cap B_N(x,r)$, $\alpha = (x,r)$, r is sufficiently small and we identify U_α with $\pi(U_\alpha)$) form a family of local trivializations of bundle $\pi : X \longrightarrow X/G$ see [1] II 5.2, II 5.8 .

Suppose that the action of G on X is free and Y is a left G-space . Consider the associated G-bundle $p : Y \times_G X \longrightarrow X/G$. If $\psi : G \times U \longrightarrow \pi^{-1}(U)$ is a local trivialization of the bundle then $\tilde{\psi} : Y \times U \approx (Y \times_G G) \times U \longrightarrow Y \times_G (G \times U) \longrightarrow Y \times_G \pi^{-1}(U) \approx p^{-1}(U)$

$$\tilde{\psi}(y,u) = [y, \psi(e,u)]$$

is a local trivialization of p . Moreover , there is a one-to-one correspondence between G-maps $f : X \longrightarrow Y$ and cross-sections s_f of p given by $s_f(\pi(x)) = [f(x), x]$ see [1] II 2.4 , II 2.6 .

Now we assume that G is a compact connected Lie group , G acts freely on X and X/G is triangulable ; $X/G = |K|$, K - triangulation of X/G . Let Y be an n-simple G-space and let L be a subcomplex of K . Consider a partial cross-section $s : K^n \cup L \longrightarrow Y \times_G X$. Since G is connected , the bundle of coefficients associated with the bundle p is trivial . So the obstruction z(s) to extending s on $K^{n+1} \cup L$ lies in $C^{n+1}(K, L, \pi_n(Y))$. From the above facts it follows :

<u>0.1 Lemma.</u> Let $f : \pi^{-1}(K^n \cup L) \longrightarrow Y$ be a G-map and let $s_f : K^n \cup L \longrightarrow Y \times_G X$ be the partial cross-section of p corresponding to f . Then $z(s_f)(\sigma) = [f \cdot \psi_{e/\partial\sigma}] \in \pi_n(Y)$ where is an n+1 - simplex contained in U , $\psi : G \times U \longrightarrow \pi^{-1}(U)$ is a local trivialization of π and $\psi_e(x) = \psi(e, x)$.

1. Let V be a real orthogonal representation of $G = S^1$, $O(V) = \{H \subsetneq S^1 ; \exists x \in V \ G_x = H\}$ and $\Omega \subset V \oplus R$ an open bounded invariant subset . We denote by $C_G(\Omega, \partial\Omega)$ the space of S^1-equivariant maps $f : (\bar{\Omega}, \partial\Omega) \longrightarrow (V, V \setminus \{0\})$ with the standard metric $|f_1 - f_2| = \sup_{x \in \Omega} |f_1(x) - f_2(x)|$. We say that $f_0, f_1 \in C_G(\Omega, \partial\Omega)$ are G-homotopic if there exists a G-homotopy $h : (\Omega \times [0,1], \partial\Omega \times [0,1]) \longrightarrow (V, V \setminus \{0\})$ (i.e. $h(gx, t) = g \cdot h(x, t)$) such that $h(\cdot, 0) = f_0$ $h(\cdot, 1) = f_1$.

For maps $f \in C_G(\Omega, \partial\Omega)$ we can define the S^1-degree $\mathrm{Deg}(f, \Omega) \in Z_2 \oplus (\bigoplus_{H \in O(V)} Z)$. We denote by $\deg_H(f, \Omega) \in Z$ the H-coordinate of $\mathrm{Deg}(f, \Omega)$ and by $\deg_{S^1}(f, \Omega) \in Z_2$ its first coordinate .

<u>1.1. Theorem.</u> Let $\Omega, \Omega_0, \Omega_1, \Omega_2 \subset V \oplus R$ be open bounded invariant subsets and $f \in C_G(\Omega, \partial\Omega)$. Then the following properties hold:

a) If $\deg_H(f,\Omega) \neq 0$, then $f^{-1}(0) \cap \Omega^H \neq \phi$.

b) If $f^{-1}(0) \subset \Omega_0 \subset \Omega$, then $\mathrm{Deg}(f,\Omega) = \mathrm{Deg}(f,\Omega_0)$.

c) If $f^{-1}(0) \subset \Omega_1 \cup \Omega_2 \subset \Omega$ and $\Omega_1 \cap \Omega_2 = \phi$, then $\mathrm{Deg}(f,\Omega) = \mathrm{Deg}(f,\Omega_1) + \mathrm{Deg}(f,\Omega_2)$.

d) If $h : (\bar{\Omega} \times [0,1], \partial\Omega \times [0,1]) \longrightarrow (V, V \setminus \{0\})$ is a G-homotopy then $\mathrm{Deg}(h_0,\Omega) = \mathrm{Deg}(h_1,\Omega)$.

e) Suppose W is another representation of $G=S^1$ and let U be an open bounded invariant subset of W such that $0 \in U$. Define $F : U \times \Omega \longrightarrow W \oplus V$ by $F(x,y) = (x, f(y))$. Then $\mathrm{Deg}(F, U\times\Omega) = \mathrm{Deg}(f,\Omega)$.

f) Let $H \in O(V) \cup \{S^1\}$ and $f^H : (\bar{\Omega}^H, \partial\Omega^H) \longrightarrow (V^H, V^H \setminus \{0\})$ denotes the restriction of f. Then $\deg_H(f,\Omega) = \deg_H(f^H, \Omega^H)$.

g) If $\Omega_H = \phi$, then $\deg_H(f,\Omega) = 0$.

The properties a-e have been proved in ([3] Theorem 1.2). The properties f,g follow immediately from ([3] Definition 3.6, 3.7).

Let $K = \bigcap_{H \in O(V)} H$. Assume now that $\bar{\Omega}_K = \bar{\Omega}$. We recall the definition of $\deg_K(f,\Omega)$.

Suppose $a \in V \oplus R$ and $G_a \neq S^1$. Let v be a tangent vector to $M = Ga$ at a and $N_a = \{x \in V \oplus R, \langle x.v \rangle = 0\}$ the normal space to Ga at a. Note that $a \in N_a$ and $V \oplus R = N_a \oplus \mathrm{span}\{v\}$. Let $A : N_a \longrightarrow V$ be a linear isomorphism. Define $A^\wedge : V \oplus R \longrightarrow V \oplus R$ by $N_a \oplus \mathrm{span}\{v\} \ni (x, \lambda v) \longrightarrow (A(x), \lambda) \in V \oplus R$ and $\mathrm{sgn}\, A = \mathrm{sgn}(\det A^\wedge)$. Recall the notation: $B_N(a,r) = \{x \in N_a, |x-a| < r\}$ $B(r) = \{x \in V, |x| < r\}$.

<u>1.2. Definition.</u> Let $a \in V \oplus R$ and $G_a \neq S^1$. We say that a continuous map $\varphi : B(2) \longrightarrow V \oplus R$ is a <u>slice map at a</u> if:

a) $\varphi(0) = a$

b) there exists $r>0$ such that $B_N(a,r)$ is a slice and $\varphi(B(2)) \subset B_N(a,r)$

c) there exists a linear isomorphism $A : N_a \longrightarrow V$ such that $\varphi(x) = a + A^{-1}(x)$ and $\operatorname{sgn} A = +1$.

__1.3 Definition.__ We say that an open invariant subset $\Omega_0 \subset \Omega$ is __elementary__ if there exists a finite family $\Omega_1,\dots,\Omega_r$ of open invariant subsets of Ω such that :

a) $\Omega_0 \subset \Omega_1 \cup \dots \cup \Omega_r$

b) $\Omega_i \cap \Omega_j = \emptyset$ for $i \neq j$

c) for each i , $1 \leq i \leq r$, there is a slice map $\varphi_i : B(2) \longrightarrow$ $\longrightarrow \Omega$ such that $\Omega_i \subset G\cdot\varphi_i(B(1))$.

__1.4. Definition.__ We say that $f \in C_G(\Omega,\partial\Omega)$ is an __elementary map__ if there exists an elementary subset $\Omega_0 \subset \Omega$ such that $f^{-1}(0) \subset \Omega_0$.

In ([3] Proposition 2.11) the following lemma has been proved :

__1.5 Lemma.__ The set of all elementary maps is an open and dense subset of $C_G(\Omega,\partial\Omega)$.

Assume now that $f \in C_G(\Omega,\partial\Omega)$ is an elementary map and $\{\Omega_i\}$ $\{\varphi_i\}$ satisfy the conditions of Definition 1.3 . Let $U_i = \varphi_i^{-1}(\Omega_i)$ and $F_i = f \circ \varphi_i : U_i \longrightarrow V$. Clearly $F_i^{-1}(0)$ is a compact subset of U_i , thus the Brouwer degree $\deg(F_i,U_i)$ is well defined . Define

__1.6 Definition.__ $\deg_K(f,\Omega) = \sum_i \deg(F_i,U_i)$

Let now $f \in C_G(\Omega,\partial\Omega)$ be any equivariant map and $\varepsilon = \min_{x\in\partial\Omega}|f(x)|$. By Lemma 1.5 , there exists an elementary map $f_e \in C_G(\Omega,\partial\Omega)$ such that $|f-f_e| < \varepsilon$. Define

<u>1.7 Definition.</u> $\deg_K(f,\Omega) = \deg_K(f_e,\Omega)$.

<u>Remark.</u> We would like to point out that in Definitions 1.2 - - 1.7 and in Lemma 1.5 the set Ω satisfied the condition $\bar{\Omega}_K = \bar{\Omega}$.

For the general case see ([3] Section 3.) .

Let $B = \{x \in V \oplus R\ ,\ |x| < 1\}$. Now we define a new invariant for $f \in C_G(B,\partial B)$. It will be needed in Section 3.4 .

<u>1.8. Definition.</u> Let V be a real representation of $G = S^1$. Define :

$$A(V) = \begin{cases} \{0\} \oplus (\bigoplus_{H \in O(V)} Z) & \text{if } \dim V^G = 1 \text{ or } 2 \\ Z \oplus (\bigoplus_{H \in O(V)} Z) & \text{if } \dim V^G = 3 \\ Z_2 \oplus (\bigoplus_{H \in O(V)} Z) & \text{if } \dim V^G \geqslant 4 \end{cases}$$

For given $f \in C_G(B,\partial B)$, by the same letter f we denote the induced map $f/|f| : S(V \oplus R) \longrightarrow S(V)$, and the same for the restriction map $f^G \in C(B^G,\partial B^G)$. Let $[f^G]$ denote the homotopy class of the map $f^G : S(V \oplus R)^G \longrightarrow S(V)^G$.

<u>1.9 Definition.</u> Let $f \in C_G(B,\partial B)$ and $H \in O(V) \cup \{S^1\}$. Define

$$d_H(f,B) = \begin{cases} \deg_H(f,B) & \text{if } H \neq S^1 \\ 0 & \text{if } H = S^1 \text{ and } \dim V^G = 1 \text{ or } 2 \\ [f^G] & \text{if } H = S^1 \text{ and } \dim V^G \geqslant 3 \end{cases}$$

and

$$D(f,B) = \{d_H(f,B)\} \in A(V) .$$

2. Connection of $\deg_H(f,\Omega)$ with the obstruction theory.

Throughout this section we shall make the following assumptions.

2.1 Let V be an orthogonal representation of S^1, $\dim V = n+1$; $\Omega \subset V \oplus R$ an open bounded invariant connected subset such that $\bar{\Omega}$ is a smooth S^1-manifold with boundary, and $\bar{\Omega}_H = \bar{\Omega}$ where (H) is the main orbit type on V.

Under the above assumptions the orbit space $\bar{\Omega}/S^1$ is a smooth manifold, hence $\bar{\Omega}/S^1$ is triangulable. Let K denote a triangulation of $\bar{\Omega}/S^1$ such that each simplex $\sigma \in K$ is contained in a chart (U,ψ) of the S^1-bundle $\bar{\Omega} \longrightarrow \bar{\Omega}/S^1$ (see Section 0.). Note that $\dim K = n+1$. Let σ be any $n+1$-simplex contained in a chart (U,ψ), $U = \pi(B_N(a,r) \cap \bar{\Omega})$ and $\varphi : B(2) \longrightarrow B_N(a,r)$ a slice map. We define $\psi_e : U \longrightarrow B_N(a,r)$ $\psi_e(x) = \psi(e,x)$ and $\tau = \varphi^{-1}(\psi_e(\sigma)) = \varphi^{-1}(\pi^{-1}(\sigma)) \subset B(1)$. Since τ is $n+1$-simplex there exists an orientation-preserving homeomorphism $h: B(1) \longrightarrow \tau$. Let σ^+ be the simplex oriented by the homeomorphism $\psi_e^{-1} \circ \varphi \circ h$.

Suppose $f \in C_G(\Omega, \partial\Omega)$ is an equivariant map such that $f^{-1}(0) \cap \pi^{-1}(|K^n|) = \emptyset$ (K^n denotes n-skeleton). From definition 1.6. and Theorem 1.1c we have

$$\deg_H(f,\Omega) = \sum_\sigma \deg_H(f, \pi^{-1}(\sigma)) = \sum_\sigma \deg(f\circ\varphi, \tau) =$$
$$= \sum_\sigma \deg(f\circ\varphi\circ h, B) = \sum_\sigma \deg(f\circ\psi_e\circ\psi_e^{-1}\circ\varphi\circ h, B)$$

where σ runs over all $n+1$-simplexes of K. Let $[f\circ\psi_{e/\sigma^+}] \in \pi_n(V\setminus\{0\})$ denote the homotopy class of the map $f\circ\psi_{e/\partial\sigma} : \sigma^+ \longrightarrow V\setminus\{0\}$. Identifying $\pi_n(V\setminus\{0\}) \approx Z$ by

$$\pi_n(V\setminus\{0\}) = [\partial B, V\setminus\{0\}] \approx [(B,\partial B);(V,V\setminus\{0\})] \xrightarrow{\deg} Z$$

we obtain

<u>2.2 Lemma.</u> Let $\Omega \subset V \oplus R$ satisfy the assumptions 2.1 and $f \in C_G(\Omega,\partial\Omega)$ be an equivariant map such that $f^{-1}(0)\cap\pi^{-1}(K^n) = \emptyset$ Then

$$\deg_H(f,\Omega) = \sum_{\sigma^+}[f\circ\psi_{e/\partial\sigma^+}] \in \pi_n(V\setminus\{0\})$$

We shall denote : $V_0 = V\setminus\{0\}$, $V_0 \times_G \Omega = E \longrightarrow K$ - the bundle associated with the bundle $\pi : \Omega \longrightarrow K$ and $L = \partial K = \partial\Omega/S^1$. Any map $f \in C_G(\Omega,\partial\Omega)$ induces a partial cross-section $s_f : L \longrightarrow E$. Since the fibre of the bundle p , V_0 is n-1 - connected , there exists an extension $s_f^\wedge : K^n \cup L \longrightarrow E$ of s_f.
The cross-section $s_f^\wedge$ induces a cocycle $z(s_f^\wedge)$ $C^{n+1}(K,L;\pi_n(V_0))$

<u>2.3 Lemma.</u> Let $\Omega \subset V \oplus R$ satisfy 2.1 and $f \in C_G(\Omega,\partial\Omega)$. Then $\deg_H(f,\Omega) = \sum z(s_f^\wedge)(\sigma^+)$.

<u>Proof.</u> Let $f^\wedge : \pi^{-1}(K^n) \longrightarrow V_0$ be the equivariant map corresponding to $s_f^\wedge$ and $\bar{f} : (\Omega,\partial\Omega) \longrightarrow (V,V\setminus\{0\})$ be an extension of $f^\wedge$. Since f and $\bar{f}$ are equal on $\partial\Omega$, $\deg_H(f,\Omega) = \deg_H(\bar{f},\Omega)$. From Theorem 1.1c we have $\deg_H(\bar{f},\Omega) = \sum_{\sigma^+}\deg_H(\bar{f},\pi^{-1}(\sigma^+))$. We shall show that $\deg_H(\bar{f},\pi^{-1}(\sigma^+)) = z(s_f^\wedge)(\sigma^+)$. Let (U,ψ) be a chart of the bundle π which contain σ . From Lemma 2.2 and 0.1 it follows that $\deg_H(\bar{f},\pi^{-1}(\sigma^+)) = [\bar{f}\circ\psi_{e/\partial\sigma^+}] = [f^\wedge\circ\psi_{e/\partial\sigma^+}] = z(s_f^\wedge)(\sigma^+)$.

Consider a homomorphism $\sum : C^{n+1}(K,L;\pi_n(V_0)) \longrightarrow \pi_n(V_0)$ $\sum(z) = \sum z(\sigma^+)$. Since $|K|$ is a compact connected and orientable manifold with boundary $|L|$, the homomorphism $\sum$ induces an isomorphism $\sum^* : H^{n+1}(K,L;\pi_n(V_0)) \longrightarrow \pi_n(V_0)$.

Let $f \in C_G(\Omega,\partial\Omega)$ and s_f , $s_f^\wedge$ denote partial cross-sections as above . We denote by c_f the cohomology class of $z(s_f^\wedge)$; $c_f \in H^{n+1}(K,L;\pi_n(V_0))$. From the obstruction theory it is known that c_f is independent of the choice of extension $s_f^\wedge$. The cohomology class c_f is called the first obstruction . It is wellknown that there exists an extension of s_f on $K^{n+1} = K$ if and only if $c_f = 0$. From the above considerations we deduce :

<u>2.4 Corollary.</u> $\sum^* c_f = \deg_H(f, \Omega)$.

<u>2.5 Corollary.</u> $\deg_H(f,\Omega) = 0$ if and only if $c_f = 0$.

<u>2.6 Theorem.</u> Let $\Omega \subset V \oplus R$ satisfy 2.1 and $f \in C_G(\Omega,\partial\Omega)$. If $\deg_H(f,\Omega) = 0$ then there exists $f_o \in C_G(\Omega,\partial\Omega)$ such that $f_o(x) = f(x)$ for $x \in \partial\Omega$ and $f_o(\bar{\Omega}) \subset V \setminus \{0\}$.

3. The group structure on $[S(W), S(V)]_G$.

Let W,V be orthogonal representations of a compact Lie group G ; S(W) , S(V) the unit spheres in W and V respectively ; $x_o \in S(W)^G$, $y_o \in S(V)^G$ fixed points . Let $[S(W),S(V)]_G$ denote the set of G-homotopy classes of G-maps $f : S(W) \longrightarrow S(V)$ and let $[S(W), x_o ; S(V), y_o]_G$ denote the set of G-homotopy classes (rel. x_o) of G-maps $f : S(W) \longrightarrow S(V)$ with $f(x_o) = y_o$.

Suppose $\dim W^G \geqslant 1$ and $\dim V^G \geqslant 2$. Let $L = \mathrm{span}\{x_o\} \subset W$ and let $W_1 = L^\perp$ be the orthogonal complement of L in W . We may identify S(W) with a non-reduced suspension $\Sigma S(W_1) = [0,1] \times S(W_1)/\sim$. Under this identification $x_o = [0,x]$, $-x_o = [1,x]$ for $x \in S(W_1)$. Let $[f_1]$, $[f_2] \in [S(W), S(V)]_G$. We can choose $f_1, f_2 : S(W) \longrightarrow S(V)$ in such a way that $f_1(-x_o) = y_o$ and $f_2(x_o) = y_o$. Define $f_3 : S(W) \longrightarrow S(V)$

$$3.1 \qquad f_3[t,x] = \begin{cases} f_1[2t,x] & \text{for } 0 \leqslant t \leqslant 1/2 \ , \ x \in S(W_1) \\ f_2[2t-1,x] & \text{for } 1/2 \leqslant t \leqslant 1 \ , \ x \in S(W_1) \end{cases}$$

Now we define a group structure on $[S(W),S(V)]_G$ by

$$[f_1] + [f_2] = [f_3]$$

The following lemma shows that the operation "+" is well defined

<u>3.2 Lemma.</u> If $f_1, f_2 : S(W) \longrightarrow S(V)$ are G-homotopic and $f_1(x_o) = f_2(x_o)$, then they are G-homotopic (rel. x_o) .

The proof of this lemma is given in [4] . The standard computations show that the operation "+" yields a group structure on $[S(W), S(V)]_G$. The G-homotopy class of the constant map $f = y_o$ is the neutral element .

Now consider the case $\dim W^G \geqslant 2$, $\dim V^G = 1$. Let $f : (S(W), x_o) \longrightarrow (S(V), y_o)$ be a G-map . Observe that $f(S(W)^G) = y_o$. Therefore we can define in the same way a group structure on $[S(W), x_o ; S(V), y_o]_G$.

The following lemma will be needed in the next section .

<u>3.3 Lemma.</u> If $\dim W^G \geqslant 2$ and $\dim V^G = 1$, then there exists a bijection

$$\Psi : Z_2 \times [S(W), x_o ; S(V), y_o]_G \longrightarrow [S(W), S(V)]_G .$$

<u>Proof.</u> Let $L = \operatorname{span}\{y_o\}$, $V_1 = L^{\perp}$ and $\psi : S(V) \longrightarrow S(V)$ be given by $\psi(\lambda, y) = (-\lambda, y)$, where $(\lambda, y) \in L \oplus V_1 = V$. Define $\Psi(0 \times [f]) = [f]$, $\Psi(1 \times [f]) = [\psi \circ f]$.

In the remainder of this section we assume that $W = V \oplus R$ and $G = S^1$. Let B denote the unit ball in $V \oplus R$. Consider an S^1-map $f : S(V \oplus R) \longrightarrow S(V)$. Let $f^{\wedge} : B \longrightarrow V$ denote an extension of f . For $f^{\wedge}$ the degree $D(f^{\wedge}, B) \in A(V)$ has been defined in Definition 1.9 . It is easily seen that $D(f^{\wedge}, B)$ is independend of the choice of the extension $f^{\wedge}$. Therefore we can define .

<u>3.4 Definition.</u> $D : [S(V \oplus R), S(V)]_G \longrightarrow A(V)$

$$D[f] = D(f^{\wedge}, B)$$

We shall denote : $d_H[f] = d_H(f^{\wedge}, B)$.

3.5 Theorem. i) If $\dim V^G \geq 2$ then

$D : [S(V\oplus R), S(V)]_G \longrightarrow A(V)$ is a group homomorphism .

ii) If $\dim V^G = 1$ then

$D : [S(V\oplus R), x_o \,;\, S(V), y_o]_G \longrightarrow A(V)$ is a group homomorphism .

Proof. i) Let $[f_1]$, $[f_2] \in [S(V+R), S(V)]_G$. We have to prove that $d_H([f_1] + [f_2]) = d_H[f_1] + d_H[f_2]$ for $H \in O(V) \cup \{S^1\}$. It is evident for $H = S^1$ (see Definition 1.9, 3.4) . Assume now that $H \in O(V)$. We identify B with $[0,1] \times B(W_1) \,/\!\sim$ where $W_1 = \mathrm{span}\{x_o\}^{\perp}$, $B(W_1)$ - the unit ball in W_1 . Let f_1, f_2, f_3 be as in 3.1 . We denote by $\hat{f}_1, \hat{f}_2 : B \longrightarrow V$ the S^1-extensions of f_1, f_2 , respectively . Define $\hat{f}_3 : B \longrightarrow V$ by

$$\hat{f}_3[t,x] = \begin{cases} \hat{f}_1[2t,x] & \text{for } 0 \leq t \leq 1/2 \ , \ x \in B(W_1) \\ \hat{f}_2[2t-1,x] & \text{for } 1/2 \leq t \leq 1 \ , \ x \in B(W_1) \end{cases}$$

Consider the sets :

$$\Omega_1 = \{[t,x] \in B \,;\, 0 < t < 1/2 \ , \ x \in \mathrm{int}\, B(W_1)\}$$

$$\Omega_2 = \{[t,x] \in B \,;\, 1/2 < t < 1 \ , \ x \in \mathrm{int}\, B(W_1)\}$$

From 1.1c, 1.9 , 3.4 we have $d_H([f_1]+[f_2]) = d_H[f_3] = d_H(\hat{f}_3, B)$ $= \deg_H(\hat{f}_3, \Omega_1) + \deg_H(\hat{f}_3, \Omega_2)$. Let us define two maps $\grave{f}_1 , \grave{f}_2 : B \longrightarrow V$

$$\grave{f}_1[t,x] = \begin{cases} \hat{f}_1[2t,x] & \text{for } 0 \leq t \leq 1/2 \ , \ x \in B(W_1) \\ y_o & \text{for } 1/2 \leq t \leq 1 \ , \ x \in B(W_1) \end{cases}$$

$$\grave{f}_2[t,x] = \begin{cases} y_o & \text{for } 0 \leq t \leq 1/2 \ , \ x \in B(W_1) \\ \hat{f}_2[2t-1,x] & \text{for } 1/2 \leq t \leq 1 \ , \ x \in B(W_1) \end{cases}$$

Observe that $\hat{f}_i , \grave{f}_i : (B, \partial B) \longrightarrow (V, V\setminus\{0\})$ are S^1-homotopic , the maps $\grave{f}_1$, $\hat{f}_3$ are equal on Ω_1 and $\grave{f}_2$, $\hat{f}_3$ are equal on Ω_2 . Therefore from 1.1 b,d it follows that $\deg_H(\hat{f}_3, \Omega_1) +$ $+ \deg_H(\hat{f}_3, \Omega_2) = \deg_H(\grave{f}_1, \Omega_1) + \deg_H(\grave{f}_2, \Omega_2) = \deg_H(\grave{f}_1, B) + \deg_H(\grave{f}_2, B)$ $= \deg_H(\hat{f}_1, B) + \deg_H(\hat{f}_2, B) = d_H[f_1] + d_H[f_2]$. This proves i) . The same proof works in the case ii) .

4. A classification of S^1-maps $S(V\oplus R) \longrightarrow S(V)$.

In this section V denotes an orthogonal representation of the group $G = S^1$. Let x_o, y_o be fixed points of $S(V\oplus R)^G$, $S(V)^G$ respectively. We now formulate our main result.

4.1 Theorem. i) If $\dim V^G \geqslant 2$ then $D : [S(V\oplus R), S(V)]_G \longrightarrow A(V)$ is a group isomorphism.

ii) If $\dim V^G = 1$ then $D : [S(V\oplus R), x_o ; S(V), y_o]_G \longrightarrow A(V)$ is a group isomorphism.

The proof of Theorem 4.1 is based on the following two lemmas.

4.2 Lemma. Let $f : (B, \partial B) \longrightarrow (V, V\setminus\{0\})$ be an S^1-map. If $D(f,B) = 0$ then there exists an S^1-map $f_o : (B,\partial B) \longrightarrow (V, V\setminus\{0\})$ such that $f(x) = f_o(x)$ for $x \in \partial B$ and $f_o(B) \subset V\setminus\{0\}$.

Proof. Choose an ordering $H_1, H_2, \ldots, H_k$ for the set $\{S^1\} \cup O(V)$ such that if $H_i \subset H_j$ then $j \leqslant i$. Let $B_i = \bigcup_{j\leqslant i} B^{H_j}$. We will define S^1-maps $f_i : (B,\partial B) \longrightarrow (V, V\setminus\{0\})$, $i=1, \ldots, k$ such that $f_i(B_i) \subset V\setminus\{0\}$ and $f_i(x) = f(x)$ for $x \in \partial B$.

Suppose first that $i=1$. Since $d_{S^1}(f,B) = [f/_{\partial B^G}] = 0$, there exists an extension $\bar{f} : B^G \longrightarrow V^G\setminus\{0\}$ of $f/_{\partial B^G}$. Define $f_1 : \partial B \cup B_1 \longrightarrow V\setminus\{0\}$ by $f_1(x) = f(x)$ for $x \in \partial B$ and $f_1(x) = \bar{f}(x)$ for $x \in B^G = B_1$. The map f_1 can be extended on B by Tietze-Gleason Theorem.

Assume the map f_i is defined. We will define f_{i+1}. Let $H = H_{i+1}$. Since $f_i(B_i) \subset V\setminus\{0\}$, we can choose an open invariant connected subset Ω of $(V\oplus R)^H$ such that $f_i^{-1}(0)_H \subset \Omega \subset \bar{\Omega} \subset B_H$ and $\bar{\Omega}$ is a smooth S^1-manifold with boundary. Denote by $g : (\Omega, \partial\Omega) \longrightarrow (V^H, V^H\setminus\{0\})$ the restriction of f_i. Theorem 1.1 b,f

implies that $\deg_H(g,\Omega) = \deg_H(f_i^H, B^H) = \deg_H(f_i, B) = d_H(f_i, B) = 0$.
From Theorem 2.6 it follows that there exists an S^1-map $g_0 : (\Omega, \partial\Omega) \longrightarrow (V^H, V^H \setminus \{0\})$ such that $g_0(x) = g(x)$ for $x \in \partial\Omega$ and $g_0(\Omega) \subset V^H \setminus \{0\}$. We define $f_{i+1} : \partial B \cup B_{i+1} \longrightarrow V \setminus \{0\}$ by $f_{i+1}(x) = f_i(x)$ for $x \in \partial B \cup (B_{i+1} \setminus \Omega)$ and $f_{i+1}(x) = g_0(x)$ for $x \in \Omega$.
The map f_{i+1} extends on B by Tietze-Gleason Theorem.

Observe that $B_k = B$ and $f_k(B) \subset V \setminus \{0\}$. We put $f_o = f_k$ and the proof is completed.

__4.3. Lemma.__ i) Let $K \in O(V)$. Then there exists a G-map $f : S(V \oplus R) \longrightarrow S(V)$ such that

$$d_H[f] = \begin{cases} 1 & \text{for } H = K \\ 0 & \text{for } H \neq K \end{cases}$$

ii) If $\dim V^G \geq 3$, then there exists a G-map $f : S(V \oplus R) \longrightarrow S(V)$ such that

$$d_H\ f = \begin{cases} 1 & \text{for } H = S^1 \\ 0 & \text{for } H \neq S^1 \end{cases}$$

__Remark.__ If $\dim V^G = 1$ or 2 then $d_{S^1}[f] = 0$ for any G-map f (see Definition 1.9).

__Proof.__ i) Choose $x_0 \in \operatorname{int} B_K$, $r > 0$ and $y_0 \in S(V)^G$ such that $|x_0| + r < 1$, $|x_0| + (3/2)\cdot r > 1$ and $D(x_0, 2r) \subset (V \oplus R)_K$ is a slice in the space $(V \oplus R)^K$ at the point x_0 ($D(x_0, 2r)$ denotes a disc in the space $(N_{x_0})^K$). Let $x_1 = x_0 + (3/2) r (x_0/|x_0|)$, $U_{2r} = G \cdot D(x_0, 2r)$. We shall define a G-map $F : V^K \oplus R \longrightarrow V^K$ such that $F(x) = y_0$ for $x \in V^K \oplus R \setminus U_{2r}$, $F^{-1}(0) = Gx_0 \cup Gx_1$ and

$$d_H(F_{/B^K}, B^K) = \begin{cases} 1 & \text{for } H = K \\ 0 & \text{for } H \neq K \end{cases}$$

It is easy to construct a map $g : D(x_0, 2r) \longrightarrow V^K$ such that $g(\partial D(x_0, 2r)) = y_0$, $g^{-1}(0) = \{x_0, x_1\}$ and $\deg(g, D(x_0, r)) = 1$

(Brouwer degree). Define $\bar{f}(z\cdot x) = y_0$ for $z\cdot x \notin U_{2r}$ and $\bar{f}(z\cdot x) = z\cdot g(x)$ for $x \in D(x_0, 2r)$, $z \in S^1$. We now extend $\bar{f}$ on $V\oplus R$ by the formula $\bar{f}(x,y) = \bar{f}(x) + y$ where $(x,y) \in (V\oplus R)^K \oplus (V^K)^{\perp} = V\oplus R$ and we define a G-map $f : S(V\oplus R) \longrightarrow S(V)$ by $f(x) = \bar{f}(x)/|\bar{f}(x)|$. From Theorem 1.1 and Definition 1.9, 3.4, it follows immediately that

$$d_H[f] = \begin{cases} 1 & \text{if } H = K \\ 0 & \text{if } H \neq K \end{cases}$$

ii) Choose a map $f^G : S(V\oplus R)^G \longrightarrow S(V)^G$ such that $[f^G] = 1 \in [S(V\oplus R)^G, S(V)^G]$. Let $\bar{f} : B^G \longrightarrow V^G$ denote an extension of f^G. We extend $\bar{f}$ on B by formula $\bar{f}(x,y) = \bar{f}(x) + y$ where $(x,y) \in (V\oplus R)^G \oplus (V^G)^{\perp}$. The G-map $f : S(V\oplus R) \longrightarrow S(V)$ is given by $f(x) = \bar{f}(x)/|\bar{f}(x)|$. This ends the proof.

<u>Proof of Theorem 4.1</u>. (Mono). Suppose that $D[f] = 0$. From Lemma 4.2, it follows that there exists a G-extension $\bar{f} : (B, \partial B) \longrightarrow (V, V\setminus\{0\})$ of f such that $\bar{f}(B) \subset V\setminus\{0\}$. We define a G-homotopy $H : S(V\oplus R)\times[0,1] \longrightarrow S(V)$ by $H(x,t) = \bar{f}(tx)/|\bar{f}(tx)|$ It is easy to check that the homotopy H joins the map f and the constant map $H(\cdot,0) = \bar{f}(0)/|\bar{f}(0)|$. Therefore we have $[f] = 0$.

(Epi). It follows immediately from Lemma 4.3

<u>Corollary 4.4</u>

$$S\ V{+}R\ ,S\ V\ \ _G = \begin{cases} Z_2 \oplus \left(\bigoplus_{H\in O(V)} Z\right) & \text{if } \dim V^G = 1 \\ \bigoplus_{H\in O(V)} Z & \text{if } \dim V^G = 2 \\ Z \oplus \left(\bigoplus_{H\in O(V)} Z\right) & \text{if } \dim V^G = 3 \\ Z_2 \oplus \left(\bigoplus_{H\in O(V)} Z\right) & \text{if } \dim V^G \geqslant 4 \end{cases}$$

In this paper we have studied $Deg(f,\Omega)$ of an S^1-maps. Nevertheless we are able to define an analogous invariant of T^n-equivariant maps, T^n - the torus. Moreover, the statement of

Theorem 4.1 extends on this case in following manner

$$[S(V\oplus R), S(V)]_{T^n} = [S(V\oplus R)^{T^n}, S(V)^{T^n}] \oplus (\bigoplus_H Z)$$

where the last sum is taken over all isotropy groups H (on S(V)) with one dimensional orbits .

Institute of Mathematics
University of Gdańsk
Wita Stwosza 57
80-952 Gdańsk

References

[1] G.E. Bredon . Introduction to Compact Transformation Group , Academic Press , New York and London 1972 .

[2] T. tom Dieck , Transformation Groups and Representation Theory, Lect. Notes in Math. 766 , Springer , Heidelberg-New York,1979.

[3] G.Dylawerski,K.Gęba,J.Jodel,W.Marzantowicz , An S^1-Equivariant Degree An The Fuller Index, Preprint No 64 , University of Gdańsk , 1987 .

[4] R.L. Rubinsztein , On the equivariant homotopy of spheres , Dissertationes Mathematicae , No 134 , Warszawa 1976 .

[5] G.B. Segal , Equivariant stable homotopy theory , Actes , Congres Inten. Math. Nice 1970 , Tome 2 , p. 59-63 .

[6] N.Steenrod , The topology of fibre bundles , Princeton University Press , 1951 .

ON CERTAIN SIEGEL MODULAR VARIETIES OF GENUS TWO AND LEVELS ABOVE TWO

Ronnie Lee* and Steven H. Weintraub**

In our previous work, we have studied spaces M_Λ^* which are moduli spaces of stable curves (i.e. Riemann surfaces) of genus 2 with level Λ structure, for two particular subgroups Λ of $PSp_4(\mathbf{Z})$.

In general, M_Λ^* is a three-dimensional complex projective variety. It is usually non-singular, though for some choices of Λ it has finite quotient singularities. (In Satake's language, it is then a V-manifold). It is the Igusa compactification of the variety $M_\Lambda = \mathbf{S}_2/\Lambda$, the quotient of $\mathbf{S}_2$, the Siegel space of degree 2, under the action of Λ. Further, $M_\Lambda^* \supset M_\Lambda \supset M_\Lambda^0$, where M_Λ^0 (a Zariski open set in M_Λ^*) is the moduli space of non-singular curves of genus 2 with level Λ structure.

We follow our previous notation and write the complement $M_\Lambda^* - M_\Lambda^0 = \partial_\Lambda \cup \Theta_\Lambda$, where ∂_Λ and Θ_Λ are unions of components (each a complex surface) and are themselves moduli spaces for the two kinds of singular but stable curves of genus 2 (see [LW_1], section 8.4).

In our papers [LW_1], [LW_2] we considered the case $\Lambda = \Gamma(2)$, the principal congruence subgroup of level 2. (In this case a level Λ structure is more commonly known as a level 2 structure.)

* Partially supported by the National Science Foundation.

** Partially supported by the National Science Foundation and the Sonderforschungsbereich fur Geometrie und Analysis (SFB 170).

In [LW_3] we considered the case $\Lambda = \Gamma$, where Γ is a certain subgroup of $\Gamma(2)$. We define Γ precisely in (1.1) below. Here we just observe that $\Gamma(2) \supset \Gamma \supset \Gamma(4)$, and $[\Gamma(2): \Gamma] = 2^6$, $[\Gamma(2): \Gamma(4)] = 2^9$. The quotient $\Gamma(2)/\Gamma(4)$ is an elementary abelian 2-group, and hence so is $\Gamma(2)/\Gamma$.

We wish to investigate the topology of these spaces M_Λ^* for various Λ.

In [LW_1], [LW_2] we proved the following theorem in case $\Lambda = \Gamma(2)$ (see also [G]).

Theorem 0.1. a) $H_i(M_\Lambda^*) = 0$ for i odd.

b) The map $H_i(\partial_\Lambda \cup \Theta_\Lambda) \longrightarrow H_i(M_\Lambda^*)$ is an epimorphism for $i < 6$.

c) $H_4(M_\Lambda^*)$ has a basis consisting of algebraic cycles (so, in particular, in the Hodge decomposition of $H^*(M_\Lambda^*)$, $H^{p,q} = 0$ unless $p = q$).

d) The integral homology of M_Λ^* is torsion-free.

In [LW_3] we proved the same theorem in case $\Lambda = \Gamma$.

In addition to this qualitative ("soft") information, in the above-mentioned papers we have the following quantitative ("hard") information.

Theorem 0.2. a) In case $\Lambda = \Gamma(2)$, rank $H_4(M_\Lambda^*) = 16$.

b) In case $\Lambda = \Gamma$, rank $H_4(M_\Lambda^*) = 79$.

(Of course, by Poincare duality, rank H_4 = rank H_2).

Our main result in this paper is the determination of the homology of M_Λ^* (at least up to 2-torsion) for all $\Gamma \subset \Lambda \subset \Gamma(2)$.

The line of argument is given to us by the following theorem:

Theorem 0.3. a) - d) The conclusions of theorem 0.1 a) - d) are valid for any $\Gamma \subset \Lambda \subset \Gamma(2)$, except that homology must be taken with coefficients in $\mathbb{Z}[\frac{1}{2}]$, rather than in $\mathbb{Z}$.

e) For any such Λ there is an exact sequence (with (co)homology having coefficients in $\mathbb{Z}[\frac{1}{2}]$)

$$0 \to H^1(M^0_\Lambda) \to H_4(\partial_\Lambda \cup \Theta_\Lambda) \to H_4(M^*_\Lambda) \to 0.$$

Proof: Recall the following general fact: If a finite group G acts on a space M, and if $\mathbb{F}$ is any field of characteristic 0 or prime to the order of G, then (see [B, theorem III.2.4]):

$$H_*(M/G: \mathbb{F}) = H_*(M: \mathbb{F})^G$$

where $(\ )^G$ denotes the elements fixed under the action of G.

Here we have the 2-group Λ/Γ acting on M^*_Γ with quotient M^*_Λ. Let us for the moment take coefficients in $\mathbb{Q}$. Then a) holds immediately, as does b), since the quotient $(\partial_\Gamma \cup \Theta_\Gamma)/\Lambda$ is $\partial_\Lambda \cup \Theta_\Lambda$. Furthermore, c) holds as well, as the required basis of algebraic cycles for $H_4(M^*_\Lambda)$ will be images in M^*_Λ of those elements of the basis in M^*_Γ whose fundamental classes are fixed under the action of Λ/Γ. As the reader will see, if in the arguments in this paper we replace $\mathbb{Q}$ by any field $\mathbb{F}$ of characteristic not equal to two, we obtain the same dimensions for all spaces, so in particular

$$\dim H_*(M^*_\Lambda: \mathbb{F}) = \dim H_*(M^*_\Lambda: \mathbb{Q})$$

and so the homology of M^*_Λ has no odd torsion and d) holds.

Of course, it is the proof of part e) that requires work. Consider the exact sequence of the pair $(M^*_\Gamma, \partial_\Gamma \cup \Theta_\Gamma)$, again with coefficients in $\mathbb{Q}$ (or $\mathbb{F}$, char $\mathbb{F} \neq 2$)

$$H_5(\partial_\Gamma \cup \Theta_\Gamma) \to H_5(M^*_\Gamma, \partial_\Gamma \cup \Theta_\Gamma) \to H_4(\partial_\Gamma \cup \Theta_\Gamma) \to H_4(M^*_\Gamma)$$

The first of these groups is zero, as $\partial_\Gamma \cup \Theta_\Gamma$ is a union of complex surfaces, and the second is isomorphic to $H^1(M^*_\Gamma - (\partial_\Gamma \cup \Theta_\Gamma))$ $= H^1(M^0_\Gamma)$ by Alexander duality. Thus we have

$$0 \to H^1(M^0_\Gamma) \to H_4(\partial_\Gamma \cup \Theta_\Gamma) \to H_4(M^*_\Gamma)$$

In fact, as we showed in [LW3], this last map is an epimorphism (and indeed, this is part of 0.1 b)). This follows from the computation of $\dim H_4(M^*_\Gamma) = 79$, $\dim H_1(M^0_\Gamma) = 27$, and $\dim H_4(\partial_\Gamma \cup \Theta_\Gamma)$ $= 106$ there. We shall see below how to make these last two computations (cf. 1.16 and 3.1, and 1.12, 2.1 and 2.2).

Now let Λ/Γ act on this short exact sequence. We again obtain a short exact sequence

$$0 \to H^1(M^0_\Gamma)^{\Lambda/\Gamma} \to H_4(\partial_\Gamma \cup \Theta_\Gamma)^{\Lambda/\Gamma} \to H_4(M^*_\Gamma)^{\Lambda/\Gamma} \to 0$$

which is nothing other than the sequence

$$0 \to H^1(M^0_\Lambda) \to H_4(\partial_\Lambda \cup \Theta_\Lambda) \to H_4(M^*_\Lambda) \to 0$$

as desired.

Corollary 0.4. $\dim H_4(M^*_\Lambda) = \dim H_4(\partial_\Lambda \cup \Theta_\Lambda) - \dim H^1(M^0_\Lambda)$, where (co)homology is taken in an arbitrary field $\mathbb{F}$ of characteristic not equal to two (and these numbers are independent of the choice of $\mathbb{F}$.) (Note that by Poincare duality this is also $\dim H_2(M^*_\Lambda)$ as M^*_Λ is an $\mathbb{F}$-homology manifold.)

This corollary tells us how to compute $H_4(M^*_\Lambda)$ - compute the two terms on the right-hand side and subtract. In principle, this is the approach we follow.

In practice, as there are very many subgroups Λ, we choose a somewhat different line of attack. Namely, what we actually determine is the action of $G = \Gamma(2)/\Gamma$ on the space $\mathbf{R} = H_4(M^*_\Gamma)$.

Knowing $\mathbb{R}$ as a representation space of G of course tells us $\dim H_4(M_\Gamma^*)^{\Lambda/\Gamma} = \dim H_4(M_\Lambda^*)$ for any $\Gamma \subset \Lambda \subset \Gamma(2)$. (It turns out as well that the final result is much easier to state in terms of $\mathbb{R}$.) How do we compute $\mathbb{R}$ as a representation of G? It turns out that we may do so by computing $\dim H_4(M_\Lambda^*)$ for a relatively small number of subgroups Λ (of a kind we call "un-twisted").

Thus our paper is arranged as follows: In section 1 we establish our notation, recall some basic results of [LW_3], and establish precisely what we what we need to compute. In section 2 and 3 we compute the two terms on the right-hand side of 0.4 for certain Λ. In section 4 we assemble this information to obtain our main result, theorem 4.2, which gives the action of G on $\mathbb{R}$. As a specific application we then give $\dim H_4(M_\Lambda^*)$ for the extreme cases Γ of index 2 in Λ, and Λ of index 2 in $\Gamma(2)$.

All (co)homology henceforth is to be understood as having coefficients in $\mathbb{Q}$. As we have remarked, this gives us information complete except for 2-torsion. We discuss this question in 4.7. In many cases we can show that the (co)homology of M_Λ^* is torsion-free.

As this manuscript is being photo-offset from typescript, the reader will be able to appreciate the marvelous typing job done on it by Nell Castleberry, to whom the author extends his deepest thanks.

1. The situation at level Γ.

We begin by establishing notation and recalling some of the results of [LW3].

Definition 1.1. Let $\Gamma(n) = \{M \in PSp_4(\mathbb{Z}) \mid M \equiv I \bmod n\}$.

$$\text{Let } \Gamma = \{M \in PSp_4(\mathbb{Z}) \mid M \equiv I + 2 \begin{matrix} 0 & & \\ a & & b \\ & 0 & \\ c & & -a \end{matrix} \bmod 4\}.$$

(Observe that $\Gamma(4) \subset \Gamma \subset \Gamma(2)$, and $[\Gamma: \Gamma(4)] = 2^3$, $[\Gamma(2): \Gamma] = 2^6$.)

Definition 1.2. Let $\mathbf{S}_2 = \{M \in M_2(\mathbb{C}) \mid M = {}^tM$ and $\mathrm{Im}(M)$ is positive definite $\}$.

$$\text{Let } \Theta^0 \subset \mathbf{S}_2 = \{\left(\begin{smallmatrix} z & 0 \\ 0 & w \end{smallmatrix}\right)\} \mid \mathrm{Im}(z) > 0, \mathrm{Im}(w) > 0\}$$

together with the union of its translates under the action of $PSp_4(\mathbb{Z})$. We call Θ^0 the Humbert surface in $\mathbf{S}_2$ and set $\mathbf{S}_2^0 = \mathbf{S}_2 - \Theta^0$.

Definition 1.3. For any subgroup Λ of $PSp_4(\mathbb{Z})$, set

$$M_\Lambda = S_2/\Lambda, \quad M_\Lambda^0 = S_2^0/\Lambda, \quad \Theta_\Lambda^0 = \Theta^0/\Lambda.$$

For any $\Lambda \subset \Gamma(2)$, Λ acts freely on $\mathbf{S}_2^0$. Also, M_2^0 is the moduli space of non-singular Riemann surfaces R of genus 2 with a level Λ structure (i.e. a choice of symplectic basis for $H_1(R: \mathbb{Z})$ modulo the action of Λ.)

Definition 1.4. Let M_Λ^* be the Igusa compactification of M_Λ. Set $\partial_\Lambda = M_\Lambda^* - M_\Lambda$, and let Θ_Λ be the closure of Θ_Λ^0 in M_Λ^*.

The space M_Λ^* is the moduli space of stable (in the sense of Mumford [M]) Riemann surfaces of genus 2 with level Λ structure. We call Θ_Λ the Humbert surface in M_Λ^*, and, by abuse of language, ∂_Λ the boundary of M_Λ^* (even though, as a projective variety, this space has no boundary in the topological sense). Of course, both the boundary and the Humbert surface are the union of many irreducible components.

We now recall our results on the structure of M_Γ^*. We begin with M_Γ^0.

First we exhibit a 4-fold cover $f: \mathbb{P}^1 \to \mathbb{P}^1$, branched over 3 points, each of whose inverse images has cardinality two.

Lemma 1.5. Let $X = \{\mathrm{Im}\ z > 0\}/\Gamma_1(4)$ and $X_0 = \{\mathrm{Im}\ z > 0\}/\Gamma_1(2)$, where $\Gamma_1(n)$ is the principal congruence subgroup of level n of $PSL_2(\mathbb{Z})$. Then X is a 4-fold cover of X_0 with group $\Gamma_1(2)/\Gamma_1(4) = (\mathbb{Z}/2) + (\mathbb{Z}/2)$. Also, we may identify X with $\mathbb{P}^1 - \{\pm i, 0, \infty, \pm 1\}$ and X_0 with $\mathbb{P}^1 - \{0, 1, \infty\}$, and the covering projection $f: X \to X_0$ with the function

$$z = f(w) = ((w^2 + 1)/(w^2-1))^2.$$

Furthermore, f extends to a branched cover of $\bar{X} = \mathbb{P}^1$ to $\bar{X}_0 = \mathbb{P}^1$ with $f(\pm i) = 0$, $f(0) = f(\infty) = 1$, and $f(\pm 1) = \infty$.

Proof. [LW3], 2.1 and 2.2.

The group of the cover $(\mathbb{Z}/2) + (\mathbb{Z}/2)$ is generated by the loops around 0 and 1 in the base, each of which has order 2. We denote them by p and 2 respectively. Of course, the loop around ∞ is then $r = pq$, the other non-trivial element of this group.

Notation 1.6. Let V denote the Klein 4-group (isomorphic to $(\mathbb{Z}/2) + (\mathbb{Z}/2)$) with elements $\{1, p, q, r\}$ which appears in 1.5.

Theorem 1.7. Let $Z_0 = \{(x_1, x_2, x_3) \in X_0 \mid x_i$ not all distinct$\}$

and $Z = \{(x_1, x_2, x_3) \in X \mid (f(x_1), f(x_2), f(x_3)) \in Z_0\}$.

Then $M^0_{\Gamma(2)} = X_0 \times X_0 \times X_0 - Z_0$, and $M^0_\Gamma = X \times X \times X - Z$,

where the map $M^0_\Gamma \to M^0_{\Gamma(2)}$ is covering projection $f \times f \times f$.

Thus M^0_Γ is a 2^6-fold cover of $M^0_{\Gamma(2)}$ with group $G = V_1 \times V_2 \times V_3$, with each V_i naturally isomorphic to V.

Proof. This is theorem 3.9 of [LW3].

We let V_i have the non-trivial elements $p_i, q_i, r_i = p_i q_i$, where the isomorphism with V is the obvious one suggested by the notation. We shall frequently identify V_i with V via the isomorphism. Of course G is generated by $\{p_i, q_i\}$, $i = 1,2,3$.

In order to do our computations, we need specific matrices representing the generators (and hence elements) of G. These we also obtained in the proof of this theorem in [LW3], and we quote the result here. (In [LW3] we denoted p_1, say, by σ_{p_1}, as it is given by the effect of a "Dehn twist" around a lift of a loop representing p_i, but for simplicity of notation we drop the σ here.)

Lemma 1.8.

$$p_1 = I + 2\begin{pmatrix} 0&1&1&0\\ 0&0&0&0\\ 0&0&0&0\\ 0&1&1&0 \end{pmatrix} \qquad q_1 = I + 2\begin{pmatrix} 1&1&1&0\\ 0&0&0&0\\ 1&1&1&0\\ 1&1&1&0 \end{pmatrix}$$

$$p_2 = I + 2\begin{pmatrix} 0&1&1&1\\ 0&1&1&1\\ 0&0&0&0\\ 0&1&1&1 \end{pmatrix} \qquad q_2 = I + 2\begin{pmatrix} 0&1&1&1\\ 1&0&1&1\\ 1&1&0&1\\ 1&1&1&0 \end{pmatrix}$$

$$p_3 = I + 2\begin{pmatrix} 0&0&1&1\\ 0&0&1&1\\ 0&0&0&0\\ 0&0&0&0 \end{pmatrix} \qquad q_3 = I + 2\begin{pmatrix} 1&0&1&1\\ 1&0&1&1\\ 1&0&1&1\\ 0&0&0&0 \end{pmatrix}$$

We now make a useful observation, which will reduce the number of cases which we have to consider. Recall that the automorphism group of V is Σ_3, which acts by permuting the non-trivial elements.

Lemma 1.9. Let $\Sigma_3 \times \Sigma_3$ operate on $G = V_1 \times V_2 \times V_3$ as follows: An element of G is a word in $p_1, \ldots, r_3$. The first factor operates on such a word by permuting the symbols p, q, and r (as above). The second factor acts by permuting the subscripts 1, 2, and 3.

If Λ_1 and Λ_2 are two subgroups of G equivalent under this action, then $M^*_{\Lambda_1}$ and $M^*_{\Lambda_2}$ are equivalent as complex algebraic varieties.

Proof. The action of Σ_3 on V may be realized by the unique automorphism of $\bar{X}_0 = \mathbb{P}^1$ permuting 0, 1, and ∞ as specified, giving an action of the first factor on $X_0 \times X_0 \times X_0$, and the second factor acts on this product by permuting the factors. These are clearly automorphisms of M^0_Γ, which extend to M^*_Γ, as their effect on $\partial_\Gamma \cup \Theta_\Gamma$ is to permute components. This automorphism of M^*_Γ then descends to an equivalence between $M^*_{\Lambda_1}$ and $M^*_{\Lambda_2}$.

We shall denote this $\Sigma_3 \times \Sigma_3$ subgroup of the automorphism group of G by $\mathbb{A}(G)$.

Now we consider $M^*_\Lambda - M^0_\Lambda = \partial_\Lambda \cup \Theta_\Lambda$. Each of ∂_Λ and Θ_Λ is a union of irreducible components which are complex surfaces. ∂_Λ is a union of complex surfaces $D_\Lambda(\ell)$, where the indexing set is

$$\{\pm\ell \mid \ell \text{ a non-zero primitive vector in } \mathbb{Z}^4\}/\text{action of } \Lambda \qquad \textbf{(1.10)}$$

where Λ acts on ℓ by ordinary matrix multiplication, $(\ell)\lambda = \ell\lambda$. (In case $\Lambda = \Gamma(n)$, the principal congruence subgroup of level n, this set is just the set of non-zero primitive vectors in $(\mathbb{Z}/n)^4$, taken up to sign.)

Similarly, Θ_Λ is a union of complex surfaces $H_\Lambda(\Delta)$ where the indexing set is

$$\{\Delta = \{\delta, \delta^\perp\} \mid \Delta = \pm\, \ell_1 \wedge \ell_2,\ \delta^\perp = \pm\, \ell_1' \wedge \ell_2',\ \text{with } \delta \text{ and } \delta^\perp \text{ mutually orthogonal \underline{anisotropic} subspaces of } \mathbb{Z}^4,\ \delta + \delta^\perp = \mathbb{Z}^4\}/\text{action of } \Lambda \,. \qquad \textbf{(1.11)}$$

We refer to such a Δ as a anisotropic pair. Again Λ acts by matrix multiplication, and in case $\Lambda = \Gamma(n)$ this is the set of pairs of such subspaces in $(\mathbb{Z}/n)^4$.

We call the components $D(\ell)$ of ∂_Λ boundary components and the components $H(\Delta)$ Humbert surfaces. The justification for all these remarks can be found in [LW_2], for $\Lambda = \Gamma(2)$, and in [LW_3], for $\Lambda = \Gamma$, but the set-up holds generally. The set in (1.5) indexes one type of vertex in the Tits building for Λ. We do not need to consider the Tits building here, (or the "Tits building with scaffolding" of [LW3]) for it contains the further information on how the various components of ∂_Λ (or $\partial_\Lambda \cup \Theta_\Lambda$) intersect, and that is superfluous here.

Lemma 1.12. $H_4(\partial_\Lambda \cup \Theta_\Lambda)$ is the free abelian group on the generators $\{[D(\ell)]\}$ of (1.5) and $\{[H(\Delta)]\}$ of (1.6), where [ ] denotes the fundamental homology class.

Proof. As we are dealing with complex surfaces, which have a canonical orientation, [ ] is well defined. Since the intersection of two irreducible components is a complex subvariety (perhaps singular or empty) it has real codimension at least two so the lemma is immediate from the Mayer-Vietoris sequence.

Lemma 1.13. As a representation space of G, $H_4(\partial_\Gamma \cup \Theta_\Gamma)$ is isomorphic to $\underset{\ell}{+}\mathbb{Z}[\ell] + \underset{\Delta}{+}\mathbb{Z}[\Delta]$, where $\gamma \in G$ acts on the latter by $\ell \longrightarrow \ell\gamma$, $\Delta \longrightarrow \Delta\gamma$.

Proof. The action of γ takes $D(\ell)$ to $D(\ell\gamma)$ and $H(\Delta)$ to $H(\Delta\gamma)$, so it induces a map γ_* on $H_4(\partial_\Gamma \cup \Theta_\Gamma)$ by $\gamma_*([D(\ell)]) = \pm[D(\ell\gamma)]$, $\gamma_*([H(\Delta)]) = \pm[H(\Delta\gamma)]$. Since the action of γ is a complex automorphism, it preserves the canonical orientations, so both signs are +, and the lemma follows.

We rephrase theorem 1.7 in a way that will be slightly more useful.

Definition 1.14. Let $\bar{Z}_0 = \{(x_1,x_2,x_3) \in \bar{X}_0 \times \bar{X}_0 \times \bar{X}_0 | (x_1,x_2,x_3) \in Z_0$ or $x_i = 0, 1,$ or ∞ for some $i\}$.

Let $\bar{Z} = \{(x_1,x_2,x_3) \in \bar{X} \times \bar{X} \times \bar{X} | (f(x_1),f(x_2),f(x_3)) \in \bar{Z}_0\}$.

Theorem 1.15. $M_\Gamma^0 = \bar{X} \times \bar{X} \times \bar{X} - \bar{Z}$.

Proof. This is immediate - compared with the description of M_Γ^0 in 1.7, we are adding and then subtracting all points (x_1,x_2,x_3) with $(f(x_1),f(x_2),f(x_3))$ not in X_0.

This latter description has the advantage that $\bar{X} \times \bar{X} \times \bar{X}$ is a compact manifold (in fact $\mathbb{P}^1 \times \mathbb{P}^1 \times \mathbb{P}^1$) and $\bar{Z}$ is a union (not disjoint) of irreducible components, each of which is compact. Typical components are $\{(x,y,0)\}$ or $\{(x,x,y)\}$, so each component of $\bar{Z}$ is $\mathbb{P}^1 \times \mathbb{P}^1$, and different components intersect in a $\mathbb{P}^1$ or $\mathbb{P}^0$, or not at all.

The group $G = \Gamma(2)/\Gamma$ has an obvious action (by covering translations on each factor $\bar{X}$) extending its previously defined action on M^0_Γ. For any $\Gamma(2) \supset \Lambda \supset \Gamma$, we will let $\bar{Z}_\Lambda$ denote the quotient of $\bar{Z}_\Gamma = \bar{Z}$ under the action of the subgroup Λ/Γ of G.

Proposition 1.16. For any subgroup Λ of $\Gamma(2)$, $H^1(M^0_\Lambda)$ is a free abelian group of rank 3 less than the number of irreducible components of $\bar{Z}_\Lambda$.

Proof. By Alexander duality, $H^1(M^0) = H_5(\bar{X} \times \bar{X} \times \bar{X}, \bar{Z})$. We then have the exact sequence of the pair:

$$H_5(\bar{X} \times \bar{X} \times \bar{X}) \longrightarrow H_5(\bar{X} \times \bar{X} \times \bar{X}, \bar{Z}) \longrightarrow H_4(\bar{Z}) \longrightarrow H_4(\bar{X} \times \bar{X} \times \bar{X})$$

The first group above is obviously zero; the last map is obviously onto a free abelian group of rank 3, and, by the same argument as in 1.12, $H_4(\bar{Z})$ has rank equal to the number of components of $\bar{Z}$, so the proposition follows for $\Lambda = \Gamma$.

Now consider a group Λ with Λ/Γ non-trivial. Certainly H^1 is free abelian, so we need only compute its rank. Thus let us take homology with coefficients in $\mathbb{Q}$.

Then we have

$$0 \longrightarrow H_5(\bar{X} \times \bar{X} \times \bar{X}, \bar{Z})^{\Lambda/\Gamma} \longrightarrow H_4(\bar{Z})^{\Lambda/\Gamma} \longrightarrow H_4(\bar{X} \times \bar{X} \times \bar{X})^{\Lambda/\Gamma} \longrightarrow 0$$

Now, by the argument of 0.3, we may identify the first two terms with $H^1(M^0_\Lambda)$ and $H_4(\bar{Z}_\Lambda)$ respectively. Furthermore, Λ/Γ acts trivially on the homology of $\bar{X} \times \bar{X} \times \bar{X}$ (as each generator $p_1,\dots,q_3$ of G does) so the last term has rank three and the proposition follows.

We single out a special class of subgroups of G.

Definition 1.17. A subgroup of G is called untwisted if it has a set of generators which are a subset of $\{p_1,p_2,p_3,q_1,q_2,q_3,r_1,r_2,r_3\}$.

We call $\Gamma \subset \Lambda \subset \Gamma(2)$ untwisted if $\Lambda/\Gamma \subset G$ is.

Such subgroups are naturally distinguished. Also, we will see that in order to determine the action of G on the homology of M_Γ^*, it suffices to consider relatively few subgroups, all of which are untwisted.

2. The action on lines and anisotropic planes.

By 1.13, the action of Λ on the fundamental classes of the components of the boundary (resp. the Humbert surface) of M_Γ^* is given by the action of Λ/Γ on lines ℓ (resp. anisotropic pairs Δ). In this section we determine these actions for certain (enough) subgroups Λ.

These are long computations, so rather than working them out in full we indicate how to do them in one or two illustrative cases.

The boundary components (resp. Humbert components) at level Λ are in 1-1 correspondence with the Λ-equivalence classes of lines (resp. anistropic pairs), so it is this number we need to compute. Of course, for $\Lambda = \Gamma$ or $\Gamma(2)$ we already know the answer; it is for the intermediate levels that work must be done.

Proposition 2.1. For the following untwisted groups Λ, the number of equivalence classes of lines at level Λ is as stated:

$[\Lambda:\Gamma]$	Generators of Λ/Γ	Number of equivalence classes
1	-	54
2	p_1	40
4	p_1, p_2	32
4	p_1, q_1	33
4	p_1, q_2	29
8	p_1, p_2, p_3	28
8	p_1, p_2, q_1	25
8	p_1, p_2, q_3	23
64	$p_1, p_2, p_3, q_1, q_2, q_3$	15

Proof. To determine the action of Λ/Γ on lines at level Γ we of course must know the latter. They are given by [LW_3], theorem 3.6 (and there are 54 of them). Recall they arise as follows:

There are 15 lines at level 2, given by (a_1,a_2,b_1,b_2) where a_i and b_i are defined mod 2 and not all are zero. Each line at level 2 is covered by 8 lines at level 4.

For example, $(1,0,0,0)$ is covered by

$$\{(1,0,0,0),\ (1,0,0,2),\ (1,0,2,0),\ (1,0,2,2),\ (1,2,0,0),\ (1,2,0,2),\ (1,2,2,0),\ (1,2,2,2)\} \mod 4,$$

and $(0,1,0,0)$ is covered by

$$\{(0,1,0,0),\ (0,1,0,2),\ (0,1,2,0),\ (0,1,2,2),\ (2,1,0,0),\ (2,1,0,2),\ (2,1,2,0),\ (2,1,2,2)\} \mod 4,$$

and $(1,1,0,0)$ is covered by

$$\{(1,1,0,0),\ (1,1,0,2),\ (1,-1,0,0,),\ (1,-1,0,2),\ (1,1,2,0),\ (1,1,2,2,),\ (1,-1,2,0),\ (1,-1,2,2)\} \mod 4.$$

(Note we have resolved the ambiguity in sign by choosing some entry to be $+1 \mod 4$.)

Now on lines over $(1,0,0,0)$, $\Gamma/\Gamma(4)$ acts trivially, so there are 8 equivalence classes of such lines at level Γ. On lines over $(0,1,0,0)$, $\Gamma/\Gamma(4)$ acts by interchanging the first and second, third and fourth, fifth and sixth, and seventh and eighth, so there are 4 equivalence classes of such lines at level Γ. On lines over $(1,1,0,0)$, $\Gamma/\Gamma(4)$ acts transitively on the first four, and transitively on the last four, so there are 2 equivalence classes of such lines at level Γ.

Now let us consider the action of a typical element p_1 of G. On the equivalence classes of lines over $(1,0,0,0)$ it acts as follows: Interchanging $(1,0,0,0)$ and $(1,2,2,0)$, $(1,0,0,2)$ and $(1,2,2,2,)$, $(1,0,2,0)$ and $(1,2,0,0)$, $(1,0,2,2,)$ and $(1,2,0,2)$. Thus p_1 has 4 orbits on these. (On the other hand, as the reader may check, there are also 8 lines at level Γ covering the line $(0,0,1,0)$ at level 2 and p_1 acts trivially on these, so there are 8 orbits of p_1 there.)

On the four equivalence classes of lines over $(0,1,0,0)$ p_1 acts trivially , giving 4 orbits. Also, p_1 acts by interchanging the two equivalence classes of lines over $(1,1,0,0)$, giving 1 orbit here. Then adding the number of orbits over each of the 15 lines at level 2 gives 40 orbits for Λ with Λ/Γ generated by p_1, giving the second line of the table.

From this point on the computation is routine.

Proposition 2.2. For the following untwisted groups Λ, the number of equivalence classes of anisotropic pairs at level Λ is as stated:

$[\Lambda:\Gamma]$	Generators of Λ/Γ	Number of equivalence classes
1	-	52
2	p_1	36
4	p_1, p_2	26
4	p_1, q_1	28
4	p_1, q_2	25
8	p_1, p_2, p_3	22
8	p_1, p_2, q_1	20
8	p_1, p_2, q_3	18
64	$p_1, p_2, p_3, q_1, q_2, q_3$	10

Proof. This is entirely analogous to the proof of the preceding proposition (only slightly more complicated as there are more choices of representatives). By theorem 3.7 of [LW_3] we know all the anisotropic pairs of level Γ. (Each of the 10 anistropic pairs at level 2 is covered by 16 at level 4, and by either 4 or 16 at level Γ. There are a total of 52 of these at level Γ.) For example, the pair $\Delta = \{\delta, \delta^{\perp}\}$ at level 2, with $\delta = (1,1,0,0)\wedge(0,0,1,0)$, is covered by the four equivalence classes of pairs at level Γ whose representatives we may take to be δ and its orthogonal complement $\delta^{\perp}$, with

$$\delta = (1,1,0,0)\wedge(0,0,1,0),\ (1,1,0,0)\wedge(0,0,1,2),$$
$$(1,1,0,0)\wedge(0,2,1,0),\ (1,1,0,0)\wedge(0,2,1,2) \mod 4.$$

Then p_1 acting on the first one of these planes sends it to $(1,-1,2,0)\wedge(0,0,1,0)$. But adding twice the second vector to the first, we see that this is the same plane as $(1,-1,0,0)\wedge(0,0,1,0)$. Also, in the proof of 2.1, we observed that the line $(1,-1,0,0)$ is equivalent to $(1,1,0,0)$, so we conclude that this plane is equivalent to $(1,1,0,0)\wedge(0,0,1,0)$, i.e. p_1 acts trivially on $(1,1,0,0)\wedge(0,0,1,0)$.

Otherwise the computation is straightforward.

3. The action on excised components.

We see from 1.16 that we must count the number of components of $\bar{Z}_\Lambda$, i.e. the number of orbits of components of $\bar{Z} = \bar{Z}_\Gamma$ under the action of Λ/Γ.

Proposition 3.1. For each untwisted subgroup Λ of $\Gamma(2)$, the number of irreducible components of Λ_Δ is as follows:

$[\Lambda : \Gamma]$	Generators of Λ/Γ	Number of components
1	-	30
2	p_1	24
4	p_1, p_2	20
4	p_1, q_1	21
4	p_1, q_2	19
8	p_1, p_2, p_3	18
8	p_1, p_2, q_1	17
8	p_1, p_2, q_3	16
64	$p_1, p_2, p_3, q_1, q_2, q_3$	12

Proof. Again there are many cases and we shall merely indicate a few.

Recall $f: \bar{X} \to \bar{X}_0$ is a branched covering with group $(\mathbb{Z}/2) + (\mathbb{Z}/2)$ generated by p and q, with each of $f^{-1}(0)$, $f^{-1}(1)$, and $f^{-1}(\infty)$ having cardinality two. The following schematic represents this cover:

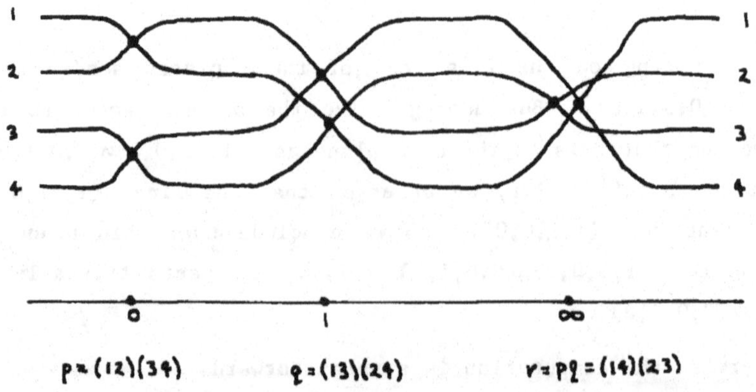

Let $\bar{X}_p$ (resp. $\bar{X}_q$) denote the quotient of $\bar{X}$ by the action of p (resp. of q). Then $\bar{X}_p$ (resp. $\bar{X}_q$) is a 2-fold branched cover of X_0, branched over 1 and ∞ (resp. 0 and ∞). We continue to let f denote the covering projection. Then $f^{-1}(0)$, $f^{-1}(1)$, $f^{-1}(\infty)$ have cardinality 2, 1, 1 in X_p (resp. cardinality 1, 2, 1 in X_2).

An irreducible component of $\bar{\Delta}_\Lambda$ projects onto one of the following types of components in $\bar{\Delta}_0$: $(*,x,y)$, $(x,*,y)$, $(x,y,*)$, (x,x,y), (x,y,x), or (y,x,x), where $* = 0$, 1, or ∞ and x and y are arbitrary. Thus we have six kinds of components, and we will gather the number of each kind into a 6-tuple, whose sum is the number of components of $\bar{\Delta}_\Lambda$. For example, when $\Lambda = \Gamma(2)$, $\bar{\Delta}_\Lambda = \bar{\Delta}_0$ has the 6-tuple (3,3,3,1,1,1,) and so has 12 components.

a) The case $\Lambda = \Gamma$. The 6-tuple is (6,6,6,4,4,4,), as follows: Here the covering space is $\bar{X} \times \bar{X} \times \bar{X}$. $F^{-1}(*,x,y)$ has 6 components as $f^{-1}(0) \cup f^{-1}(1) \cup f^{-1}(\infty)$ has cardinality 6, giving the first entry, and the second and third are identical. The fourth entry is the number of components of X , the inverse of the diagonal in $\bar{X} \times \bar{X}$. But this inverse consists of $\{(x_1,x_2) \mid x_1$ and x_2 differ by a covering translation$\}$, and so has 4 components (as the group of covering translations has 4 elements), and the fifth and sixth entries are identical.

b) The case Λ/Γ generated by p_1. Now the 6-tuple is (4,6,6,2,2,4). Here the covering space is $\bar{X}_p \times \bar{X} \times \bar{X}$. $F^{-1}(*,x,y)$ has 4 components as $f^{-1}(0) \cup f^{-1} \cup f^{-1}(\infty)$ has cardinality 4 in $\bar{X}_p$, giving the first entry. The second and third are as in a). The fourth and fifth entries are the number of components of the inverse image of the diagonal in $\bar{X}_p \times \bar{X}$. But this inverse image is the quotient of X (as in a)) by the group generated by $p \times id: \bar{X} \times \bar{X} \longrightarrow \bar{X} \times \bar{X}$, and this quotient has two components. The sixth entry is as in a).

c) The case Λ/Γ generated by p_1 and p_2. The 6-tuple is (4,4,6,2,2,2) and the covering space is $\bar{X}_p \times \bar{X}_p \times \bar{X}$. In particular note that the fourth entry is 2 by the same argument as in a).

d) The case Λ/Γ generated by p_1 and q_1. Now the 6-tuple is $(3,6,6,1,1,4)$. Here the covering space is $\bar{X}_0 \times \bar{X} \times \bar{X}$, and the argument is similar to b).

e) The case Λ/Γ generated by p_1 and q_2. The 6-tuple is $(4,4,6,1,2,2)$ and the covering space is $\bar{X}_p \times \bar{X}_q \times \bar{X}$. The only (subtle) difference between this and case c) is the following: The inverse image of the diagonal of $\bar{X}_0 \times \bar{X}_0$ in $\bar{X} \times \bar{X}$ has 4 components. Under the action of the group generated by p_1 and p_2 they are identified to <u>two</u> components in $\bar{X}_p \times \bar{X}_p$ (i.e. this group acts on the 4 components with p_1 and p_2 each acting non-trivially but giving the same identification), while here, under the action of the group generated by p_1 and q_2 they are identified to <u>one</u> component in $\bar{X}_p \times \bar{X}_q$ (i.e. this group acts on the 4 components with p_1 and p_2 each acting non-trivially but giving different identifications).

The remaining cases are similar.

4. The representation of G on the homology of M_Γ^*.

In this section we obtain our main result. We use the calculations of section 2 and 3, which give $\dim H_4(M_\Lambda^*)$ for some Λ, to decompose $H_4(M_\Gamma^*)$ into a sum of irreducible representations of $G = \Gamma(2)/\Gamma$.

First we assemble some information.

Proposition 4.1. For the following untwisted subgroups $\Gamma \subset \Lambda \subset \Gamma(2)$, $\dim H_4(M_\Lambda^*: \mathbb{Q})$ is as stated:

$[\Lambda: \Gamma]$	Generators of Λ/Γ	Dimension
1	–	79
2	p_1	55
4	p_1, p_2	41
4	p_1, q_1	43
4	p_1, q_2	38
8	p_1, p_2, p_3	35
8	p_1, p_2, q_1	31
8	p_1, p_2, q_3	28
64	$p_1, p_2, p_3, q_1, q_2, q_3$	16

Proof. Immediate from 0.4, 2.1, 2.2, 1.16 and 3.1.

Now G has 64 irreducible representations, all 1-dimensional, which are obtained by letting each of the six generators p_1, p_2, p_3, q_1, q_2, q_3 act by multiplication by ± 1. We will denote an irreducible representation of G by $e = (e_1, \ldots, e_6)$ where each e_i is + or – according as the corresponding generator acts by +1 or –1.

Let $\mathbb{R} = H_4(M_\Gamma^*)$ regarded as a representation space of G. Let $\mathbb{R}(e)$ be the subspace on which G acts by the representation e. Our problem is to determine $\dim \mathbb{R}(e)$ = the multiplicity of e in $\mathbb{R}$. The answer is this:

Theorem 4.2. The multiplicities of the irreducible representations of G in its action on $H_4(M_\Gamma^*)$ are given by the following table:

e_1,e_2,e_3 / e_4,e_5,e_6	+++	++−	+−+	−++	+−−	−+−	−−+	−−−
+++	16	3	3	3	3	3	3	1
++−	3	3	0	0	0	0	0	0
+−+	3	0	3	0	0	0	0	0
−++	3	0	0	3	0	0	0	0
+−−	3	0	0	0	3	1	1	0
−+−	3	0	0	0	1	3	1	0
−−+	3	0	0	0	1	1	3	0
−−−	1	0	0	0	0	0	0	1

(Thus for example, the multiplicity of the representation where p_1,p_2,p_3 (resp. q_1,q_2,q_3) act by $(+1,-1,-1)$ (resp. $(-1,+1,-1)$) is the intersection of the column labelled +−− and the row labelled −+− and is 1. (Note in this representation (r_1,r_2,r_3) act by $(-1,-1,+1)$.)

Remark 4.3. The reader will observe that the multiplicity of each non-trivial representation is one less than a power of two. Why this should be so, or what it means, is a complete mystery to us.

Gathering the irreducible representation of G into $A(G)$-equivalence classes, we may rephrase the theorem as follows. (Note that when we compare representations of different V_i's, we are using their identification to V.)

Theorem 4.4. As a representation space of $G = V_1 \times V_2 \times V_3$, $R = H_4(M_T^*)$ decomposes as follows:

Type of irreducible	No. of irreducibles of this type	Multiplicity of each in R	Total Dimension in R
Trivial	1	16	16
V_i acts non-trivially for one value of i	9	3	27

V_i acts non-trivially for two values of i - both act the same way	9	3	27
V_i acts non-trivially for two values of i - they act differently	18	0	0
V_i acts non-trivially for all values of i - all act the same way	3	1	3
V_i acts non-trivially for all i - two act same, one different	18	0	0
V_i acts non-trivially for all i - all act differently	6	1	6

Proof. Since all representations of a given type are $\mathbf{A}(G)$-equivalent, they occur with the same multiplicity, so we must determine this common value for each type. Let the multiplicities of these types be $m_0,\ldots,m_6$ (i.e. m_0 is the multiplicity of the trivial representation, m_1 the multiplicity of each irreducible representation in which V_i acts non-trivially for one value of i, etc).

It is easy to check that the number of each type of irreducible appearing in $\mathbf{R}$ is as claimed. Thus by counting dimensions we obtain the equation

$$m_0 + 9m_1 + 9m_2 + 18m_3 + 3m_3 + 81m_5 + 6m_6 = \dim \mathbf{R} = 79.$$

Now consider the action of p_1 on $\mathbf{R}$. By 4.1, the dimension of the subspace of $\mathbf{R}$ on which p_1 acts trivially is 55. This subspace is a sum of copies of 32 of the 64 irreducible representations of G, and it is easy to see that the number of these of type 0 is 1, of type 1 is 7, of type 2 is 5, etc.

Proceeding in this fashion for all the subgroups Λ given in 4.1 yields the linear system

$$\begin{pmatrix} 1 & 9 & 9 & 18 & 3 & 18 & 6 \\ 1 & 7 & 5 & 10 & 1 & 6 & 2 \\ 1 & 5 & 3 & 4 & 1 & 2 & 0 \\ 1 & 6 & 3 & 6 & 0 & 0 & 0 \\ 1 & 5 & 2 & 5 & 0 & 2 & 1 \\ 1 & 3 & 3 & 0 & 1 & 0 & 0 \\ 1 & 4 & 1 & 2 & 0 & 0 & 0 \\ 1 & 3 & 1 & 2 & 0 & 1 & 0 \\ 1 & 0 & 0 & 0 & 0 & 0 & 0 \end{pmatrix} \begin{pmatrix} m_0 \\ m_1 \\ m_2 \\ m_3 \\ m_4 \\ m_5 \\ m_6 \end{pmatrix} = \begin{pmatrix} 79 \\ 55 \\ 41 \\ 43 \\ 38 \\ 35 \\ 31 \\ 28 \\ 16 \end{pmatrix}$$

This (consistent) system has rank 7, and hence a unique solution, $(m_0, m_1, \ldots, m_6) = (16,3,3,0,1,0,1)$, yielding the theorem.

From this theorem we may of course determine $\dim H_4(M_\Lambda^*)$ for any $\Gamma \subset \Lambda \subset \Gamma(2)$. There are very many such Λ (even up to **A**(G)-equivalence) so we content ourselves with listing the extreme cases.

Corollary 4.5. Let $\Gamma \subset \Lambda \subset \Gamma(2)$ be any subgroup with $[\Lambda: \Gamma] = 2$. Then Λ is a **A**(G)-equivalent to one of the following, and $\dim H_4(M_\Lambda^*)$ is as stated:

Generator of Λ/Γ	Dimension of $H_4(M_\Lambda^*)$
p_1	55
p_1p_2	51
p_1q_2	45
$p_1p_2p_3$	59
$p_1p_2q_3$	43
$p_1p_2q_1q_3$	41

Corollary 4.6. Let $\Gamma \subset \Lambda \subset \Gamma(2)$ be any subgroup with $[\Gamma(2): \Lambda] = 2$, so Λ is the kernel of a homomorphism $\phi_\Lambda: \Gamma(2)/\Gamma \to \{\pm 1\}$. Then Λ is a **A**(G)-equivalent to one of the following, and $\dim H_4(M_\Lambda^*)$ is as stated:

Generators not in $Ker(\Phi_\Lambda)$	Dimension of $H_4(M^*_\Lambda)$
p_1	19
p_1, p_2	19
p_1, q_2	16
p_1, p_2, p_3	17
p_1, p_2, q_3	16
p_1, p_2, q_1, q_3	17

(Note that here $H_4(M^*_\Lambda)$ will be a sum of two types of irreducible representations of G, the trivial one and one other. The six cases of this corollary correspond, in order, to the six non-trivial types of irreducibles in theorem 4.4.)

We close by considering the question of torsion in the homology of M^*_Λ. As we have seen, the only possible torsion is 2-torsion.

Theorem 4.7. Suppose Λ is untwisted. Then $H_*(M^*_\Lambda)$ is torsion free.

Proof. If Λ is untwisted, then Λ/Γ may be written as a product $W_1 \times W_2 \times W_3$ with $W_i \subset V_i$. (The different W_i need not be isomorphic.)

From theorem 1.15, we see that M^*_Γ is rational, and indeed, this theorem shows that M^0_Γ, a Zariski open set in M^*_Γ, is isomorphic to a Zariski open set in $\bar{X} \times \bar{X} \times \bar{X} = \mathbb{P}^1 \times \mathbb{P}^1 \times \mathbb{P}^1$. But then M^0_Λ is isomorphic to a Zariski open set in $(\bar{X}/W_1) \times (\bar{X}/W_2) \times (\bar{X}/W_3)$, which is itself isomorphic to $\mathbb{P}^1 \times \mathbb{P}^1 \times \mathbb{P}^1$, so M^*_Λ is rational.

Then by [AM, proposition 1], $H_*(M^*_\Lambda)$ is torsion-free.

References

[AM] Artin, M. and Mumford, D. Some elementary examples of unirational varieties which are not rational, Proc. Lond. Math. Soc. 25 (1972), 75-95.

[B] Bredon, G. Introduction to compact transformation groups, Academic Press, New York, 1972.

[G] van der Geer, G. On the geometry of a Siegel modular threefold, Math. Ann. 260 (1982), 317-350.

[LW_1] Lee, R. and Weintraub, S. H. Cohomology of a Siegel modular variety of degree two, in Group Actions on Manifolds, R. Schultz, ed., Amer. Math. Soc., Providence, RI, 1985, 433-488.

[LW_2] ________ Cohomology of $Sp_4(\mathbb{Z})$ and related groups and spaces, Topology 24 (1985), 291-310.

[LW_3] ________ Moduli spaces of Riemann surfaces of genus two with level structures, to appear in Trans. Amer. Math. Soc.

[M] Mumford, D. Stability of projective varieties, L'Enseignement Math. 23 (1977), 39-110.

Yale University
Louisiana State University and Universität Göttingen

THE RO(G)-GRADED EQUIVARIANT ORDINARY COHOMOLOGY OF COMPLEX PROJECTIVE SPACES WITH LINEAR $\mathbb{Z}/p$ ACTIONS

L. Gaunce Lewis, Jr.

INTRODUCTION. If X is a CW complex with cells only in even dimensions and R is a ring, then, by an elementary result in cellular cohomology theory, the ordinary cohomology $H^*(X;R)$ of X with R coefficients is a free, $\mathbb{Z}$-graded R-module. Since this result is quite useful in the study of well-behaved complex manifolds like projective spaces or Grassmannians, it would be nice to be able to generalize it to equivariant ordinary cohomology. The result does generalize in the following sense. Let G be a finite group, X be a G-CW complex (in the sense of [MAT, LMSM]), and R be a ring-valued contravariant coefficient system [ILL]. Then the G-equivariant ordinary Bredon cohomology $H^*(X;R)$ of X with R coefficients may be regarded as a coefficient system. If the cells of X are all even dimensional, then $H^*(X;R)$ is a free module over R in the sense appropriate to coefficient systems. Unfortunately, this theorem does not apply to complex projective spaces or complex Grassmannians with any reasonable nontrivial G-action because these spaces do not have the right kind of G-CW structure. In fact, if G is $\mathbb{Z}/p$, for any prime p, and η is a nontrivial irreducible complex G-representation, then the theorem does not apply to S^η, the one-point compactification of η. Moreover, the $\mathbb{Z}$-graded Bredon cohomology of S^η with coefficients in the Burnside ring coefficient system is quite obviously not free over the coefficient system.

The purpose of this paper is to provide an equivariant generalization of the "freeness" theorem which does apply to an interesting class of G-spaces and to use this result to describe the equivariant ordinary cohomology of complex projective spaces with linear $\mathbb{Z}/p$ actions. These results are obtained by regarding equivariant ordinary cohomology as a Mackey functor-valued theory graded on the real representation ring RO(G) of G [LMM, LMSM]. To obtain such a theory, we take the Burnside ring Mackey functor as our coefficient ring. Instead of using cells of the form $G/H \times e^n$, where H runs over the subgroups of G, we use the unit disks of real G-representations as cells. Our main theorem, Theorem 2.6, then has roughly the following form.

THEOREM. Let G be $\mathbb{Z}/p$ and let X be a G-CW complex constructed from the unit disks of real G-representations. If these disks are all even dimensional and are attached in the proper order, then the equivariant ordinary cohomology $\underline{H}_G^* X$ of X is a free RO(G)-graded module over the equivariant ordinary cohomology of a point.

To show that this theorem is not without applications, we prove in Theorem 3.1 that if V is a complex G-representation and P(V) is the associated complex projective space with the induced linear G-action, then P(V) has the required type of cell structure. Theorems 4.3 and 4.9, which describe the ring structure of $\underline{H}_G^* P(V)$, follow from the freeness of $\underline{H}_G^* P(V)$. As a sample of these results, assume that $p = 2$ and V is a complex G-representation consisting of countably many copies of both the (complex) one-dimensional sign representation λ and the one dimensional trivial representation 1. Then P(V) is the classifying space for G-equivariant complex line bundles. As an RO(G)-graded ring, $\underline{H}_G^*P(V)$ is generated by an element c in dimension λ and an element C in dimension $1+\lambda$. The second generator is a polynomial generator; the first satisfies the single relation

$$c^2 = \epsilon^2 c + \xi C,$$

where ϵ and ξ are elements in the cohomology of a point. If, instead, V contains an equal, but finite, number of copies of λ and 1, then the only change in $\underline{H}_G^*P(V)$ is that the polynomial generator C is truncated in the appropriate dimension. If the number of copies of 1 in V is different from the number of copies of λ in V, or if p is odd, then the ring structure of $\underline{H}_G^*P(V)$ is more complex.

Equivariant ordinary Bredon cohomology with Burnside ring coefficients is just the part of RO(G)-graded equivariant ordinary cohomology with Burnside ring coefficients that is indexed on the trivial representations. All of the generators of $\underline{H}_G^*P(V)$ occur in dimensions corresponding to nontrivial representations of G. This behavior of the generators offers a partial explanation of the difficulties encountered in trying to compute Bredon cohomology. All that can been seen of $\underline{H}_G^*P(V)$ with $\mathbb{Z}$-graded Bredon cohomology is some junk connected to the RO(G)-graded cohomology of a point whose presence in $\underline{H}_G^*P(V)$ is forced by the unseen generators in the nontrivial dimensions.

Using $\underline{H}_G^*P(V)$, it is possible to give an alternative proof of the homotopy rigidity of linear $\mathbb{Z}/p$ actions on complex projective spaces [LIU]. Moreover, the "freeness" theorem should apply to complex Grassmannians with linear $\mathbb{Z}/p$ actions, and it should be possible to compute the ring structure of the equivariant ordinary cohomology of these spaces. Of course, it would be nice to extend the main theorem to groups other than $\mathbb{Z}/p$. Unfortunately, the obvious generalization of this theorem fails for groups other than $\mathbb{Z}/p$. The counterexamples have some interesting connections with the equivariant Hurewicz theorem [LE1]. All of these topics are being investigated.

All of the results in this paper depend on the observation that equivariant cohomology theories are Mackey functor-valued. Therefore, the first section of this paper contains a discussion of Mackey functors for the group $\mathbb{Z}/p$. In the second section, we discuss the RO(G)-graded cohomology of a point, precisely define what we mean by a G-CW complex, and prove our "freeness" theorem. The G-cell structure of complex projective spaces with linear $\mathbb{Z}/p$ actions is discussed in section 3. There the cohomology of these spaces is shown to be free over the cohomology of a point. Section 4 is devoted to the multiplicative structure of the cohomology of a point. The multiplicative structure of the cohomology of complex projective spaces is discussed in section 5. The results stated in this section are proved in section 6. The results on the cohomology of a point stated in sections 2 and 4 are proved in the appendix.

A few comments on notational conventions are necessary. Hereafter, all homology and cohomology is reduced. If X is a G-space and we wish to work with the unreduced cohomolgy of X, then we take the reduced cohomology of X^+, the disjoint union of X and a G-trivial basepoint. In particular, instead of speaking of the cohomology of a point, hereafter we speak of the cohomology of S^0, which always has trivial G action. If V is a G-representation, then SV and DV are the unit sphere and unit disk of V with respect to some G-invariant norm. The one-point compactification of V is denoted S^V and the point at infinity is taken as the basepoint. If X is a based G-space, then $\Sigma^V X$ denotes the smash product of X and S^V. Unless otherwise noted, all spaces, maps, homotopies, etc., are G-spaces, G-maps, and G-homotopies, etc. We will shift back and forth between real and complex G-representations; in general, real representations will be used for grading our cohomology groups and complex representations will be used in discussions of the structure of projective spaces. If the virtual representation α is represented by the difference $V - W$ of representations V and W, then $|\alpha| = \dim V - \dim W$ is the real virtual dimension of α and $\alpha^G = V^G - W^G$ is the fixed virtual representation associated to α. The trivial virtual representation of real dimension n is denoted by n. Recall that the set of irreducible complex representations of G forms a group under tensor product. If η is an irreducible complex representation, then η^{-1} denotes the inverse of η in this group. The tensor product of η and any representation V is denoted ηV. Many of our formulas contain terms of the form A/p, where A is some integer-valued espression. The claim that A is divisible by p is implicitly included in the use of such a term.

I would like to thank Tammo tom Dieck, Sonderforschungsbereich 170, and the Mathematisches Institut at Göttingen for their hospitality during the initial stages of this work. I would especially like to thank Tammo tom Dieck for suggesting the problem which led to this paper and for invaluable comments, especially on the main theorem, Theorem 2.6.

Equivariant cohomology theories graded on RO(G) are not universally familiar objects, so a few remarks about what this paper assumes of its readers seem appropriate. Equivariant ordinary cohomology with Burnside ring coefficients assigns to each virtual representation α in RO(G) a contravariant functor $\underline{H}_G^\alpha$ from the homotopy category of based G-spaces to the category of Mackey functors. It also assigns a suspension natural isomorphism

$$\underline{H}_G^{\alpha+V}(\Sigma^V X) \cong \underline{H}_G^\alpha(X)$$

to each pair (α, V) consisting of a virtual representation α and an actual representation V. The isomorphisms associated to the three pairs (α, V), (α, W), and $(\alpha, V+W)$ are required to satisfy a coherence condition. The functors $\underline{H}_G^\alpha$ are required to be exact in the sense that they convert cofibre sequences into long exact sequences. The dimension axiom requires that $\underline{H}_G^0 S^0$ be the Burnside ring Mackey functor and that $\underline{H}_G^n S^0$ be zero if $n \in \mathbb{Z}$ and $n \neq 0$. If α is a nontrivial virtual representation, then $\underline{H}_G^\alpha S^0$ need not be zero, but it is uniquely determined by the axioms. Note that because $\underline{H}_G^* S^0$ is nonzero in dimensions other than zero, the assertion that the cohomology of certain spaces is free over the cohomology of S^0 is very different from the assertion that the cohomology is free over the coefficient ring. Our cohomology theory is ring valued; that is, any pair of elements drawn from $\underline{H}_G^\alpha X$

and $\underline{H}_G^{\beta}X$ have a cup product which is in $\underline{H}_G^{\alpha+\beta}X$. We will also work with RO(G)-graded, Mackey functor-valued, reduced equivariant ordinary homology with Burnside ring coefficients. This homology theory satisfies the obvious analogs of the cohomology axioms. Also, it has a Hurewicz map, which we use to convert various space level maps into homology classes. Finally, we assume that S^0 and the free orbit G^+ satisfy equivariant Spanier-Whitehead duality [WIR, LMSM]; that is, for any α in RO(G) there are isomorphisms

$$\underline{H}_G^{\alpha}S^0 \cong \underline{H}_{-\alpha}^G S^0 \quad \text{and} \quad \underline{H}_G^{\alpha}G^+ \cong \underline{H}_{-\alpha}^G G^+.$$

The proofs of all our results flow from these basic assumptions. In fact, most of the proofs are simple long exact sequence arguments which would be left to the reader in a paper dealing with a $\mathbb{Z}$-graded, abelian group-valued, nonequivariant cohomology. One of the points of this paper is that these simple techniques work perfectly well in RO(G)-graded, Mackey functor-valued, equivariant cohomology theories and yield useful results. The one serious demand made of the reader is a willingness to work with Mackey functors. When the group is $\mathbb{Z}/p$, these are really very simple objects. Section one is intended as a tutorial on them.

1. MACKEY FUNCTORS FOR $\mathbb{Z}/p$. Since the language of Mackey functors pervades this paper, this section contains a brief introduction to Mackey functors for the groups $\mathbb{Z}/p$. For any finite group G, a G-Mackey functor M is a contravariant additive functor from the Burnside category B(G) of G to the category Ab of abelian groups [DRE, LE2, LIN]. However, since we are only concerned with $G = \mathbb{Z}/p$, rather than describing B(G) in detail, we simply note that a $\mathbb{Z}/p$-Mackey functor M is determined by two abelian groups, M(G/G) and M(G/e); two maps, a restriction map

$$\rho : M(G/G) \to M(G/e)$$

and a transfer map

$$\tau : M(G/e) \to M(G/G);$$

and an action of G on M(G/e). The trace of this action and the composite $\rho\tau$ are required to be equal by the definition of the composition of maps in B(G); that is, if $x \in M(G/e)$, then

$$\rho\tau(x) = \sum_{g \in G} gx.$$

The abelian groups M(G/G) and M(G/e) are the values of the Mackey functor M at the trivial orbit and the free orbit; or, if one prefers to think in terms of subgroups instead of orbits, the values of M at the group and at the trivial subgroup. For convenience, we abbreviate G/G to 1 and write M(e) for M(G/e). Frequently the G-action on M(e) is trivial; in these cases the composite $\rho\tau$ is just multiplication by p.

A map $f : M \to N$ between Mackey functors consists of homomorphisms

$$f(1) : M(1) \to N(1) \quad \text{and} \quad f(e) : M(e) \to N(e)$$

which commute with ρ and τ in the obvious sense. The map f(e) must also be G-equivariant. The category $\mathfrak{M}$ of Mackey functors is a complete and cocomplete abelian category. The limit or colimit of a diagram in $\mathfrak{M}$ is formed by taking the limit or colimit of the corresponding two diagrams consisting of the abelian groups associated to G/G and to G/e. The maps ρ and τ and the group action on the limit or colimit are the obvious induced maps and action.

We will describe Mackey functors diagramatically in the form

$$\begin{array}{c} M(1) \\ \rho \big\downarrow \quad \big\uparrow \tau \\ M(e) \\ \circlearrowleft \\ \theta \end{array}$$

where M(1) and M(e) will be replaced by the appropriate abelian groups, ρ and τ may be replaced by explicit descriptions of the restriction and transfer maps, and θ may be replaced by an explicit description of the group action. If ρ or τ is replaced by a number (usually 0, 1, or p), then the map is just multiplication by that number. If θ is omitted or replaced by 1, then the group action on M(e) is trivial. If $p = 2$ and θ is replaced by -1, then the generator of $G = \mathbb{Z}/2$ acts by multiplication by -1.

EXAMPLES 1.1 The following Mackey functors and maps appear repeatedly in our cohomology computations.

(a) The Burnside ring Mackey functor A is given by

$$\begin{array}{c} \mathbb{Z}\oplus\mathbb{Z} \\ (1,p) \big\downarrow \quad \big\uparrow (0,1) \\ \mathbb{Z} \end{array}$$

where the notation (1,p) means that the restriction map ρ is the identity on the first component and multiplication by p on the second. Similarly, (0,1) means that the transfer map is the inclusion into the second factor. For any Mackey functor M, there is a one-to-one correspondence between maps $f : A \to M$ and elements of M(1). The correspondence relates the map f to the element f(1)((1,0)) of M(1). It follows from this correspondence that A is a projective Mackey functor.

(b) The d-twisted Burnside ring Mackey functor A[d] is given by

$$\begin{array}{c} \mathbb{Z}\oplus\mathbb{Z} \\ (d,p) \big\downarrow \quad \big\uparrow (0,1) \\ \mathbb{Z} \end{array}$$

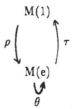

where d $\in \mathbb{Z}$. Note that $A = A[1]$. If $d \equiv \pm d' \bmod p$, then there is an isomorphism $f: A[d] \cong A[d']$ of Mackey functors. The map f(e) is the identity and if $d' = \pm d + np$, then

$$f(1)(1,0) = (\pm 1,n) \in \mathbb{Z}\oplus\mathbb{Z}$$
$$f(1)(0,1) = (0,1).$$

If $d \equiv 0 \bmod p$, then A[d] decomposes as the sum of two other Mackey functors; thus A[d] is only of interest when $d \not\equiv 0 \bmod p$. In this case, it is a projective Mackey functor. An alternative $\mathbb{Z}$-basis for A[d](1) will be used in some of our cohomology calculations. To distinguish the two bases, we denote (1,0) and (0,1) in the present basis by μ and τ respectively. Select integers a and b such that $ad + bp = 1$. The alternative $\mathbb{Z}$-basis consists of $\sigma = a\mu + b\tau$ and $\kappa = p\mu - d\tau$. Note that $\rho(\sigma) = 1$, $\rho(\kappa) = 0$, and $\tau(1) = \tau$. In fact, κ generates the kernel of ρ, and τ generates the image of the map τ for which it is named. Of course, σ depends on the choice of a and b; in our applications, these choices will always be specified.

(c) If C is any abelian group, then we use $\langle C\rangle$ to denote the Mackey functor described by the diagram

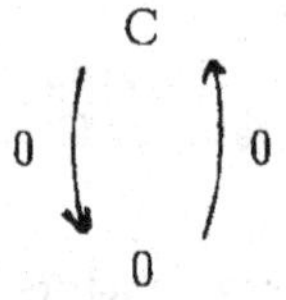

(d) If d_1 and d_2 are integers prime to p, then there is an isomorphism

$$g_{12}: A[d_1] \oplus \langle\mathbb{Z}\rangle \to A[d_2] \oplus \langle\mathbb{Z}\rangle.$$

Let μ_i and τ_i be the standard generators for $A[d_i]$, and let z_1 and z_2 be generators of $\langle\mathbb{Z}\rangle(1)$ in the domain and range of g_{12}. Select integers a_i and b_i such that $a_i d_i + b_i p = 1$, for $i = 1$ or 2. The map $g_{12}(e): \mathbb{Z} \to \mathbb{Z}$ is the identity map, and the map $g_{12}(1)$ is given by

$$g_{12}(1)(\mu_1) = d_1(a_2\mu_2 + b_2\tau_2) + (b_1 + b_2 - b_1b_2p)z_2$$
$$g_{12}(1)(\tau_1) = \tau_2$$

and

$$g_{12}(1)(z_1) = p\mu_2 - d_2\tau_2 - a_1d_2z_2.$$

The inverse of g_{12} is just g_{21}. The existence of this isomorphism will explain an apparent inconsistency in our description of the equivariant cohomology of projective spaces.

(e) Associated to an abelian group B with a G-action, we have the Mackey functors L(B) and R(B) given by

L(B)

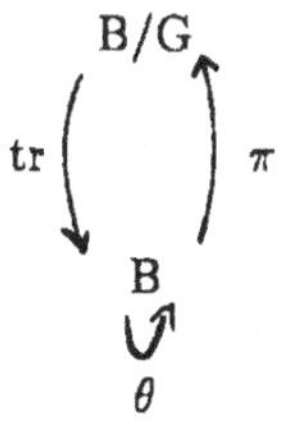

R(B)

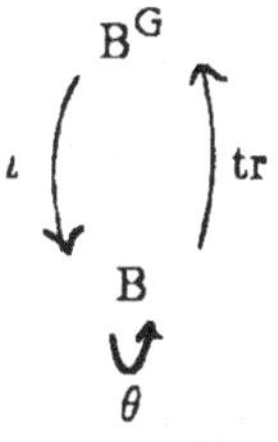

Here, $\iota : B^G \to B$ is the inclusion of the fixed point subgroup and $\pi : B \to B/G$ is the projection onto the orbit group. The two maps tr are variants of the trace map. The map $tr : B \to B^G$ takes $x \in B$ to $\sum_{g \in G} gx \in B^G$. If $x \in B$ and $[x]$ is the associated equivalence class in B/G, then $tr : B/G \to B$ is given by

$$tr([x]) = \sum_{g \in G} gx \ \in \ B.$$

These two constructions give functors from the category of $\mathbb{Z}[G]$-modules to the category of Mackey functors. These functors are the left and right adjoints to the obvious forgetful functor from the category of Mackey functors to the category of $\mathbb{Z}[G]$-modules. We will encounter these functors most often when B is $\mathbb{Z}$ with the trivial action or, if $p = 2$, with the sign action. Denote the resulting Mackey functors by L, R, L_-, and R_-. These functors are described by the diagrams

L

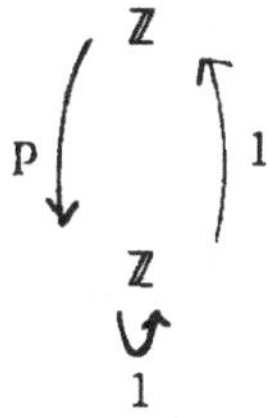

R

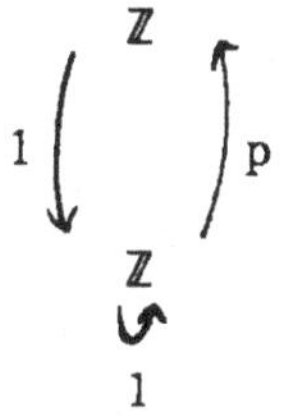

L_-

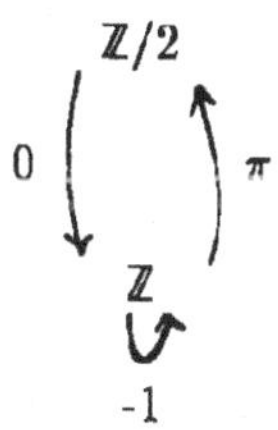

R_-

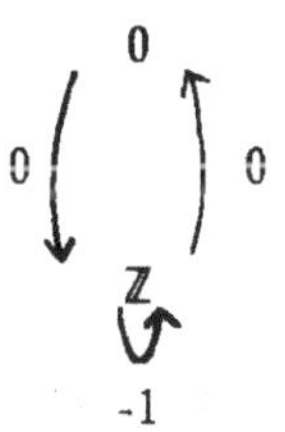

If C is any abelian group, there is an obvious permutation action of G on C^p, the direct sum of p copies of C. Unless otherwise indicated, this action is assumed when we refer to $L(C^p)$ or $R(C^p)$. These two functors are isomorphic and are described by the diagram

$$\begin{array}{c} C \\ \Delta \big\downarrow \big\uparrow \nabla \\ C^{p} \\ \circlearrowleft \\ \theta \end{array}$$

where Δ is the diagonal map. ∇ is the folding map, and θ is the permutation action.

(f) If M is a Mackey functor, then $L(M(e)^{p}) \cong R(M(e)^{p})$ is denoted M_G. There are two reasonable choices of a G action on $M(e)^{p}$, the permutation action or the composite of the permutation action and the given action of G on each factor M(e). These actions yield isomorphic $\mathbb{Z}[G]$-modules, so the choice is not important. The simple permutation action is always assumed here. The assignment of M_G to M is a special case of an important construction in induction theory [DRE, LE2] that assigns a Mackey functor M_b to each object b of B(G) and each Mackey functor M.

The restriction map $\rho: M(1) \to M(e) \cong M_G(1)$ and the diagonal map $\Delta: M(e) \to M(e)^{p} \cong M_G(e)$ form a natural transformation ρ from M to M_G. Similarly, $\tau: M_G(1) \cong M(e) \to M(1)$ and the folding map $\nabla: M_G(e) \cong M(e)^{p} \to M(e)$ form a natural transformation $\tau: M_G \to M$. The Mackey functor $A_G = L(\mathbb{Z}^{p})$ is characterized by the fact that, for any Mackey functor M, there is a one-to-one correspondence between maps $f: A_G \to M$ and elements of M(e). This correspondence relates the map f to the element $f(e)((1,0,0, \ldots ,0))$ of M(e). It follows that A_G is a projective Mackey functor.

(g) If Y is a G-space, M is a Mackey functor, $\alpha \in RO(G)$, and $H_G^{\alpha}(Y;M)$ and $H_{\alpha}^{G}(Y;M)$ denote the abelian group-valued equivariant ordinary cohomology and homology of Y with coefficients M in dimension α, then the Mackey functor valued cohomology $\underline{H}_G^{\alpha}(Y;M)$ and homology $\underline{H}_{\alpha}^{G}(Y;M)$ are described by the diagrams

$$\begin{array}{c} H_G^{\alpha}(Y;M) \\ \pi^{*} \big\downarrow \big\uparrow \pi_{!} \\ H_G^{\alpha}(G \times Y;M) \\ \circlearrowleft \\ \theta \end{array} \qquad \text{and} \qquad \begin{array}{c} H_{\alpha}^{G}(Y;M) \\ \pi^{!} \big\downarrow \big\uparrow \pi_{*} \\ H_{\alpha}^{G}(G \times Y;M) \\ \circlearrowleft \\ \theta \end{array}$$

where the maps $\pi^{*}$and π_{*} are induced by the projection $\pi: G \times Y \to Y$, and the maps $\pi_{!}$ and $\pi^{!}$ are the transfer maps arising from regarding the projection π as a covering space. The group $H_G^{\alpha}(G \times Y;M)$ is isomorphic to the nonequivariant cohomology group $H^{|\alpha|}(Y;M(e))$. If α is represented by the difference $V - W$ of representations V and W, then, under this isomorphism, the action of an element g of

G on $H_G^\alpha(G \times Y; M)$ may be described as the composite of multiplication by the degrees of the maps $g: S^V \to S^V$ and $g: S^W \to S^W$ and the actions of g on $H^{|\alpha|}(Y; M(e))$ induced by the action of g on M(e) and the action of g^{-1} on Y. Similar remarks apply in homology. If no coefficient Mackey functor M is indicated in equivariant cohomology or homology, then Burnside ring coefficients are intended.

(h) For any Mackey functor M and any abelian group B, the Mackey functor $M \otimes B$ has value $M(G/H) \otimes B$ for the orbit G/H and the obvious restriction, transfer, and action by G. If M^* is an RO(G)-graded G-Mackey functor and B^* is a $\mathbb{Z}$-graded abelian group, then $M^* \otimes B^*$ is the RO(G)-graded G-Mackey functor defined by

$$(M^* \otimes B^*)^\alpha = \sum_{\beta+n=\alpha} M^\beta \otimes B^n.$$

If a CW complex Y with cells only in even dimensions is regarded as a G-space by assigning it the trivial G-action, then there is an isomorphism of RO(G)-graded Mackey functors

$$\underline{H}_G^* Y \cong \underline{H}_G^* S^0 \otimes \underline{H}^*(Y; \mathbb{Z})$$

which preserves cup products.

For any finite group G, there is a box product operation $\Box$ on the category $\mathfrak{M}$ of G-Mackey functors which behaves like the tensor product on the category of abelian groups. In particular, $\mathfrak{M}$ is a symmetric monoidal closed category under the box product. The Burnside ring Mackey functor A is the unit for $\Box$. If $G = \mathbb{Z}/p$, then the box product $M \Box N$ of Mackey functors M and N is described by the diagram

$$\begin{array}{c} \big[M(1) \otimes N(1) \oplus M(e) \otimes N(e)\big]/\approx \\ \rho \downarrow \quad \uparrow \tau \\ M(e) \otimes N(e) \\ \circlearrowleft \\ \theta \end{array}$$

The equivalence relation $\approx$ is given by

$$x \otimes \tau y \approx \rho x \otimes y \quad \text{for } x \in M(1) \text{ and } y \in N(e)$$

$$\tau v \otimes w \approx v \otimes \rho w \quad \text{for } v \in M(e) \text{ and } w \in N(1).$$

The action θ of G on $M(e) \otimes N(e)$ is just the tensor product of the actions of G on M(e) and N(e). The map τ is derived from the inclusion of $M(e) \otimes N(e)$ as a summand of the direct sum used to define $M \Box N(1)$. The map ρ is induced by $\rho \otimes \rho$ on the first summand and the trace map of the action θ on the second.

EXAMPLES 1.2(a) For any integers d_1 and d_2, there is an isomorphism

$$A[d_1] \Box A[d_2] \cong A[d_1 d_2]$$

of Mackey functors.

(b) If B is a $\mathbb{Z}[G]$-module and M is any Mackey functor, then there is an isomorphism

$$L(B) \square M \;\cong\; L(B \otimes M(e)).$$

(c) For any Mackey functor M, the product $R \square M$ is described by the diagram

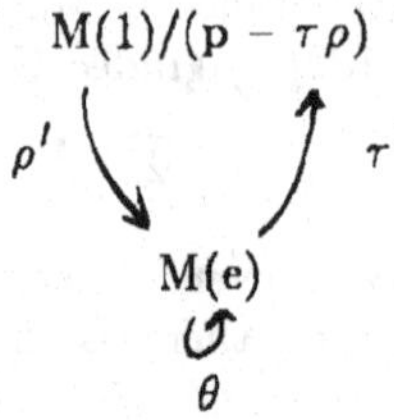

where $M(1)/(p - \tau\rho)$ is the cokernel of the difference between the multiplication by p map and the composite $\tau\rho$. The maps ρ' and τ' are induced by the restriction and transfer maps for M. In particular, if $M = R(B)$ for some $\mathbb{Z}[G]$-module B, then $R \square R(B) \cong R(B)$. Also, for any abelian group C, $R \square \langle C \rangle \cong \langle C/pC \rangle$.

(d) If $p = 2$, then for any Mackey functor M, the product $R_- \square M$ is described by the diagram

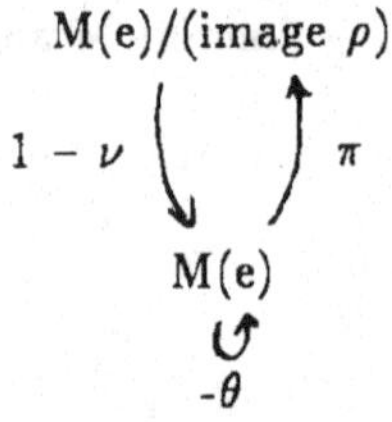

Here $\pi: M(e) \to M(e)/(\text{image } \rho)$ is the projection onto the cokernel of the restriction map and $\nu: M(e) \to M(e)$ describes the action of the nontrivial element of G on M(e). The action $-\theta$ is the composite of the given action θ of G on M(e) and the sign action of G on M(e). In particular, $R_- \square R_- \cong L$.

(e) For any abelian group C and any Mackey functor M,

$$\langle C \rangle \square M \;\cong\; \langle C \otimes (M(1)/\text{image } \tau) \rangle.$$

A Mackey functor ring (or Green functor [DRE, LE2]) is a Mackey functor S together with a multiplication map $\mu : S \square S \to S$ and a unit map $\eta : A \to S$ making the appropriate diagrams commute. A module over S is just a Mackey functor M together with an action map $\xi : S \square M \to M$ making the appropriate diagrams commute. Since the Burnside ring Mackey functor A is the unit for $\square$, it is a Mackey functor ring whose multiplication is the isomorphism $A \square A \to A$ and whose unit is the identity map A → A. Every Mackey functor is a module over A with action map the isomorphism $A \square M \to M$. Note that if S is a Mackey functor ring and R is a ring, then the Mackey functor $S \otimes R$ of Examples 1.1(h) is a Mackey functor ring. Similar remarks apply in the graded case. The cohomology of any G-space Y with coefficients a Mackey functor ring S is an RO(G)-graded Mackey functor ring whose multiplication is given by maps

$$\underline{H}_G^\alpha(Y;S) \square \underline{H}_G^\beta(Y;S) \to \underline{H}_G^{\alpha+\beta}(Y;S),$$

for α and β in RO(G).

The following result characterizes maps out of box products and allows us to describe a Mackey functor ring S in terms of S(1) and S(e). This is the approach to Mackey functor rings used in our discussion of the ring structure of the cohomology of complex projective spaces.

PROPOSITION 1.3 For any Mackey functors M, N and P, there is a one-to-one correspondence between maps $h : M \square N \to P$ and pairs $H = (H_1, H_e)$ of maps

$$H_1 : M(1) \otimes N(1) \to P(1)$$

$$H_e : M(e) \otimes N(e) \to P(e)$$

such that, for $x \in M(1)$, $y \in N(1)$, $z \in M(e)$, and $w \in N(e)$,

$$\begin{aligned} H_e(\rho x \otimes \rho y) &= \rho H_1(x \otimes y) \\ H_1(\tau z \otimes y) &= \tau H_e(z \otimes \rho y) \\ H_1(x \otimes \tau w) &= \tau H_e(\rho x \otimes w). \end{aligned}$$

The second and third of these equations are called the Frobenius relations.

PROOF. The maps H_e and h are related by $H_e = h(e)$. Given h, H_1 is derived in an obvious way from h(1) using the definition of $M \square N$. Given H_1 and H_e, h(1) is constructed from the maps H_1 and τH_e on the two summands used to define $M \square N(1)$.

It follows easily from the proposition that, if S is a Mackey functor ring, then S(1) and S(e) are rings, $\rho : S(1) \to S(e)$ is a ring homomorphism, and $\tau : S(e) \to S(1)$ is an S(1)-module map when S(e) is considered an S(1)-module via ρ. Moreover, if M is a Mackey functor module over S, then M(1) is an S(1)-module and M(e) is an S(e)-module. If we regard M(e) as an S(1)-module via $\rho : S(1) \to S(e)$, then the maps $\rho : M(1) \to M(e)$ and $\tau : M(e) \to M(e)$ are S(1)-module maps.

2. $H_G^* S^0$ AND SPACES WITH FREE COHOMOLOGY.

Here, we recall Stong's unpublished description of the additive structure of the RO(G)-graded equivariant ordinary cohomology of S^0. We use this to show that if X is a generalized G-cell complex constructed from suitable even-dimensional cells, then $\underline{H}_G^* X$ and $\underline{H}_*^G X$ are free over $\underline{H}_G^* S^0$. The additive structure of the cohomology $\underline{H}_G^* G^+$ of the free orbit is also described. This is used to show that $\underline{H}_G^* X$ and $\underline{H}_*^G X$ are projective over $\underline{H}_G^* S^0$

when X is constructed from a slightly more general class of even-dimensional cells.

Since $\mathbb{Z}/2$ has only one nontrivial irreducible representation, $\underline{H}_G^* S^0$ is very easy to describe when $G = \mathbb{Z}/2$.

THEOREM 2.1. If $G = \mathbb{Z}/2$ and $\alpha \in RO(G)$, then

$$\underline{H}_G^\alpha S^0 = \begin{cases} A, & \text{if } |\alpha| = |\alpha^G| = 0, \\ R, & \text{if } |\alpha| = 0,\ |\alpha^G| < 0, \text{ and } |\alpha^G| \text{ is even}, \\ R_-, & \text{if } |\alpha| = 0,\ |\alpha^G| \leq 1, \text{ and } |\alpha^G| \text{ is odd}, \\ L, & \text{if } |\alpha| = 0,\ |\alpha^G| > 0, \text{ and } |\alpha^G| \text{ is even}, \\ L_-, & \text{if } |\alpha| = 0,\ |\alpha^G| > 1, \text{ and } |\alpha^G| \text{ is odd}, \\ \langle \mathbb{Z} \rangle, & \text{if } |\alpha| \neq 0 \text{ and } |\alpha^G| = 0, \\ \langle \mathbb{Z}/2 \rangle, & \text{if } |\alpha| > 0,\ |\alpha^G| < 0, \text{ and } |\alpha^G| \text{ is even}, \\ \langle \mathbb{Z}/2 \rangle, & \text{if } |\alpha| < 0,\ |\alpha^G| > 1, \text{ and } |\alpha^G| \text{ is odd}, \\ 0, & \text{otherwise}. \end{cases}$$

The most effective way to visualize $\underline{H}_G^* S^0$ is to display $\underline{H}_G^\alpha S^0$ for various α on a coordinate plane in which the horizontal and vertical coordinates specify $|\alpha^G|$ and $|\alpha|$, respectively. In such a plot, given as Table 2.2 below, the zero values of $\underline{H}_G^* S^0$ are indicated by blanks. The only values in this plot with odd horizontal coordinate are the R_- and L_- on the horizontal axis and the $\langle \mathbb{Z}/2 \rangle$ in the fourth quadrant.

	⋮		⋮		⋮		⋮							
⋯	$\langle \mathbb{Z}/2 \rangle$		$\langle \mathbb{Z}/2 \rangle$		$\langle \mathbb{Z}/2 \rangle$		$\langle \mathbb{Z} \rangle$							
⋯	$\langle \mathbb{Z}/2 \rangle$		$\langle \mathbb{Z}/2 \rangle$		$\langle \mathbb{Z}/2 \rangle$		$\langle \mathbb{Z} \rangle$							
⋯	$\langle \mathbb{Z}/2 \rangle$		$\langle \mathbb{Z}/2 \rangle$		$\langle \mathbb{Z}/2 \rangle$		$\langle \mathbb{Z} \rangle$							
⋯	$\langle \mathbb{Z}/2 \rangle$		$\langle \mathbb{Z}/2 \rangle$		$\langle \mathbb{Z}/2 \rangle$		$\langle \mathbb{Z} \rangle$							
⋯	R	R_-	R	R_-	R	R_-	A	R_-	L	L_-	L	L_-	L	⋯
							$\langle \mathbb{Z} \rangle$			$\langle \mathbb{Z}/2 \rangle$		$\langle \mathbb{Z}/2 \rangle$		⋯
							$\langle \mathbb{Z} \rangle$			$\langle \mathbb{Z}/2 \rangle$		$\langle \mathbb{Z}/2 \rangle$		⋯
							$\langle \mathbb{Z} \rangle$			$\langle \mathbb{Z}/2 \rangle$		$\langle \mathbb{Z}/2 \rangle$		⋯
							$\langle \mathbb{Z} \rangle$			$\langle \mathbb{Z}/2 \rangle$		$\langle \mathbb{Z}/2 \rangle$		⋯
							⋮			⋮		⋮		

TABLE 2.2. $H_G^* S^0$ for $p = 2$.

Even though the representation ring of G is much more complicated when $p \neq 2$, $\underline{H}_G^\alpha S^0$ is completely determined by the integers $|\alpha|$ and $|\alpha^G|$ except in the special case where $|\alpha| = |\alpha^G| = 0$. In this special case, $\underline{H}_G^\alpha S^0$ is A[d] for some integer d which depends on α. Unfortunately, because of the isomorphism described in Examples 1.1(b), d is only determined up to a multiple of p. The major source of unpleasantness in the description of the multiplicative structure of the equivariant cohomology of a point and of complex projective spaces is this lack of a canonical choice for d. To explain the relation between α and d, we introduce several relatives of the representation ring. Let R(G) be the complex representation ring of G and RSO(G) be the ring of SO-isomorphism classes of SO-representations of G. Since any real representation of G is also an SO-representation, the difference between RO(G) and RSO(G) is that, in RSO(G), equivalences between representations are required to preserve underlying nonequivariant orientations on the representation spaces. The difference between R(G) and RSO(G) is that elements of RSO(G) may contain an odd number of copies of the trivial one-dimensional real representation of G. Let $R_0(G)$, $RO_0(G)$, and $RSO_0(G)$ denote the subrings of R(G), RO(G), and RSO(G) containing those virtual representations α with $|\alpha| = |\alpha^G| = 0$. Note that $R_0(G) = RSO_0(G)$. Let $\tilde{R}_0(G)$ be the free abelian monoid generated by the formal differences $\phi - \eta$ of complex isomorphism classes of nontrivial irreducible complex representations. Note that $R_0(G)$ is the quotient of $\tilde{R}_0(G)$ obtained by allowing the obvious cancellations and that $RO_0(G)$ is the quotient of $R_0(G)$ obtained by identifying conjugate representations. Let λ be the irreducible complex representation which sends the standard generator of $\mathbb{Z}/p$ to $e^{2\pi i/p}$. The monoid $\tilde{R}_0(G)$ is generated by elements of the form $\lambda^m - \lambda^n$, where $1 \leq m, n \leq p-1$. Define a homomorphism from $\tilde{R}_0(G)$ to $\mathbb{Z}$, regarded as a monoid under multiplication, by sending the generator $\lambda^m - \lambda^n$ to $m(n^{-1})$, where n^{-1} denotes the unique integer such that $1 \leq n^{-1} \leq p-1$ and $n(n^{-1}) \equiv 1 \bmod p$. Define functions from $RSO_0(G)$ and $RO_0(G)$ into $\mathbb{Z}$ by composing this homomorphism with sections of the projections from $\tilde{R}_0(G)$ to $RSO_0(G)$ or $RO_0(G)$. Let d_α denote the integer assigned to the virtual representation α by either map. The sections can not be chosen to be homomorphisms, so the assignment of d_α to α will not be a homomorphism from $RSO_0(G)$ or $RO_0(G)$ to the multiplicative monoid $\mathbb{Z}$. However, the assignment of d_α to α does give a homomorphism from $R_0(G)$ to the group of units $(\mathbb{Z}/p)^*$ of $\mathbb{Z}/p$ and a homomorphism from $RO_0(G)$ to the quotient $(\mathbb{Z}/p)^*/\{\pm 1\}$ of $(\mathbb{Z}/p)^*$. For later convenience, we select our sections so that d_0 is 1.

Stong's description of the additive structure of $\underline{H}_G^* S^0$ can now be translated into the Mackey functor language of section one.

THEOREM 2.3. If p is odd, then

$$\underline{H}_G^\alpha S^0 = \begin{cases} A[d_\alpha] & \text{if } |\alpha| = |\alpha^G| = 0 \\ R & \text{if } |\alpha| = 0 \text{ and } |\alpha^G| < 0 \\ L & \text{if } |\alpha| = 0 \text{ and } |\alpha^G| > 0 \\ \langle \mathbb{Z} \rangle & \text{if } |\alpha| \neq 0 \text{ and } |\alpha^G| = 0 \\ \langle \mathbb{Z}/p \rangle & \text{if } |\alpha| > 0,\ |\alpha^G| < 0, \text{ and } |\alpha^G| \text{ is an even integer} \\ \langle \mathbb{Z}/p \rangle & \text{if } |\alpha| < 0,\ |\alpha^G| > 1, \text{ and } |\alpha^G| \text{ is an odd integer} \\ 0 & \text{otherwise} \end{cases}$$

As in the case $p = 2$, $\underline{H}_G^* S^0$ is best visualized by plotting it on a coordinate plane whose horizontal and vertical axes specify $|\alpha^G|$ and $|\alpha|$ respectively. In this plot, given as Table 2.4 below, the zero values of $\underline{H}_G^* S^0$ are indicated by blanks. The vertical and horizontal coordinates of all the nonzero values, except the $\langle \mathbb{Z}/p\rangle$ values in the fourth quadrant, are even. Notice in the plots for both the odd primes and 2 that the vanishing of $\underline{H}_G^* S^0$ on the vertical line $|\alpha^G| = 1$ (for $|\alpha| \neq 0$ if $p = 2$) is unlike its behavior on the vertical lines corresponding to the other odd positive values for $|\alpha^G|$. These unusual zeroes for $\underline{H}_G^\alpha S^0$ are the key to our freeness and projectivity results. When $G = \mathbb{Z}/p^n$ for $n > 1$, the corresponding values are not zero, so our techniques do not extend to these groups.

Hereafter, we will often describe elements in $\underline{H}_G^* S^0$ by their position in these plots. For example, we may refer to the torsion in the fourth quadrant or the copies of $\langle \mathbb{Z}\rangle$ on the positive vertical axis.

	⋮	⋮	⋮	⋮				
⋯	$\langle \mathbb{Z}/p\rangle$	$\langle \mathbb{Z}/p\rangle$	$\langle \mathbb{Z}/p\rangle$	$\langle \mathbb{Z}\rangle$				
⋯	$\langle \mathbb{Z}/p\rangle$	$\langle \mathbb{Z}/p\rangle$	$\langle \mathbb{Z}/p\rangle$	$\langle \mathbb{Z}\rangle$				
⋯	$\langle \mathbb{Z}/p\rangle$	$\langle \mathbb{Z}/p\rangle$	$\langle \mathbb{Z}/p\rangle$	$\langle \mathbb{Z}\rangle$				
⋯	R	R	R	$A[d_\alpha]$	L	L	L	⋯
						$\langle \mathbb{Z}/p\rangle$	$\langle \mathbb{Z}/p\rangle$	⋯
				$\langle \mathbb{Z}\rangle$				
						$\langle \mathbb{Z}/p\rangle$	$\langle \mathbb{Z}/p\rangle$	⋯
				$\langle \mathbb{Z}\rangle$				
						$\langle \mathbb{Z}/p\rangle$	$\langle \mathbb{Z}/p\rangle$	⋯
				$\langle \mathbb{Z}\rangle$				
				⋮		⋮	⋮	

TABLE 2.4. $\underline{H}_G^* S^0$ for p odd.

Recall, from Examples 1.1(f), the new Mackey functor M_G which can be derived from any Mackey functor M, and the observation that $A_G = L(\mathbb{Z}^p) = R(\mathbb{Z}^p)$. It is easy to check that $\underline{H}_G^\alpha G^+$ is $\underline{H}_G^\alpha(S^0)_G$, and from this, to compute $\underline{H}_G^* G^+$.

COROLLARY 2.5. For any prime p,

$$\underline{H}_G^* G^+ = \begin{cases} A_G & \text{if } |\alpha| = 0 \\ 0 & \text{otherwise} \end{cases}$$

Proposition 4.12 tells us that $\underline{H}_G^* G^+$ is an RO(G)-graded projective module over $\underline{H}_G^* S^0$, and that any map

$$f: \underline{H}_G^* G^+ \to M^*$$

of RO(G)-graded modules over $\underline{H}_G^* S^0$ is completely determined by the image of $(1,0,0,\ldots,0) \in \mathbb{Z}^p = \underline{H}_G^0(G^+)(e)$ in $M^0(e)$.

A generalized G-cell complex X is a G-space X together with an increasing sequence of subspaces X_n of X such that X_0 is a single orbit, $X = \cup X_n$, X has the colimit (or weak) topology from the X_n, and X_{n+1} is formed from X_n by attaching G-cells. We will allow two types of G-cells. If V is a G-representation and DV and SV are the unit disk and sphere of V, then the first type of allowed cell is a copy of DV attached to X_n by a G-map from SV to X_n. The second type of cell is a copy of $G \times e^m$, where e^m is the unit m-disk with trivial G action, attached to X_n by a G-map from $G \times S^{m-1}$ to X_n. For each n, we let J_{n+1} denote the set of cells added to X_n to form X_{n+1}. Regard a cell DV of the first type as even-dimensional if $|V|$ and $|V^G|$ are even. Regard a cell $G \times e^m$ as even dimensional if m is even.

THEOREM 2.6. Let X be a generalized G-cell complex with only even-dimensional cells.

(a) Assume that $X_0 = *$ and all the cells of X are of the first type; that is, disks DV of G-representations V. Assume also that $|V^G| \geq |W^G|$ whenever $DV \in J_n$, $DW \in J_k$, $1 \leq k < n$, and $|V| > |W|$. Then $\underline{H}_G^* X^+$ is a free RO(G)-graded module over $\underline{H}_G^* S^0$ with one generator in dimension 0 and one generator in dimension V for each $DV \in J_n$, $n \geq 1$. The homology $\underline{H}_*^G X^+$ of X is also a free RO(G)-graded module over $\underline{H}_G^* S^0$ with generators in the same dimensions.

(b) If X contains cells of both types and all the cells of X of the first type satisfy the condition in part (a), then $\underline{H}_G^* X^+$ is a projective RO(G)-graded module over $\underline{H}_G^* S^0$. Moreover, $\underline{H}_G^* X^+$ is the sum of one copy of $\underline{H}_G^* X_0^+$, which is $\underline{H}_G^* S^0$ or $\underline{H}_G^* G^+$, in dimension 0, one copy of $\underline{H}_G^* S^0$ in dimension V for each $DV \in J_n$, and one copy of $\underline{H}_G^* G^+$ in dimension 2k for each $G \times e^{2k} \in J_n$, $n \geq 1$. The homology $\underline{H}_*^G X^+$ of X is also a projective RO(G)-graded module over $\underline{H}_G^* S^0$ and decomposes into the same summands.

PROOF. Abusing notation, we let J_{n+1} denote both the set of cells to be added to X_n and the space consisting of the disjoint union of those cells. Let ∂J_{n+1} denote the space consisting of the disjoint union of the boundaries of the cells in J_{n+1}. Associated to the cofibre sequence

$$X_n^+ \to X_{n+1}^+ \to J_{n+1}/\partial J_{n+1},$$

we have the long exact sequences

$$\cdots \to \underline{H}_\alpha^G X_{n+1}^+ \to \underline{H}_\alpha^G(J_{n+1}/\partial J_{n+1}) \xrightarrow{\partial} \underline{H}_{\alpha-1}^G X_n^+ \to \cdots$$

and

$$\cdots \to \underline{H}_G^\alpha X_{n+1}^+ \to \underline{H}_G^\alpha X_n^+ \xrightarrow{\partial} \underline{H}_G^{\alpha+1}(J_{n+1}/\partial J_{n+1}) \to \cdots.$$

The space $J_{n+1}/\partial J_{n+1}$ is a wedge of one copy of S^V for each $DV \in J_{n+1}$ and one copy of $G^+ \wedge S^{2k}$ for each $G \times e^{2k} \in J_{n+1}$. Thus, $\underline{H}_G^*(J_{n+1}/\partial J_{n+1})$ and $\underline{H}_*^G(J_{n+1}/\partial J_{n+1})$ are projective modules over $\underline{H}_G^* S^0$ with generators in dimensions corresponding to the cells added to X_n to form X_{n+1}. Moreover, if J_{n+1} contains only cells of the first type, then $\underline{H}_G^*(J_{n+1}/\partial J_{n+1})$ and $\underline{H}_*^G(J_{n+1}/\partial J_{n+1})$ are free modules over $\underline{H}_G^* S^0$. The space X_0 is either a point or the free orbit G, so $\underline{H}_G^* X_0^+$ and $\underline{H}_*^G X_0^+$ are projective, and perhaps free, modules over $\underline{H}_G^* S^0$ generated by single elements in dimension 0.

We will show inductively that the boundary maps ∂ in both long exact sequences are zero. The long exact sequences must then break up into short exact sequences which split by the projectivity of $\underline{H}_*^G(J_{n+1}/\partial J_{n+1})$ and $\underline{H}_G^* X_n^+$. Thus, by induction, $\underline{H}_G^* X_n^+$ and $\underline{H}_*^G X_n^+$ are free or projective, as appropriate, over $\underline{H}_G^* S^0$, with the indicated generators. It follows by the usual colimit argument that $\underline{H}_*^G X^+$ is free, or projective, with the appropriate generators. Since the map

$$\underline{H}_G^\alpha X_{n+1}^+ \to \underline{H}_G^\alpha X_n^+$$

is always a surjection, the appropriate $\lim^1$ term vanishes, and the cohomology of X, being the limit of the cohomologies of the X_n, is free (or projective) with the appropriate generators.

The graded Mackey functors $\underline{H}_G^*(J_{n+1}/\partial J_{n+1})$, $\underline{H}_*^G(J_{n+1}/\partial J_{n+1})$, $\underline{H}_G^* X_0^+$ and $\underline{H}_*^G X_0^+$ are sums of copies of $\underline{H}_G^* S^0$ and $\underline{H}_G^* G^+$ in various dimensions. By induction, we may assume that $\underline{H}_G^* X_n^+$ and $\underline{H}_*^G X_n^+$ are also of this form. To show that the maps ∂ are zero, it therefore suffices to show that they are zero from each summand of the domain to each summand of the range. For the cohomology sequence, the four possibilities for the summands and the map between them are:

$$\underline{H}_G^{*-2k}G^+ \cong \underline{H}_G^*(G^+\wedge S^{2k}) \rightarrow \underline{H}_G^{*+1}(G^+\wedge S^{2m}) \cong \underline{H}_G^{*+1-2m}G^+$$

$$\underline{H}_G^{*-W}S^0 \cong \underline{H}_G^*S^W \rightarrow \underline{H}_G^{*+1}(G^+\wedge S^{2m}) \cong \underline{H}_G^{*+1-2m}G^+$$

$$\underline{H}_G^{*-2k}G^+ \cong \underline{H}_G^*(G^+\wedge S^{2k}) \rightarrow \underline{H}_G^{*+1}S^V \cong \underline{H}_G^{*+1-V}S^0$$

and

$$\underline{H}_G^{*-W}S^0 \cong \underline{H}_G^*S^W \rightarrow \underline{H}_G^{*+1}S^V \cong \underline{H}_G^{*+1-V}S^0.$$

Here, we use $\underline{H}_G^*(G^+\wedge S^{2k})$ and $\underline{H}_G^*S^W$ to denote summands of $\underline{H}_G^*X_n^+$ isomorphic to $\underline{H}_G^*G^+$ in dimension 2k or $\underline{H}_G^*S^0$ in dimension W. The four maps above are all maps of RO(G)-graded modules over $\underline{H}_G^*S^0$. Any such map out of $\underline{H}_G^*S^0$ is determined by the image of $1 \in A(1) = \underline{H}_G^0(S^0)(1)$. By Proposition 4.12, such a map out of $\underline{H}_G^*G^+$ is determined by the image of $(1,0,0, \dots ,0) \in \mathbb{Z}^p = \underline{H}_G^0(G^+)(e)$. Thus, to show that the four maps are zero, it suffices to show that the groups $\underline{H}_G^{2k+1-2m}(G^+)(e)$, $\underline{H}_G^{W+1-2m}(G^+)(1)$, $\underline{H}_G^{2k+1-V}(S^0)(e)$, and $\underline{H}_G^{W+1-V}(S^0)(1)$ are zero. The integers $|2k+1-2m|$ and $|W+1-2m|$ are odd and $\underline{H}_G^\alpha G^+$ vanishes whenever $|\alpha|$ is odd, so the first two groups are zero. The integer $|2k+1-V|$ is odd and $\underline{H}_G^\alpha(S^0)(e)$ vanishes when $|\alpha|$ is odd, so the third group is zero. For the fourth group, if $|V| \leq |W|$, then $\underline{H}_G^{W+1-V}S^0$ is zero because $|W^G+1-V^G|$ is odd and $|W+1-V|$ is positive. Otherwise, $|V^G| \geq |W^G|$, and $\underline{H}_G^{W+1-V}S^0$ is zero because $|W^G+1-V^G|$ is at most one. An analogous proof shows that the map ∂ in the homology sequence is zero. Note that if $|V|>|W|$ and $|V^G|=|W^G|$, then the vanishing of $\underline{H}_G^{W+1-V}S^0$ is a result of the anomalous zeroes on the $|\alpha^G| = 1$ line in the graph of $\underline{H}_G^\alpha S^0$.

In order to compute the ring structure of the equivariant cohomology of X, we must compare it with more familiar objects, such as the nonequivariant ordinary cohomology of X and X^G. If X is a generalized G-cell complex satisfying the conditions of either part of Theorem 2.6, then so is X^G. Thus, Examples 1.1(h) describes $\underline{H}_G^*(X^G)^+$ in terms of the nonequivariant cohomology of X^G. Since the group $\underline{H}_G^*(X^+)(e)$ is just the nonequivariant ordinary cohomology of X with $\mathbb{Z}$ coefficients, the map

$$\rho \oplus i^* : \underline{H}_G^\alpha(X^+)(1) \rightarrow \underline{H}_G^\alpha(X^+)(e) \oplus \underline{H}_G^\alpha((X^G)^+)(1)$$

offers a comparison between $\underline{H}_G^\alpha(X^+)(1)$ and two more easily understood cohomology rings. This map does not detect the torsion in $\underline{H}_G^*(X^+)(1)$ coming from the fourth quadrant torsion in $\underline{H}_G^*S^0$. Moreover, the torsion in $\underline{H}_G^*((X^G)^+)(1)$ makes it hard to compute the image of $\rho \oplus i^*$. These difficulties suggest that we also consider the image of $\underline{H}_G^*(X^+)(1)/\text{torsion}$ in $(\underline{H}_G^\alpha(X^+)(e) \oplus \underline{H}_G^\alpha((X^G)^+)(1))/\text{torsion}$. Since $\underline{H}_G^*(X^+)(e)$ contains no torsion, in the range we are only collapsing out the torsion in $\underline{H}_G^*((X^G)^+)(1)$. The most useful comparison map is produced by also collapsing out the image of the transfer map τ from $\underline{H}_G^*((X^G)^+)(e)$. The quotient

$$\underline{H}_G^*((X^G)^+)(1)/(\text{torsion} \ominus \text{im } \tau)$$

consists of copies of $\mathbb{Z}$ in various dimensions; there is one $\mathbb{Z}$ in the quotient for each $A[d]$ or $\langle \mathbb{Z} \rangle$ which appears in $\underline{H}_G^*((X^G)^+)(1)$.

For many spaces, including complex projective spaces with linear actions, the cells can be ordered so that $|V| \geq |W|$ whenever $DV \in J_n$, $DW \in J_k$, and $k < n$. When the cells can be so ordered, there is no torsion in $\underline{H}_G^*(X^+)(1)$ in the dimensions of the generators of $\underline{H}_G^*X^+$ as a module over $\underline{H}_G^*S^0$. Therefore, the collapsing we have done causes a minimal loss of information. The following result describes the extent to which $\underline{H}_G^*(X^+)(1)$ is detected by $\rho \oplus i^*$.

COROLLARY 2.7. Let X be a generalized G-cell complex satisfying the conditions of either part of Theorem 2.6 and let $i: X^G \to X$ be the inclusion of the fixed point set. Then, for any $\alpha \in RO(G)$ with $|\alpha|$ even, the map

$$\rho \oplus i^* : \underline{H}_G^\alpha(X^+)(1) \to \underline{H}_G^\alpha(X^+)(e) \oplus \underline{H}_G^\alpha((X^G)^+)(1)$$

is a monomorphism. Moreover, for any $\alpha \in RO(G)$, the map

$$\rho \oplus i^* : (\underline{H}_G^\alpha(X^+)(1))/\text{torsion} \to \underline{H}_G^\alpha(X^+)(e) \oplus (\underline{H}_G^\alpha((X^G)^+)(1))/(\text{torsion} \oplus \text{im } \tau)$$

is a monomorphism.

PROOF. Since the equivariant cohomology of X is the limit of the cohomologies of the X_n, it suffices to show that the result holds for every X_n. It is easy to check the second part for X_0. Assume the second part for X_n, and let x be an element of $\underline{H}_G^\alpha(X_{n+1}^+)(1)/\text{torsion}$ vanishing under the map into

$$\underline{H}_G^\alpha(X_{n+1}^+)(e) \oplus (\underline{H}_G^\alpha((X_{n+1}^G)^+)(1))/(\text{torsion} \oplus \text{im } \tau)$$

induced by $\rho \oplus i^*$. We must show that x is zero. The group $\underline{H}_G^\alpha(X_{n+1}^+)(1)$ is the direct sum of the groups $\underline{H}_G^{\alpha}(J_{n+1}/\partial J_{n+1})(1)$ and $\underline{H}_G^{\alpha}(X_n^+)(1)$, and this decomposition is respected by the map $\rho \oplus i^*$. Thus, x is the sum of classes y and z in $\underline{H}_G^{\alpha}(J_{n+1}/\partial J_{n+1})(1)/\text{torsion}$ and $\underline{H}_G^{\alpha}(X_n^+)(1)/\text{torsion}$, respectively, which vanish under the analogous maps. By our inductive hypothesis, z is zero. Since $J_{n+1}/\partial J_{n+1}$ is a wedge of copies of S^V and $G^+ \wedge S^{2k}$ for various V and k, y vanishes by our remark about X_0. Thus, x is zero. The proof of the first part is similar. For this part, we must assume that $|\alpha|$ is even because the map $\rho \oplus i^*$ does not detect the torsion in the fourth quadrant of $\underline{H}_G^*(S^0)(1)$.

3. THE COHOMOLOGY OF COMPLEX PROJECIVE SPACES. As an application of the results from section two, we show that the cohomology of a complex projective space with a linear action is free over $\underline{H}_G^* S^0$. Let V be a finite or countably infinite dimensional complex G-representation and let $\mathbb{C}^*$ be $\mathbb{C} - \{0\}$. The complex projective space P(V) with linear G-action associated to V is the quotient G-space $(V - \{0\})/\mathbb{C}^*$. Note that if $W \subset V$, then P(W) may be regarded as a subspace of P(V). If V is infinite dimensional, then we topologize V as the colimit of its finite dimensional subspaces W; the quotient topology on P(V) is then the same as the colimit topology from the associated subspaces P(W). To describe the cohomology of P(V), we must write V as the sum $\sum_{i=0}^{n} \phi_i$ of irreducible complex representations (including possibly the trivial complex representation). Of course, if V is infinite dimensional, then $n = \infty$. Points in P(V) will be described by homogeneous coordinates of the form

$$\langle x_0, x_1, x_2, \dots, x_n \rangle, \qquad x_i \in \phi_i$$

with the conventions that not all of the x_i are zero, and if V is infinite dimensional, that all but finitely many of the x_i are zero. Each element of the group G acts on each homogeneous coordinate of P(V) by multiplication by a complex number. Therefore, if all the irreducibles in V are isomorphic, then the action of G on P(V) is trivial. Moreover, if η is any irreducible complex representation, then P(V) and $P(\eta V)$ are isomorphic G-spaces. If η and ϕ are irreducible complex representations, then $P(\eta)$ is just a point and $P(\eta \oplus \phi)$ is G-homeomorphic to the one-point compactification of either $\eta^{-1}\phi$ or $\eta\phi^{-1}$.

Since we have selected a colimit topology on P(V) when V is infinite, to show that P(V) is a generalized G-cell complex for any G-representation V, it suffices to show this when V is finite dimensional. Let V_k be the representation $\sum_{i=0}^{k-1} \phi_i$ and let W be the representation $\phi_n^{-1} V_{n-1}$. Describe points in the unit disk DW by complex coordinates $(x_0, x_1, \dots, x_{n-1})$, with $x_i \in \phi_n^{-1}\phi_i$. Define a map f: DW $\rightarrow$ P(V) by

$$f((x_0, x_1, \dots, x_{n-1})) = \langle x_0, x_1, x_2, \dots, x_{n-1}, 1 - \sum_{i=0}^{n-1} |x_i|^2 \rangle.$$

The tensor product with ϕ_n^{-1} is inserted in the definition of W to ensure that the map f is equivariant. The image of SW in P(V) lies in the subspace $P(V_{n-1})$ of P(V), and f is a homeomorphism from DW − SW to its image in P(V). Thus P(V) is formed from $P(V_{n-1})$ by adjoining the G-cell DW along the map $f|SW : SW \to P(V_{n-1})$. Working backwards through the sequence of representations V_k, we conclude that P(V) is a generalized G-cell complex with cells the unit disks of the representations $\phi_k^{-1} V_k$ for $1 \leq k \leq n$.

In order to show that the equivariant cohomology of P(V) is free over $\underline{H}_G^* S^0$, we must show that the cells of P(V) can be attached in an order satisfying the condition in Theorem 2.6(a). This proper ordering of cells is derived from a careful ordering of the set Φ of irreducible summands of V. Since the remainder of our discussion focuses on Φ, we write $P(\Phi)$ for P(V). An ordering $\phi_0, \phi_1, \phi_2, \ldots$ of the elements of Φ is said to be proper if the number of irreducibles in the set $\{\phi_i\}_{0 \leq i \leq k-1}$ isomorphic to ϕ_k is a nondecreasing function of k. For example, if ϕ and η are distinct complex irreducibles and Φ consists of two copies of ϕ and one of η, then η, ϕ, ϕ and ϕ, η, ϕ are proper orderings of Φ, but ϕ, ϕ, η is not. The dimension of the fixed subrepresentation of the representation $\phi_k^{-1} \sum_{i=0}^{k-1} \phi_i$ is the number of irreducibles in the set $\{\phi_i\}_{0 \leq i \leq k-1}$ isomorphic to ϕ_k. Thus, if Φ is properly ordered, then the cell structure described above satisfies the conditions of Theorem 2.6.(a).

PROPOSITION 3.1. If $\phi_0, \phi_1, \phi_2, \ldots$ is any ordering of the elements of a set Φ of irreducible representations, then $P(\Phi)$ is a generalized G-cell complex with cells the unit disks of the G-representations $\phi_k^{-1} \sum_{i=0}^{k-1} \phi_i$, for $k \geq 1$. Moreover, $\underline{H}_G^* P(\Phi)^+$ and $\underline{H}_*^G P(\Phi)^+$ are free RO(G)-graded modules over $\underline{H}_G^* S^0$. If the ordering of Φ is proper, then the homology and cohomology of $P(\Phi)$ are each generated by one element in dimension zero and one in each of the dimensions $\phi_k^{-1} \sum_{i=0}^{k-1} \phi_i$, for $k \geq 1$.

The G-fixed subspace of $P(\Phi)$ is a disjoint union of complex projective spaces, one for each isomorphism class of irreducibles in Φ. The (complex) dimension of the complex projective space in $P(\Phi)^G$ associated to the irreducible ϕ is one less than the multiplicity of ϕ in Φ. Thus, the effect of properly ordering the irreducibles is that the maximal dimension of the components of the G-fixed subspace of $P(\{\phi_i\}_{0 \leq i \leq k})$ increases as slowly as possible with increasing k.

REMARKS 3.2. Our description of the cohomology of $P(\Phi)$ contains one apparent anomaly. Suppose that ζ, η, and ϕ are distinct complex irreducible representations and $\Phi = \{\zeta, \eta, \phi\}$. If we assign the proper ordering ζ, η, ϕ to Φ, then we find that the generators of $\underline{H}_G^* P(\Phi)^+$ are in dimensions 0, $\eta^{-1}\zeta$, and $\phi^{-1}(\zeta \oplus \eta)$. However, if we select the proper ordering ϕ, ζ, η, we find that the generators are in dimensions 0, $\zeta^{-1}\phi$, and $\eta^{-1}(\phi \oplus \zeta)$. In particular, the cohomology in dimension $\eta^{-1}\zeta$ must be $A \oplus \langle \mathbb{Z} \rangle \oplus \langle \mathbb{Z} \rangle$ if we use the first set of generators, and $A[d] \oplus \langle \mathbb{Z} \rangle \oplus \langle \mathbb{Z} \rangle$ if we use the second, where d is the integer associated to the element $\eta^{-1}\zeta - \zeta^{-1}\phi$ of $RO_0(G)$. There is no contradiction in these two claims about the cohomology in dimension

$\eta^{-1}\zeta$ because these two Mackey functors are isomorphic by Examples 1.1.(d). The apparent difficulties in all the other dimensions are resolved in exactly the same way.

This example illustrates the latitude that one has in selecting the dimensions of the generators of the cohomology of $P(\Phi)$ for almost any set Φ of irreducibles. This latitude is necessary because, for most Φ, there are many proper orderings and a choice of a proper ordering corresponds to a selection of the dimensions of the generators.

It would be nice to have some simple cohomology invariants of $P(\Phi)$ which could be used for problems like comparing the cohomology of projective spaces with different G-actions. The fact that the dimensions for the cohomology generators don't provide such an invariant is a disappointment. However, one invariant related to the dimensions of the generators is readily available. Select a proper ordering of Φ and plot the dimensions α of the resulting set of generators of $\underline{H}_G^* P(\Phi)^+$ on a coordinate plane whose horizontal and vertical axes indicate $|\alpha^G|$ and $|\alpha|$, respectively. The dimensions lie on a stair-step pattern whose foot is at the origin. This plot is an invariant of $P(\Phi)$. The height of the steps in the plot decreases, or remains constant, as one goes up the steps (that is, moves in the direction of increasing $|\alpha^G|$ and $|\alpha|$). The height remains constant only if irreducible types appearing in Φ have equal multiplicity. The step-like structure of the plot reflects a filtration on Φ which plays an important role in our discussion of the ring structure of $\underline{H}_G^* P(\Phi)^+$. An increasing filtration

$$\emptyset = \Phi(0),\ \Phi(1),\ \Phi(2),\ \ldots,\ \Phi(r),\ \ldots$$

of the set Φ is said to be proper if, for every r and every complex irreducible ϕ, the number of irreducibles in $\Phi(r)$ isomorphic to ϕ is the lesser of r and the number of irreducibles in Φ isomorphic to ϕ. Any two proper filtrations of Φ differ only by an interchange of isomorphic irreducible complex representations, so there is essentially only one proper filtration of Φ. The steps in the plot of the dimensions of the generators are in a one-to-one correspondence with the stages in the filtration of Φ. The height of the step corresponding to filtration level r is the number of elements in $\Phi(r) - \Phi(r-1)$.

4. CUP PRODUCTS IN $\underline{H}_G^* S^0$. Here we describe the multiplicative structure of $\underline{H}_G^* S^0$. We begin with the case $p = 2$, which is due to Stong.

DEFINITIONS 4.1. Let ζ be the real one-dimensional sign representation of $G = \mathbb{Z}/2$. The identity element 1 in $A(1) = \underline{H}_G^0(S^0)(1)$ is the identity element of the RO(G)-graded Mackey functor ring $\underline{H}_G^* S^0$. Let $\kappa \in \underline{H}_G^0(S^0)(1)$ be $2 - \tau\rho(1)$. Observe that $\kappa^2 = 2\kappa$. Let $\epsilon \in \underline{H}_G^{\zeta}(S^0)(1)$ be the Euler class; that is, the image of $1 \in \underline{H}_G^0(S^0)(1)$ under the map induced by the inclusion $S^0 \subset S^{\zeta}$. Select a nonequivariant identication of S^ζ with S^1 and let $\iota_{1-\zeta} \in \underline{H}_G^{1-\zeta}(S^0)(e) \cong \underline{H}_G^1(S^\zeta)(e)$ and $\iota_{\zeta-1} \in \underline{H}_G^{\zeta-1}(S^0)(e) \cong \underline{H}_G^{\zeta}(S^1)(e)$ be the images of $\rho(1) \in \underline{H}_G^0(S^0)(e) \cong \underline{H}_G^1(S^1)(e)$ under the maps induced by this identification. Let $\xi \in \underline{H}_G^{2\zeta-2}(S^0)(1)$ be the unique element with $\rho(\xi) = \iota_{\zeta-1}^2$. The elements 1 and κ generate the abelian group $\underline{H}_G^0(S^0)(1)$ and the Mackey functor $\underline{H}_G^0 S^0$. Each of the elements ϵ^m, ξ^m, and $\epsilon^m\xi^n$, for m, n $\geq$ 1, generates the abelian group $\underline{H}_G^\alpha(S^0)(1)$ and the Mackey functor $\underline{H}_G^\alpha S^0$ in the appropriate dimension α. For m $\geq$ 1, the element $\iota_{1-\zeta}^m$ or $\iota_{\zeta-1}^m$ generates the abelian group $\underline{H}_G^\alpha(S^0)(e)$ in the appropriate dimension and $\iota_{1-\zeta}^m$ generates the Mackey functor $\underline{H}_G^* S^0$ in the appropriate dimension. For m $\geq$ 2, $\tau(\iota_{1-\zeta}^m)$ generates the abelian group $\underline{H}_G^*(S^0)(1)$ in the appropriate dimension.

LEMMA 4.2. The class $\kappa \in \underline{H}_G^0(S^0)(1)$ and, for n $\geq$ 1, the classes

$$\tau(\iota_{1-\zeta}^{2n+1}) \in \underline{H}_G^{(2n+1)(1-\zeta)}(S^0)(1)$$

are infinitely divisible by $\epsilon \in \underline{H}_G^\zeta(S^0)(1)$; that is, for m $\geq$ 1, there are unique elements

$$\epsilon^{-m}\kappa \in \underline{H}_G^{-m\zeta}(S^0)(1)$$

and

$$\epsilon^{-m}\tau(\iota_{1-\zeta}^{2n+1}) \in \underline{H}_G^{2n+1-(2n+m+1)\zeta}(S^0)(1)$$

such that

$$\epsilon^m(\epsilon^{-m}\kappa) = \kappa \quad \text{and} \quad \epsilon^m(\epsilon^{-m}\tau(\iota_{1-\zeta}^{2n+1})) = \tau(\iota_{1-\zeta}^{2n+1}).$$

Moreover, each of the elements $\epsilon^{-m}\kappa$ or $\epsilon^{-m}\tau(\iota^{2n+1})$ generates the abelian group $\underline{H}_G^*(S^0)(1)$ and the Mackey functor $\underline{H}_G^* S^0$ in its dimension.

THEOREM 4.3. The elements

$$\epsilon \in \underline{H}_G^\zeta(S^0)(1)$$
$$\iota_{1-\zeta} \in \underline{H}_G^{1-\zeta}(S^0)(e)$$
$$\iota_{\zeta-1} \in \underline{H}_G^{\zeta-1}(S^0)(e)$$
$$\xi \in \underline{H}_G^{2\zeta-2}(S^0)(1)$$
$$\epsilon^{-m}\kappa \in \underline{H}_G^{-m\zeta}(S^0)(1), \qquad \text{for m} \geq 1,$$

and

$$\epsilon^{-m}\tau(\iota_{1-\zeta}^{2n+1}) \in \underline{H}_G^{2n+1-(2n+m+1)\zeta}(S^0)(1), \qquad \text{for m, n} \geq 1,$$

generate $\underline{H}_G^* S^0$ as an RO(G)-graded Mackey functor algebra over the Burnside Mackey functor ring A. The only relations among these elements, other than those forced by the Frobenius relations or the vanishing of $\underline{H}_G^* S^0$ in various dimensions, are generated by the relations

$$\begin{aligned}
\rho(\epsilon) &= 0 \\
\iota_{1-\zeta}\,\iota_{\zeta-1} &= \rho(1) \\
\tau(\iota_{1-\zeta}) &= 0 \\
\tau(\iota_{\zeta-1}^{2m+1}) &= 0, && \text{for } m \geq 0, \\
\tau(\iota_{\zeta-1}^{2m}) &= 2\xi^m, && \text{for } m \geq 1, \\
\tau(\iota_{1-\zeta}^{m})\,\tau(\iota_{1-\zeta}^{n}) &= \begin{cases} 0, & \text{if m or n is odd,} \\ 2\tau(\iota_{1-\zeta}^{m+n}), & \text{if m and n are even,} \end{cases} \\
\rho(\xi) &= \iota_{\zeta-1}^2 \\
2\epsilon\xi &= 0 \\
\rho(\epsilon^{-m}\kappa) &= 0, && \text{for } m \geq 0, \\
\epsilon(\epsilon^{-m}\kappa) &= \epsilon^{1-m}\kappa, && \text{for } m \geq 1, \\
(\epsilon^{-m}\kappa)(\epsilon^{-n}\kappa) &= 2\epsilon^{-(m+n)}\kappa, && \text{for } m, n \geq 0, \\
2\epsilon^{-m}\tau(\iota_{1-\zeta}^{2n+1}) &= 0, && \text{for } m \geq 0 \text{ and } n \geq 1, \\
\rho(\epsilon^{-m}\tau(\iota_{1-\zeta}^{2n+1})) &= 0, && \text{for } m \geq 0 \text{ and } n \geq 1, \\
\epsilon(\epsilon^{-m}\tau(\iota_{1-\zeta}^{2n+1})) &= \epsilon^{1-m}\tau(\iota_{1-\zeta}^{2n+1}), && \text{for } m, n \geq 1, \\
(\epsilon^{-m}\tau(\iota_{1-\zeta}^{2n+1}))(\epsilon^{-q}\kappa) &= 0, && \text{for } m, q \geq 0 \text{ and } n \geq 1,
\end{aligned}$$

and

$$\xi(\epsilon^{-m}\tau(\iota_{1-\zeta}^{2n+1})) = \epsilon^{-m}\tau(\iota_{1-\zeta}^{2n-1}), \qquad \text{for } m \geq 0 \text{ and } n \geq 2.$$

REMARKS 4.4. (a) The last relation indicates that, for $m \geq 0$ and $n \geq 1$, $\epsilon^{-m}\tau(\iota_{1-\zeta}^{2n+1})$ is infinitely divisible by ξ. Thus, we can think of all the elements in the fourth quadrant of the graph of $\underline{H}_G^*(S^0)$ as being derived from $\tau(\iota_{1-\zeta}^3)$ via division by powers of ϵ and ξ. One mnemonic for the effect of ϵ and ξ on the elements in the fourth quadrant is to denote the nonzero element in $\underline{H}_G^{1-m\zeta-2n(\zeta-1)}(S^0)(1)$, for $m \geq 2$ and $n \geq 1$, by $\epsilon^{-m}\xi^{-n}\omega$, where ω is regarded as a fictitious element in dimension 1. The reason for selecting a fictitious element in dimension 1, instead of the actual element in dimension $3-3\zeta$, is discussed in Remarks 4.10(b).

(b) For $p = 2$, the elements $\pm(1-\tau\rho(1))$ in A(1) are units, and $1-\tau\rho(1)$ appears in the formula describing the anticommutativity of cup products. For any G-space X, if $a \in \underline{H}_G^{i+j\zeta}X^+$ and $b \in \underline{H}_G^{m+n\zeta}X^+$, then

$$ab = (-1)^{im}(1-\tau\rho(1))^{jn}\,ba.$$

The generators $\iota_{1-\zeta}$, $\iota_{\zeta-1}$, ϵ, $\epsilon^{-n}\kappa$, and $\epsilon^{-m}\tau(\iota_{1-\zeta}^{2n+1})$ are in dimensions where the behavior of this nontrivial unit matters. Of course, since $\epsilon^{-m}\tau(\iota_{1-\zeta}^{2n+1})$ has order 2, any unit acts trivially on it. It is easy to check that

$$(1-\tau\rho(1))\,\iota_{1-\zeta} = -\iota_{1-\zeta} \quad \text{and} \quad (1-\tau\rho(1))\,\iota_{\zeta-1} = -\iota_{\zeta-1}.$$

This action of $1-\tau\rho(1)$ on $\iota_{1-\zeta}$ and $\iota_{\zeta-1}$ never affects cup products in $\underline{H}_G^* S^0$ because it is always balanced by the $(-1)^{im}$ term in the commutativity formula. However, there are algebras over $\underline{H}_G^* S^0$ where the effects of this unit on $\iota_{1-\zeta}$ and $\iota_{\zeta-1}$ are visible. The unit $1-\tau\rho(1)$ acts trivially on ϵ and $\epsilon^{-n}\kappa$. This shows up dramatically in $\underline{H}_G^* S^0$. The elements ϵ and $\epsilon^{-2n+1}\kappa$ are odd-dimensional, so our intuition about graded algebras from the nonequivariant context suggests that their squares should vanish, or at least be 2-torsion. In fact, the squares are not torsion elements, an apparent anomaly possible only because the action of $1-\tau\rho(1)$ is trivial. The overall effect of the actions of the units of A on the generators of $\underline{H}_G^* S^0$ is that $\underline{H}_G^* S^0$ is commutative in both the graded and the ungraded sense.

When p is odd, several complications in the multiplicative structure of $\underline{H}_G^* S^0$ arise from the greater complexity of RO(G). The most obvious are a host of sign problems coming from the identification of representations with their complex conjugates. Initially, we resolve these sign problems by grading $\underline{H}_G^* S^0$ on RSO(G) instead of RO(G). In Remark 4.11, we explain steps which must be taken to pass back to an RO(G)-grading. The most serious complication arises from the misbehavior of the integers d_α associated to the virtual representations α in $RSO_0(G)$. One way to deal with this complication is to avoid it. This can be done very nicely if one is only interested in $\underline{H}_G^* S^0$. Because of the intuition this approach offers, we outline it as an introduction to the odd primes case.

The stable homotopy groups $\underline{\pi}_\beta^G S^0$, for $\beta \in RSO_0(G)$, have been studied extensively by tom Dieck and Petrie [tDP], and the stable Hurewicz map

$$h: \underline{\pi}_{-\beta}^G S^0 \to \underline{H}_{-\beta}^G S^0 \cong \underline{H}_G^\beta S^0.$$

is an isomorphism [LE1] if $\beta \in RSO_0(G)$. Thus, many of their results can be applied to homology in the appropriate dimensions. They have shown that the multiplication map

$$\underline{\pi}_\beta^G S^0 \,\Box\, \underline{\pi}_\gamma^G S^0 \to \underline{\pi}_{\beta+\gamma}^G S^0$$

is an isomorphism for any $\beta \in RSO_0(G)$ and any $\gamma \in RSO(G)$. By similar reasoning, the multiplication map

$$\underline{H}_G^\beta S^0 \,\Box\, \underline{H}_G^\gamma S^0 \to \underline{H}_G^{\beta+\gamma} S^0$$

is an isomorphism under the same conditions on β and γ. Thus, to understand all of $\underline{H}_G^* S^0$, it suffices to understand the part of $\underline{H}_G^* S^0$ which tom Dieck and Petrie have already described and the part indexed on some subset of RSO(G) complementary to $RSO_0(G)$. Recall that λ is the irreducible complex representation that takes the standard generator of $\mathbb{Z}/p$ to $e^{2\pi i/p}$. Let $RSO_\lambda(G)$ be the additive subgroup of $RSO(G)$ generated by 1 and λ. As an additive group, $RSO(G)$ is the internal direct sum of $RSO_0(G)$ and $RSO_\lambda(G)$. To complete our description of $\underline{H}^*_G S^0$, it suffices to describe that part of it indexed on $RSO_\lambda(G)$. This part is almost identical to $\underline{H}^*_G S^0$ for $G = \mathbb{Z}/2$. Consider the description given above of that part of $\underline{H}^*_G S^0$ for $p = 2$ indexed on the additive subgroup of $RO(\mathbb{Z}/2)$ generated by 1 and 2ζ. Replace 2ζ by λ and all the other 2's by p's. The result is a description of the part of $\underline{H}^*_G S^0$ for p odd indexed on $RSO_\lambda(G)$. This approach describes $\underline{H}^*_G S^0$ as the graded box product of two subrings indexed on complementary subsets of $RSO(G)$. The unpleasant behavior of the integers d_α is buried in the computations of the box products.

Unfortunately, because of peculiarities in the dimensions of the algebra generators of $\underline{H}^*_G P(V)^+$, this description of $\underline{H}^*_G S^0$ as the box product of two subrings can not be used to describe the ring structure of the cohomology of complex projective spaces. Thus, we offer an alternative description of the ring structure of $\underline{H}^*_G S^0$ for p odd. In section 2, we defined a function from $RO_0(G)$ to $\mathbb{Z}$ using a section of the projection from $\tilde{R}_0(G)$ to $RO_0(G)$. Since we are now working with $RSO_0(G)$ instead of $RO_0(G)$, we define an analogous function from $RSO_0(G)$ to $\mathbb{Z}$ using a section $s\colon RSO_0(G) \to \tilde{R}_0(G)$ of the projection from $\tilde{R}_0(G)$ to $RO_0(G)$. We insist that $s(0) = 0$ and that our original section $RO_0(G) \to \tilde{R}_0(G)$ factor through s.

DEFINITIONS 4.5. (a) If $\alpha \in RSO_0(G)$ and $s(\alpha) = \sum_i \phi_i - \eta_i$, then we wish to define an equivariant map $\mu_\alpha\colon S^{\Sigma\eta_i} \to S^{\Sigma\phi_i}$ with nonequivariant degree d_α. If $\alpha = \lambda^m - \lambda^n$ with $0 < m, n < p$ and n^{-1} is the unique integer such that $1 \leq n^{-1} \leq p-1$ and $n n^{-1} \equiv 1 \bmod p$, then μ_α is the extension to one-point compactifications of the complex power map $z \to z^{m(n^{-1})}$, for $z \in \mathbb{C}$. In general, we identify $S^{\Sigma\phi_i}$ and $S^{\Sigma\eta_i}$ with $\bigwedge_i S^{\phi_i}$ and $\bigwedge_i S^{\eta_i}$, respectively, and take the smash product of the appropriate complex power maps to obtain the equivariant map μ_α from $S^{\Sigma\phi_i}$ to $S^{\Sigma\eta_i}$ with nonequivariant degree d_α. Also denote by μ_α the image of this map in $\underline{H}^\alpha_G(S^0)(1)$ under the Hurewicz map. Clearly, if the ϕ_i and the η_i were paired off in a different order, then a different map from $S^{\Sigma\phi_i}$ to $S^{\Sigma\eta_i}$ would be obtained. However, the maps coming from the two pairings would be equivariantly homotopic and so would give the same element in $\underline{H}^\alpha_G(S^0)(1)$.

(b) Let α be an element of $RSO(G)$ with $|\alpha| = 0$. Then α must be represented by a sum $\sum_i \phi_i - \eta_i$, where the ϕ_i and η_i are irreducible complex representations, some of which may be trivial. Since the ϕ_i and η_i are complex representations, they have canonical nonequivariant orientations. Combine these to produce a nonequivariant identification ι_α of $S^{\Sigma\phi_i}$ with $S^{\Sigma\eta_i}$ which is unique up to homotopy. Let ι_α also denote the image of this identification in $\underline{H}_G^\alpha(S^0)(e)$. The resulting cohomology classes ι_α are then independent of the ordering of the ϕ_i and the η_i. The abelian group $\underline{H}_G^\alpha(S^0)(e)$ is generated by ι_α. If $|\alpha^G| > 0$, then $\tau(\iota_\alpha)$ generates the abelian group $\underline{H}_G^\alpha(S^0)(1)$ and ι_α generates the Mackey functor $\underline{H}_G^\alpha S^0$.

(c) If $\alpha \in RSO_0(G)$, then in $\underline{H}_G^\alpha S^0$,

$$\rho(\mu_\alpha) = d_\alpha \iota_\alpha \quad \text{and} \quad \rho\tau(\iota_\alpha) = p\,\iota_\alpha.$$

We have already asserted that $\underline{H}_G^\alpha S^0$ is $A[d_\alpha]$. Under this identification, μ_α and $\tau(\iota_\alpha)$ become the elements μ and τ of $A[d_\alpha](1)$ and ι_α becomes $1 \in \mathbb{Z} = A[d_\alpha](e)$. There is a unique integer b_α such that $d_{-\alpha} d_\alpha + b_\alpha p = 1$. Let $\kappa_\alpha = p\,\mu_\alpha - d_\alpha \tau(\iota_\alpha)$ and $\sigma_\alpha = d_{-\alpha}\mu_\alpha + b_\alpha \tau(\iota_\alpha)$. Then, σ_α and κ_α form an alternative $\mathbb{Z}$-basis for $\underline{H}_G^\alpha(S^0)(1)$.

(d) Let β be an element of RSO(G) with $|\beta| > 0$ and $|\beta^G| = 0$. There exist an α in $RSO_0(G)$ and a G-representation V such that $V^G = 0$ and $\beta = \alpha + V$. Let $\epsilon_\beta \in \underline{H}_G^\beta(S^0)(1)$ be the image of $\mu_\alpha \in \underline{H}_G^\alpha(S^0)(1)$ under the map from $\underline{H}_G^\alpha(S^0)(1)$ to $\underline{H}_G^\beta(S^0)(1)$ induced by the inclusion $S^0 \subset S^V$. In Lemma A.11, it is shown that this Euler class ϵ_β is independent of the choice of the decomposition of β into the sum of the representation V and the element α of $RSO_0(G)$. The class ϵ_β generates the abelian group $\underline{H}_G^\beta(S^0)(1)$ and the Mackey functor $\underline{H}_G^\beta S^0$.

(e) If $|\alpha| = 0$ and $|\alpha^G| < 0$, let ξ_α be the unique element of $\underline{H}_G^\alpha(S^0)(1)$ with $\rho(\xi_\alpha) = \iota_\alpha$; this class generates the abelian group $\underline{H}_G^\alpha(S^0)(1)$ and the Mackey functor $\underline{H}_G^\alpha S^0$.

When p is odd, it is harder to pick a multiplicative basis for the torsion in the fourth quadrant of the graph of $\underline{H}_G^* S^0$. In each dimension there is a choice of $p-1$ generators, instead of a single nonzero element. Moreover, since these torsion elements are not tied by an Euler class to elements on the positive horizontal axis, there is no way to base the choice of a generator on choices already made for the axis. The following lemma justifies the procedure we employ to select multiplicative generators for the fourth quadrant.

LEMMA 4.6. Let β be an element of $RSO_0(G)$ and let α, γ, and δ be elements of RSO(G) such that

$$|\delta| = |\gamma^G| = 0,$$
$$|\alpha|, |\delta^G| < 0,$$
$$|\gamma| > 0,$$
$$|\alpha^G| \geq 3,$$

and

$$|\alpha^G| \text{ is odd.}$$

If x is any nonzero element in $\underline{H}_G^\alpha(S^0)(1)$, then $\mu_\beta x$ is a generator in $\underline{H}_G^{\alpha+\beta}(S^0)(1)$. Moreover, x and $\mu_\beta x$ are uniquely divisible by both ϵ_γ and ξ_δ.

DEFINITIONS 4.7. Select a generator in $\underline{H}_G^{3-2\lambda}(S^0)(1)$ and denote it by $\nu_{3-2\lambda}$. If $\alpha = 1 - m(\lambda - 2) - n\lambda$, for m, n ≥ 1, then let ν_α be the unique element in $\underline{H}_G^\alpha(S^0)(1)$ such that

$$\epsilon_{(n-1)\lambda}\,\xi_{(m-1)(\lambda-2)}\,\nu_\alpha \;=\; \nu_{3-2\lambda}.$$

For any $\alpha \in RSO(G)$, there are unique integers m, n, and q such that q = 0 or 1 and

$$\alpha - [q - m(\lambda - 2) - n\lambda] \;\in\; RO_0(G).$$

Denote by $<\alpha>$ the element $q - m(\lambda - 2) - n\lambda$ associated to α by these conditions. If $\alpha \in RSO(G)$ with $|\alpha| < 0$, $|\alpha^G| \geq 3$, $|\alpha^G|$ odd, and $\alpha \neq <\alpha>$, then define $\nu_\alpha \in \underline{H}_G^\alpha(S^0)(1)$ by

$$\nu_\alpha \;=\; \mu_{\alpha-<\alpha>}\,\nu_{<\alpha>}.$$

The element ν_α generates the abelian group $\underline{H}_G^\alpha(S^0)(1)$ and the Mackey functor $\underline{H}_G^\alpha S^0$.

LEMMA 4.8. If $\alpha \in RSO_0(G)$, then $\kappa_\alpha \in \underline{H}_G^\alpha(S^0)(1)$ is divisible by ϵ_β, for any $\beta \in RSO(G)$ with $|\beta| > 0$ and $|\beta^G| = 0$; that is, there is a unique element

$$\epsilon_\beta^{-1}\kappa_\alpha \;\in\; \underline{H}_G^{\alpha-\beta}(S^0)(1)$$

such that

$$\epsilon_\beta(\epsilon_\beta^{-1}\kappa_\alpha) \;=\; \kappa_\alpha.$$

The element $\epsilon_\beta^{-1}\kappa_\alpha$ generates the abelian group $\underline{H}_G^{\alpha-\beta}(S^0)(1)$ and the Mackey functor

$\underline{H}_G^{\alpha-\beta}S^0$.

THEOREM 4.9. The elements

$$\mu_\alpha \in \underline{H}_G^\alpha(S^0)(1), \qquad \text{for } \alpha = \pm(\lambda^n - \lambda), \text{ with } 1 < n < p,$$
$$\iota_\alpha \in \underline{H}_G^\alpha(S^0)(e), \qquad \text{for } \alpha = \pm(\lambda^n - \lambda), \text{ with } 1 < n < p,$$
$$\epsilon_\lambda \in \underline{H}_G^\lambda(S^0)(1)$$
$$\xi_{\lambda-2} \in \underline{H}_G^{\lambda-2}(S^0)(1)$$
$$\iota_{2-\lambda} \in \underline{H}_G^{2-\lambda}(S^0)(e)$$
$$\epsilon_{m\lambda}^{-1}\kappa_0 \in \underline{H}_G^{-m\lambda}(S^0)(1), \qquad \text{for } m \geq 1,$$

and

$$\nu_\alpha \in \underline{H}_G^\alpha(S^0)(1), \qquad \text{for } \alpha = 1 - m(\lambda - 2) - n\lambda, \text{ with } m, n \geq 1,$$

generate $\underline{H}_G^* S^0$ as an RSO(G)-graded Mackey functor algebra over the Burnside Mackey functor ring A. All of relations among the elements of $\underline{H}_G^* S^0$, other than those forced by the Frobenius relations or the vanishing of $\underline{H}_G^* S^0$ in various dimensions, are generated by the relations

$$\rho(\mu_\alpha) = d_\alpha \iota_\alpha, \qquad \text{for } \alpha \in RSO_0(G);$$

$$\mu_\alpha \mu_\beta = \mu_{\alpha+\beta} + \left[\frac{d_\alpha d_\beta - d_{\alpha+\beta}}{p}\right]\tau(\iota_{\alpha+\beta}), \qquad \text{for } \alpha, \beta \in RSO_0(G);$$

$$\rho(\epsilon_\beta) = 0, \qquad \text{for } |\beta| > 0 \text{ and } |\beta^G| = 0;$$

$$\epsilon_\alpha \epsilon_\beta = \epsilon_{\alpha+\beta}, \qquad \text{for } |\alpha|, |\beta| > 0 \text{ and } |\alpha^G| = |\beta^G| = 0;$$

$$\mu_\alpha \epsilon_\beta = \epsilon_{\alpha+\beta}, \qquad \text{for } \alpha \in RSO_0(G), |\beta| > 0, \text{ and } |\beta^G| = 0;$$

$$\rho(\xi_\alpha) = \iota_\alpha, \qquad \text{for } |\alpha| = 0 \text{ and } |\alpha^G| < 0;$$

$$\tau(\iota_\alpha) = p\,\xi_\alpha, \qquad \text{for } |\alpha| = 0 \text{ and } |\alpha^G| < 0;$$

$$\xi_\alpha \xi_\beta = \xi_{\alpha+\beta}, \qquad \text{for } |\alpha| = |\beta| = 0 \text{ and } |\alpha^G|, |\beta^G| < 0;$$

$$\mu_\alpha \xi_\beta = d_\alpha \xi_{\alpha+\beta}, \qquad \text{for } \alpha \in RSO_0(G), |\beta| = 0, \text{ and } |\beta^G| < 0;$$

$$\mathrm{p}\,\epsilon_\beta\,\xi_\alpha = 0, \qquad \text{for } |\alpha| = |\beta^{\mathrm{G}}| = 0,\ |\alpha^{\mathrm{G}}| < 0, \text{ and } |\beta| > 0;$$

$$\epsilon_\beta\,\xi_\alpha = \mathrm{d}_{\delta-\alpha}\,\epsilon_\gamma\,\xi_\delta, \qquad \text{for } |\alpha| = |\delta| = |\beta^{\mathrm{G}}| = |\gamma^{\mathrm{G}}| = 0,\ |\alpha^{\mathrm{G}}|, |\delta^{\mathrm{G}}| < 0,\ |\beta|, |\gamma| > 0, \text{ and } \alpha + \beta = \gamma + \delta;$$

$$\epsilon_\beta^{-1}\,\kappa_\alpha = \epsilon_\gamma^{-1}\,\kappa_\delta, \qquad \text{for } \alpha, \delta \in \mathrm{RSO}_0(\mathrm{G}),\ |\beta^{\mathrm{G}}| = |\gamma^{\mathrm{G}}| = 0,\ |\beta|, |\gamma| > 0, \text{ and } \alpha + \gamma = \beta + \delta;$$

$$\rho(\epsilon_\beta^{-1}\,\kappa_\alpha) = 0, \qquad \text{for } \alpha \in \mathrm{RSO}_0(\mathrm{G}),\ |\beta^{\mathrm{G}}| = 0, \text{ and } |\beta| > 0;$$

$$\mu_\gamma(\epsilon_\beta^{-1}\,\kappa_\alpha) = \epsilon_\beta^{-1}\,\kappa_{\alpha+\gamma}, \qquad \text{for } \alpha, \gamma \in \mathrm{RSO}_0(\mathrm{G}),\ |\beta^{\mathrm{G}}| = 0, \text{ and } |\beta| > 0;$$

$$\epsilon_\beta(\epsilon_\beta^{-1}\,\kappa_\alpha) = \kappa_\alpha, \qquad \text{for } \alpha \in \mathrm{RSO}_0(\mathrm{G}),\ |\beta^{\mathrm{G}}| = 0, \text{ and } |\beta| > 0;$$

$$\epsilon_\gamma(\epsilon_\beta^{-1}\,\kappa_\alpha) = \epsilon_{\beta-\gamma}^{-1}\,\kappa_\alpha, \qquad \text{for } \alpha \in \mathrm{RSO}_0(\mathrm{G}),\ |\beta^{\mathrm{G}}| = |\gamma^{\mathrm{G}}| = 0, \text{ and } |\beta| > |\gamma| > 0;$$

$$(\epsilon_\beta^{-1}\kappa_\alpha)(\epsilon_\gamma^{-1}\kappa_\delta) = \mathrm{p}\,\epsilon_{\beta+\gamma}^{-1}\,\kappa_{\alpha+\delta}, \qquad \text{for } \alpha, \delta \in \mathrm{RSO}_0(\mathrm{G}),\ |\beta^{\mathrm{G}}| = |\gamma^{\mathrm{G}}| = 0, \text{ and } |\beta|, |\gamma| > 0;$$

$$\mathrm{p}\,\nu_\alpha = 0, \qquad \text{for } |\alpha| < 0,\ |\alpha^{\mathrm{G}}| \geq 3, \text{ and } |\alpha^{\mathrm{G}}| \text{ odd};$$

$$\rho(\nu_\alpha) = 0, \qquad \text{for } |\alpha| < 0,\ |\alpha^{\mathrm{G}}| \geq 3, \text{ and } |\alpha^{\mathrm{G}}| \text{ odd};$$

$$\mu_\beta\,\nu_\alpha = \nu_{\alpha+\beta}, \qquad \text{for } \beta \in \mathrm{RSO}_0(\mathrm{G}),\ |\alpha| < 0,\ |\alpha^{\mathrm{G}}| \geq 3, \text{ and } |\alpha^{\mathrm{G}}| \text{ odd};$$

$$\epsilon_\beta \nu_\alpha = \nu_{\alpha+\beta}, \qquad \text{for } |\alpha+\beta| < 0,\ |\alpha^G| \geq 3,\ |\alpha^G| \text{ odd},\ |\beta| > 0, \text{ and } |\beta^G| = 0;$$

$$\xi_\beta \nu_\alpha = d_{<\beta>-\beta} \nu_{\alpha+\beta}, \qquad \text{for } |\alpha| < 0,\ |\alpha^G + \beta^G| \geq 3,\ |\alpha^G| \text{ odd},\ |\beta| = 0, \text{ and } |\beta^G| < 0;$$

$$(\epsilon_\beta^{-1} \kappa_\gamma) \nu_\alpha = 0, \qquad \text{for } \gamma \in RSO_0(G),\ |\alpha| < 0,\ |\alpha^G| \geq 3,\ |\alpha^G| \text{ odd},\ |\beta^G| = 0, \text{ and } |\beta| > 0;$$

$$\iota_\alpha \iota_\beta = \iota_{\alpha+\beta}, \qquad \text{for } |\alpha| = |\beta| = 0.$$

REMARKS 4.10. (a) For p odd, the only units in A(1) are ± 1. The only generators in odd dimensions are the ν_α. Since $\nu_\alpha \nu_\beta$ is zero for any α and β, no sign problems occur in commuting products in $\underline{H}_G^* S^0$. Thus, $\underline{H}_G^* S^0$ is commutative in both the graded and ungraded senses.

(b) As an alternative to using the ν_α as a basis in the fourth quadrant, one may define elements $\epsilon_\beta^{-1} \xi_\alpha^{-1} \omega$ in $\underline{H}_G^{1-\alpha-\beta}(S^0)(1)$, for $|\alpha| = |\beta^G| = 0$, $|\alpha^G| < 0$, and $|\beta| > 0$, by

$$\epsilon_\beta^{-1} \xi_\alpha^{-1} \omega \;=\; d_{\alpha - <\alpha>} \nu_{1-\alpha-\beta}.$$

Here, ω is regarded as a fictitious element in dimension 1 which is divisible by any product $\xi_\alpha \epsilon_\beta$. We employ a fictitious element because there is no canonical choice for the dimension of an actual element. The relations satisfied by the elements $\epsilon_\beta^{-1} \xi_\alpha^{-1} \omega$ are

$$\epsilon_\gamma (\epsilon_\beta^{-1} \xi_\alpha^{-1} \omega) = \epsilon_{\beta-\gamma}^{-1} \xi_\alpha^{-1} \omega, \qquad \text{for } |\alpha| = |\beta^G| = |\gamma^G| = 0,\ |\beta| > |\gamma| > 0, \text{ and } |\alpha^G| < 0;$$

$$\xi_\gamma (\epsilon_\beta^{-1} \xi_\alpha^{-1} \omega) = \epsilon_\beta^{-1} \xi_{\alpha-\gamma}^{-1} \omega, \qquad \text{for } |\alpha| = |\gamma| = |\beta^G| = 0,\ |\alpha^G| < |\gamma^G| < 0, \text{ and } |\beta| > 0;$$

$$\mu_\gamma(\epsilon_\beta^{-1}\xi_\alpha^{-1}\omega) = \epsilon_{\beta-\gamma}^{-1}\xi_\alpha^{-1}\omega, \qquad \text{for } \gamma \in RSO_0(G),\ |\alpha| = |\beta^G| = 0,\ |\alpha^G| < 0, \text{ and } |\beta| > 0;$$

$$\mu_\gamma(\epsilon_\beta^{-1}\xi_\alpha^{-1}\omega) = d_{<\gamma>-\gamma}\epsilon_\beta^{-1}\xi_{\alpha-\gamma}^{-1}\omega, \qquad \text{for } \gamma \in RSO_0(G),\ |\alpha| = |\beta^G| = 0,\ |\alpha^G| < 0, \text{ and } |\beta| > 0.$$

The one difficulty with this alternative basis is that if $\alpha + \beta = \gamma + \delta$, then $\epsilon_\beta^{-1}\xi_\alpha^{-1}\omega$ and $\epsilon_\delta^{-1}\xi_\gamma^{-1}\omega$ are in the same dimension, but they need not be equal. In fact,

$$\epsilon_\beta^{-1}\xi_\alpha^{-1}\omega = d_{\alpha-\gamma-<\alpha-\gamma>}\epsilon_\delta^{-1}\xi_\gamma^{-1}\omega.$$

(c) Observe that in the formulas for the product of μ_α with any of ϵ_β, $\epsilon_\beta^{-1}\kappa_\gamma$, or ν_β there is no d_α, but there is such a constant in the formula for the product $\mu_\alpha\xi_\beta$. On the other hand, $\sigma_\alpha\xi_\beta = \xi_{\alpha+\beta}$, but there is a $d_{-\alpha}$ in the formula for the product of σ_α with any of ϵ_β, $\epsilon_\beta^{-1}\kappa_\gamma$, or ν_β. This difference in the behavior of the elements μ_α and σ_α of $\underline{H}_G^*(S^0)(1)$ reflects the fact that there is a conjugacy class of subgroups of G associated to any well chosen element of any G-Mackey functor M for any finite group G. This association is based on the splitting of M which occurs when M is localized away from the order of G. This splitting can not be observed directly before localization, but it can be seen indirectly in the association of subgroups to well chosen elements in the Mackey functor. The elements μ_α, ϵ_β, $\epsilon_\beta^{-1}\kappa_\gamma$, and ν_β are all associated to the subgroup G of G, and products of pairs of them behave nicely. The elements σ_α and ξ_β are associated to the trivial subgroup, and their product is nice. However, the product of elements associated to two different subgroups will either be zero or involve some fudge factor like a d_α. We have introduced both μ_α and σ_α so that, when one of these elements is needed in our description of the relations in $\underline{H}_G^*P(V)^+$, we can always choose the one that will give us the simpler formula.

REMARKS 4.11. In order to explain the passage from an RSO(G) grading on $\underline{H}_G^*S^0$ to an RO(G) grading, we must first clarify what is meant by the assertion that $\underline{H}_G^*S^0$ is RO(G)-graded. The assertion does not mean that, for $\alpha \in RO(G)$, $\underline{H}_G^\alpha S^0$ can be described without reference to a choice of a representative for α. Rather it means that if $V_1 - W_1$ and $V_2 - W_2$ are two representatives for α and $\underline{H}^1$ and $\underline{H}^2$ are the values of $\underline{H}_G^\alpha S^0$ obtained using these representatives, then we can construct an isomorphism between $\underline{H}^1$ and $\underline{H}^2$ in a natural way from any isomorphism $f: V_2 \oplus W_1 \to V_1 \oplus W_2$ of representations illustrating the equivalence of $V_1 - W_1$ and $V_2 - W_2$ in RO(G). This is exactly what we mean when we say that nonequivariant homology is $\mathbb{Z}$ graded. To define the nonequivariant homology group H^nX, we must pick a standard n-simplex. Different choices of the n-simplex lead to different groups, as anyone who has been embarrassed by an orientation mistake knows all too well.

Let $\beta = V_2 \oplus W_1 - V_1 \oplus W_2$ and let $\tilde{f}$ denote the image of f in $\underline{H}_G^{\beta}(S^0)(1)$. Then the isomorphism from $\underline{H}^1$ to $\underline{H}^2$ is just multiplication by $\tilde{f}$. To provide a means of computing the effect of this isomorphism, we write $\tilde{f}$ in terms of the standard generators of $\underline{H}_G^{\beta}(S^0)(1)$. The map f induces a map f^G between the fixed point subspaces of the representations. If nonequivariant orientations are choose for their domains and ranges, then the maps f and f^G have well-defined nonequivariant degrees. It follows from Lemma A.12 that

$$\tilde{f} = (\deg f^G)\,\mu_\beta + \frac{(\deg f) - (\deg f^G) d_\beta}{p}\,\tau(\iota_\beta).$$

The structure of $\underline{H}_G^* G^+$ as an algebra over $\underline{H}_G^* S^0$ follows easily from our results on $\underline{H}_G^* S^0$ and the description of the additive structure of $\underline{H}_G^* G^+$ given in section 2.

PROPOSITION 4.12. As an RO(G)-graded module over $\underline{H}_G^* S^0$, $\underline{H}_G^* G^+$ is generated by the single element $\psi = (1, 0, 0, \ldots, 0)$ of $\underline{H}_G^0(G^+)(e) = \mathbb{Z}^p$. Moreover, for any RO(G)-graded module M^* over $\underline{H}_G^* S^0$, there is a one-to-one correspondence between maps $f: \underline{H}_G^* G^+ \to M^*$ of RO(G)-graded modules over $\underline{H}_G^* S^0$ and elements in $M^0(e)$. This correspondence associates the map f with the element $f(e)(\psi)$ of $M^0(e)$. Thus, $\underline{H}_G^* G^+$ is a projective RO(G)-graded module over $\underline{H}_G^* S^0$.

PROOF. Unless $|\alpha| = 0$, $\underline{H}_G^{\alpha}(G^+) = 0$. If $|\alpha| = 0$, then $\iota_\alpha \psi$ generates $\underline{H}_G^{\alpha} G^+$ as a module over A. Thus, ψ generates $\underline{H}_G^* G^+$ as an RO(G)-graded module over $\underline{H}_G^* S^0$, and any RO(G)-graded module map $f: \underline{H}_G^* G^+ \to M^*$ is determined by $f(\psi)$. On the other hand, recall the observation from Examples 1.1(f) that a map from A_G to any Mackey functor N can be specified by giving the image of $(1, 0, 0, \ldots, 0) \in A_G(e)$ in N(e). Let m be an element of $M^0(e)$. For each $\alpha \in RO(G)$ with $|\alpha| = 0$, $\iota_\alpha m$ is in $M^\alpha(e)$ and there is a unique map $f^\alpha : \underline{H}_G^\alpha G^+ \to M^\alpha$ of Mackey functors sending $\iota_\alpha \psi \in \underline{H}_G^\alpha(G^+)(e)$ to $\iota_\alpha m \in M^\alpha(e)$. These maps fit together to form a map $f: \underline{H}_G^* G^+ \to M^*$ of RO(G)-graded modules over $\underline{H}_G^* S^0$. The projectivity of $\underline{H}_G^* G^+$ follows immediately.

5. THE MULTIPLICATIVE STRUCTURE OF $\underline{H}_G^*P(V)^+$. We assume that there are at least two distinct isomorphism classes of irreducibles in V; otherwise, the multiplicative structure of $\underline{H}_G^*P(V)^+$ is completely described in Examples 1.1.(h). As in section 3, we take Φ to be the set of irreducible summands of the complex representation V. Let $\Phi(0)$, $\Phi(1)$, $\Phi(2)$, ... be a proper filtration of Φ. Then $\Phi(1)$ consists of exactly one representative of each of the isomorphism classes of irreducibles that appears in Φ. Let $\phi_0, \phi_1, \phi_2, \ldots, \phi_m$ be an enumeration of the elements in $\Phi(1)$, and let n_i be the number of elements of Φ isomorphic to ϕ_i (with $n_i = \infty$ allowed). Arrange the enumeration of the elements of $\Phi(1)$ so that $n_0 \geq n_1 \geq \ldots \geq n_m$. Extend the ordering of $\Phi(1)$ to Φ by selecting the unique proper ordering of Φ which is consistent with the filtration and in which, for each $r \geq 1$, the ordering of the representations in $\Phi(r+1) - \Phi(r)$ is the same as the ordering of the corresponding representations in $\Phi(1)$. If the irreducibles which appear in Φ appear with equal multiplicity, then, regarded as an ordered set, Φ is a sequence of blocks, each of which is a copy of $\Phi(1)$. If the multiplicities are not equal, then Φ is still a sequence of blocks, but each block after the first will be either a copy of $\Phi(1)$ or of an initial segment of $\Phi(1)$. The lengths of the initial segments in the sequence can not increase. We will abuse notation by writing $\phi_i \in \Phi(r+1)$ $\Phi(r)$ to mean that $\Phi(r+1) - \Phi(r)$ contains an irreducible representation isomorphic to ϕ_i. We say that two sets of irreducible representations are equivalent if they contain the same number of irreducibles in each isomorphism class. Moreover, we sometimes identify equivalent sets of irreducibles.

Corollary 2.7 will be used to derive the multiplicative structure of $\underline{H}_G^*P(V)^+$ from the multiplicative structures of $\underline{H}_G^*(P(V)^+)(e)$ and $\underline{H}_G^*((P(V)^G)^+)(1)$. The group $\underline{H}_G^\alpha(P(V)^+)(e)$ is isomorphic to the nonequivariant cohomology group $H^{|\alpha|}(P(V)^+;\mathbb{Z})$, and we will think of the restriction map ρ as a map from $\underline{H}_G^\alpha(P(V)^+)(1)$ to $\underline{H}^{|\alpha|}(P(V)^+;\mathbb{Z})$. Select an algebra generator $x \in \underline{H}^2(P(V)^+;\mathbb{Z})$ for $\underline{H}^*(P(V)^+;\mathbb{Z})$. The fixed point space of P(V) is the disjoint union of the spaces $P(n_i\phi_i) \cong P(n_i)$. Let q_i denote both the inclusion of the subspace $P(n_i)$ into P(V) and the map $\underline{H}_G^*(P(V)^+)(1) \to \underline{H}_G^*(P(n_i)^+)(1)$ induced by this inclusion. By Examples 1.1.(h), $\underline{H}_G^*P(n_i)^+$ is a truncated polynomial algebra over $\underline{H}_G^*S^0$ generated by an element x_i in $\underline{H}_G^2(P(n_i)^+)(1)$. Let

$$\tilde{q}_i : \underline{H}_G^*(P(V)^+)(1) \to \underline{H}_G^*(P(n_i)^+)(1)/(\text{torsion} \oplus \text{im } \rho)$$

denote the composition of q_i and the projection onto the quotient. If y is in $\underline{H}_G^*(P(n_i)^+)(1)$, then [y] denotes its image in $\underline{H}_G^*(P(n_i)^+)(1)/(\text{torsion} \oplus \text{im } \rho)$.

Throughout this section, we will index $\underline{H}_G^*P(V)^+$ on RSO(G) to simplify the selection of the integers d_α. The comments in Remarks 4.11 on the passage from RSO(G)-grading to RO(G)-grading for $\underline{H}_G^*S^0$ apply equally well to $\underline{H}_G^*P(V)^+$. Recall that λ is the irreducible complex representation that sends the standard generator of

$\mathbb{Z}/p$ to $e^{2\pi i/p}$ and that ζ is the real one-dimensional sign representation of $\mathbb{Z}/2$. If p is 2, then λ, regarded as a real representation, is just 2ζ.

We begin with the case $p=2$. Any complex irreducible representation is isomorphic to either the complex one-dimensional trivial representation or the complex one-dimensional sign representation λ. Since $P(V)$ and $P(\lambda V)$ are G-homeomorphic, we may assume that there are at least as many copies of the trivial representation in Φ as there are copies of the sign representation. Thus, we may take ϕ_0 to be the trivial representation and ϕ_1 to be the sign representation.

By Theorem 3.1, $\underline{H}_G^* P(V)^+$, regarded as a module over $\underline{H}_G^* S^0$, has one generator in each of the dimensions

$$2k + 2k\zeta \text{ and } 2k + 2(k+1)\zeta,$$

for $0 \leq k < n_1$, and one in each of the dimensions

$$2k + 2n_1\zeta,$$

for $n_1 \leq k < n_0$. If one assumes $n_0 = n_1$, or ignores the generators special to the case $n_0 > n_1$, then one might guess that, as an algebra, $\underline{H}_G^* P(V)^+$ had an exterior generator in dimension 2ζ and a truncated polynomial generator in dimension $2(1+\zeta)$. Except for the fact that the generator in dimension 2ζ is not quite an exterior generator and for some difficulties in the higher dimensions when $n_0 > n_1$, this guess is a good description of $\underline{H}_G^* P(V)^+$. However, in order to describe the behavior in the higher dimensions as simply as possible, we adopt a notation that does not immediately suggest this.

THEOREM 5.1. (a) As an algebra over $\underline{H}_G^* S^0$, $\underline{H}_G^* P(V)^+$ is generated by an element c of $\underline{H}_G^*(P(V)^+)(1)$ in dimension 2ζ and elements $C(k)$ of $\underline{H}_G^*(P(V)^+)(1)$ in dimensions $2k + 2\min(k, n_1)\zeta$, for $1 \leq k < n_0$.

(b) For any positive integer k, let $\bar{k}$ denote the minimum of k and n_1. Then the generators c and $C(k)$ are uniquely determined by

$$\tilde{q}_0(c) = [0]$$
$$\tilde{q}_1(c) = [\epsilon^2]$$
$$\rho(c) = x \in H^2(P(V)^+; \mathbb{Z})$$
$$\tilde{q}_0(C(k)) = \left[\epsilon^{2\bar{k}} x_0^k\right]$$
$$\tilde{q}_1(C(k)) = \left[\epsilon^{2\bar{k}} x_1^k\right]$$

and

$$\rho(C(k)) = x^{k+\bar{k}}.$$

Moreover,

$$q_0(c) = \xi x_0 \in \underline{H}_G^{2\zeta}(P(n_0)^+)(1)$$
$$q_1(c) = \epsilon^2 + \xi x_1 \in \underline{H}_G^{2\zeta}(P(n_1)^+)(1)$$
$$q_0(C(k)) = x_0^k(\epsilon^2 + \xi x_0)^{\bar{k}} \in \underline{H}_G^{2(k+\bar{k}\zeta)}(P(n_0)^+)(1)$$

and

$$q_1(C(k)) = x_1^k(\epsilon^2 + \xi x_1)^{\bar{k}} \in \underline{H}_G^{2(k+\bar{k}\zeta)}(P(n_1)^+)(1).$$

If n_i is finite, then $x_i^{n_i} = 0$ and some of the terms in the last two sums above may vanish.

(c) The generators c and C(k) satisfy the relations

$$c^2 = \epsilon^2 c + \xi C(1),$$
$$c\,C(k) = \xi C(k+1), \qquad \text{for } k \geq n_1,$$

and

$$C(j)\,C(k) = \begin{cases} C(j+k), & \text{for } j+k \leq n_1, \\ \sum_{i=0}^{\bar{j}+\bar{k}-n_1} \binom{\bar{j}+\bar{k}-n_1}{i} \epsilon^{2(\bar{j}+\bar{k}-n_1-i)} \xi^i C(j+k+i), & \text{for } j+k > n_1. \end{cases}$$

In these relations, we take $C(i)$ to be zero if $i \geq n_0$.

REMARKS 5.2. (a) By iteratively applying the third relation, we obtain

$$C(k) = (C(1))^k, \qquad \text{for } k \leq n_1,$$

so that below the dimensions where we run short of copies of the sign representation, $\underline{H}_G^* P(V)^+$ is generated by c and C(1). Moreover, in these dimensions, C(1) acts like a polynomial generator.

(b) If $n_0 = n_1$, then $\underline{H}_G^* P(V)^+$ is generated by c and C(1). The only relations satisfied by these two generators are the relation

$$c^2 = \epsilon^2 c + \xi C(1)$$

and, if $n_0 < \infty$, the relation

$$C(1)^{n_0} = 0.$$

REMARKS 5.3. Notice that the maps $\tilde{q}_0$ and $\tilde{q}_1$ behave differently on the generator c. The element $\tilde{c} = c + \epsilon^2 - \kappa c$ of $\underline{H}_G^{2\zeta} P(V)^+$ may be used as a generator in the place of c and its behavior with respect to $\tilde{q}_0$ and $\tilde{q}_1$ is exactly the reverse of the behavior of c. To understand the geometric relation between these elements, observe that c and $\tilde{c}$ can be detected in the cohomology of any subspace $P(1+\lambda)$ of $P(V)$ arising from an inclusion $1+\lambda \subset V$. The space $P(1+\lambda)$ is G-homeomorphic to S^λ, but unlike S^λ, it lacks a canonical basepoint. Either choice for the basepoint of $P(1+\lambda)$ determines a splitting of $\underline{H}_G^* P(1+\lambda)^+$ into the direct sum of one copy of

$\underline{H}_G^* S^0$ and one copy of $\underline{H}_G^* S^\lambda$. The canonical generator of $\underline{H}_G^* S^\lambda$ in dimension 2ζ is identified with c by one of the two splittings and with $\bar{c}$ by the other.

When p is 2, the multiplicative structure of $\underline{H}_G^* P(V)^+$ does not really exhibit any complexities beyond those one might experience in a $\mathbb{Z}$-graded ring. However, when p is odd, there are quirks in the multiplicative structure of $\underline{H}_G^* P(V)^+$ which are only possible because of the RSO(G)-grading. For the odd prime case, recall the stairstep diagram obtained by plotting the dimensions α of the generators of $\underline{H}_G^* P(V)^+$ in terms of $|\alpha|$ and $|\alpha^G|$. Looking at this diagram in the special case where the irreducibles appearing in V appear with equal multiplicity, one might guess that $\underline{H}_G^* P(V)^+$ was generated by two truncated polynomial generators, one in a dimension α with $|\alpha| = 2$ and $|\alpha^G| = 0$ and one in a dimension β with $|\beta| = 2m + 2$ and $|\beta^G| = 2$. Unfortunately, such a guess would badly underestimate the complexity of $\underline{H}_G^* P(V)^+$. The set of dimensions for a full set of additive generators must generate a larger additive subgroup of RSO(G) than can be accounted for by a pair of truncated polynomial generators. For example, recall that the first two additive generators of $\underline{H}_G^* P(V)^+$ are in dimensions $\phi_1^{-1}\phi_0$ and $\phi_2^{-1}(\phi_0 + \phi_1)$. If the additive generator in dimension $\phi_1^{-1}\phi_0$ were to serve as a truncated polynomial generator, then the additive generator in the next higher dimension would need to be in dimension $2\phi_1^{-1}\phi_0$ instead of $\phi_2^{-1}(\phi_0 + \phi_1)$. Any replacement of these two generators by an element and its square requires the introduction of further generators in some other dimensions inconsistent with a simple truncated polynomial structure. To provide a better feeling for the multiplicative structure of $\underline{H}_G^* P(V)^+$, we give two sets of multiplicative generators. The first is a natural set with a great deal of symmetry. It does not exhibit a preference for any one ordering of Φ. Unfortunately, this set is much too large. By selecting an ordering on Φ, we are able to construct a much smaller, but very asymmetrical, set of algebra generators.

In order to describe the effect of the maps q_i on our algebra generators, we must introduce more notation related to the integers d_α.

DEFINITIONS 5.4. (a) For any two distinct integers i and j with $0 \leq i, j \leq m$, let β_{ij} denote the irreducible representation $\phi_i^{-1}\phi_j$, and let d_{rs}^{ij} denote the integer d_α, for $\alpha = \beta_{ij} - \beta_{rs}$. Note that d_{ij}^{ij} is 1 for any pair of distinct integers i and j. For any integer i and any distinct pair of integers r and s such that $0 \leq i, r, s \leq m$, let d_{rs}^{ii} be zero. The integers d_{rs}^{ij} satisfy the relations

$$d_{rs}^{ij} d_{uv}^{rs} \equiv d_{uv}^{ij} \qquad \text{mod } p,$$

$$d_{rs}^{ij} + d_{rs}^{jk} \equiv d_{rs}^{ik} \qquad \text{mod } p,$$

and

$$d_{rs}^{ij} d_{vw}^{tu} \equiv d_{rs}^{tu} d_{vw}^{ij} \qquad \text{mod } p.$$

(b) If $\phi_i \in \Phi(r+1) - \Phi(r)$, then let $\alpha_i(r)$ denote the representation $\phi_i^{-1} \sum_{\phi \in \Phi(r)} \phi$, and let $\tilde{d}^r_{ij}$ be d_α, for $\alpha = \alpha_i(r) - \alpha_j(r)$. Note that, if $\phi_i \in \Phi(r+1) - \Phi(r)$, $\tilde{d}^r_{ii} = 1$. If either ϕ_i or ϕ_j is not in $\Phi(r+1) - \Phi(r)$, then let $\tilde{d}^r_{ij}$ be zero. If ϕ_i, ϕ_j, and ϕ_k are in $\Phi(r+1) - \Phi(r)$, then the integers $\tilde{d}^r_{ij}$ satisfy the relations

$$\tilde{d}^r_{ij}\tilde{d}^r_{jk} \equiv \tilde{d}^r_{ik} \qquad \mod p$$

and, if $i \neq j$,

$$\tilde{d}^r_{ij} \equiv (d^{ij}_{ji})^r \prod_{\substack{0 \leq k \leq m \\ k \neq i,j}} (d^{ik}_{jk})^{a_k} \qquad \mod p,$$

where a_k is the multiplicity of ϕ_k in $\Phi(r)$.

THEOREM 5.5. (a) If i and j are distinct integers with $0 \leq i, j \leq m$, then there is a unique element c_{ij} in $\underline{H}_G^{\beta_{ij}}(P(V)^+)(1)$ such that

$$\tilde{q}_k(c_{ij}) = \left[d^{kj}_{ij}\epsilon_{\beta_{ij}}\right], \qquad \text{for } 0 \leq k \leq m,$$

and

$$\rho(c_{ij}) = x.$$

If $r \geq 0$ and $\phi_j \in \Phi(r+1) - \Phi(r)$, then there is a unique element $C_j(r)$ in $\underline{H}_G^{\alpha_j(r)}(P(V)^+)(1)$ such that

$$\tilde{q}_k(C_j(r)) = \left[\tilde{d}^r_{kj}\epsilon_{\alpha_j(r)-r}x^r_k\right], \qquad \text{for } 0 \leq k \leq m,$$

and

$$\rho(C_j(r)) = x^{|\alpha_j(r)|/2}.$$

The elements c_{ij}, for $0 \leq i,j \leq m$ and $i \neq j$, and the elements $C_k(r)$, for $r \geq 1$ and $\phi_k \in \Phi(r+1) - \Phi(r)$, generate $\underline{H}_G^* P(V)^+$ as an algebra over $\underline{H}_G^* S^0$.

(b) For $0 \leq i,j,k \leq m$ and $i \neq j$,

$$q_k(c_{ij}) = d^{kj}_{ij}\epsilon_{\beta_{ij}} + \xi_{\beta_{ij}-2}x_k.$$

(c) For $r \geq 1$ and $\phi_k \in \Phi(r+1) - \Phi(r)$,

$$q_k(C_k(r)) = x_k^r \left[\prod_{\substack{\phi_i \in \Phi(r) \\ i \neq k}} (\epsilon_{\beta_{ki}} + \xi_{\beta_{ki}-2} x_k) \right].$$

If $\phi_j \in \Phi(r+1) - \Phi(r)$ and $j \neq k$, then

$$q_k(C_j(r)) = x_k^r \left(d_{jk}^{kj} \epsilon_{\beta_{jk}} + \xi_{\beta_{jk}-2} x_k \right)^r \left[\prod_{\substack{\phi_i \in \Phi(r) \\ i \neq j,k}} (d_{ji}^{ki} \epsilon_{\beta_{ji}} + \xi_{\beta_{ji}-2} x_k) \right] +$$

$$\left[\tilde{d}_{kj}^r - (d_{jk}^{kj})^r \prod_{\substack{\phi_i \in \Phi(r) \\ i \neq j,k}} (d_{ji}^{ki}) \right] \epsilon_{\alpha_j(r)-r} x_k^r .$$

If $\phi_k \notin \Phi(r+1) - \Phi(r)$, then $q_k(C_j(r))$ is zero.

(d) For $1 \leq j \leq m$, let γ_j be the representation $\phi_j^{-1} \sum_{i=0}^{j-1} \phi_i$ and let D_j be the element $\prod_{i=0}^{j-1} c_{ji}$ in $\underline{H}_G^{\gamma_j}(P(V)^+)(1)$. Then the elements D_j, for $1 \leq j \leq m$, the elements $C_0(r)$, for $r \geq 1$ and $\phi_0 \in \Phi(r+1) - \Phi(r)$, and the elements $D_j C_j(r)$, for $r \geq 1$ and $\phi_j \in \Phi(r+1) - \Phi(r)$, generate $\underline{H}_G^* P(V)^+$ as an algebra over $\underline{H}_G^* S^0$.

REMARKS 5.6. In order to simplify our indexing, we define D_0 and $C_j(0)$, for $0 \leq j \leq m$, to be $1 \in \underline{H}_G^0(P(V)^+)(1)$. We also define γ_0 and $\alpha_j(0)$ to be 0. Our second set of generators for $\underline{H}_G^* P(V)^+$ is then just the set of elements $D_j C_j(r)$, for $r \geq 0$ and $\phi_j \in \Phi(r+1) - \Phi(r)$. This set of elements of $\underline{H}_G^*(P(V)^+)(1)$ is also a set of additive generators of $\underline{H}_G^* P(V)^+$ as a module over $\underline{H}_G^* S^0$. One might hope that a set of multiplicative generators could be much smaller than a set of additive generators, but if the various irreducibles in Φ appear with very different multiplicities, then small sets of multiplicative generators do not exist.

We will order the set of generators $D_j C_j(r)$ by the dictionary order on r and then j. On the stairstep plot of the dimensions of these generators, moving in the direction of increasing order corresponds to moving up and to the right.

REMARKS 5.7. Nothing that has been said in the discussion of the odd prime case actually depends on p being odd; rather, mod 2 arithmetic is so simple that most of the technicalities necessary when p is odd are unnecessary when $p = 2$. The elements c and $\tilde{c}$ in the case $p = 2$ are c_{10} and c_{01}. The element C(j) is $C_0(j)$.

In order to describe the relations among the generators in $\underline{H}_G^*P(V)^+$ in a palatable form, we must introduce one more batch of elements in $\underline{H}_G^*(P(V)^+)(1)$.

DEFINITION 5.8. Observe that, for $1 \leq j \leq m$, κD_j is divisible by ϵ_{γ_j}. Moreover, $\rho(\epsilon_{\gamma_j}^{-1}\kappa D_j) = 0$, and

$$\tilde{q}_k(\epsilon_{\gamma_j}^{-1}\kappa D_j) = \left[p\prod_{i=0}^{j-1} d_{ji}^{ki}\right] \in \underline{H}_G^0(P(n_k\phi_k)^+)/(\text{torsion} \oplus \text{im } \tau).$$

Since $\prod_{i=0}^{j-1} d_{ji}^{ki}$ is zero if $k<j$ and 1 if $k=j$, the coefficients $p\prod_{i=0}^{j-1} d_{ji}^{ki}$ which appear in the $\tilde{q}_k(\epsilon_{\gamma_j}^{-1}\kappa D_j)$ form a matrix which is p times an upper triangular matrix with 1's on the main diagonal. Applying the obvious analog of the process for diagonalizing an upper triangular matrix to the elements $\epsilon_{\gamma_j}^{-1}\kappa D_j$ produces elements $\hat{\kappa}_j$ of $\underline{H}_G^0(P(V)^+)(1)$ characterized by the conditions

$$\rho(\hat{\kappa}_j) = 0,$$

and

$$\tilde{q}_k(\hat{\kappa}_j) = \begin{cases} [p], & \text{if } k=j, \\ 0, & \text{otherwise.} \end{cases}$$

These elements can be described inductively by the equations

$$\hat{\kappa}_m = \epsilon_{\gamma_m}^{-1}\kappa D_m$$

and, for $1 \leq j < m$,

$$\hat{\kappa}_j = \epsilon_{\gamma_j}^{-1}\kappa D_j - \sum_{k=j+1}^{m}\left(\prod_{i=0}^{j-1} d_{ji}^{ki}\right)\hat{\kappa}_k.$$

Define $\hat{\kappa}_0 \in \underline{H}_G^0(P(V)^+)(1)$ to be $\kappa - \sum_{j=1}^{m}\hat{\kappa}_j$. The equations above characterizing $\hat{\kappa}_j$ for $j \neq 0$ then also characterize $\hat{\kappa}_0$. Moreover,

$$q_k(\hat{\kappa}_j) = \begin{cases} p, & \text{if } k=j, \\ 0, & \text{otherwise.} \end{cases}$$

For $r \geq 1$ and $\phi_j \in \Phi(r+1) - \Phi(r)$, define $\hat{\kappa}_j(r) \in \underline{H}_G^{\alpha_j(r)}(P(V)^+)(1)$ to be $\hat{\kappa}_j C_j(r)$. These elements $\hat{\kappa}_j(r)$ are characterized by the equations

$$\rho(\hat{\kappa}_j(r)) = 0,$$

and

$$\tilde{q}_k(\hat{\kappa}_j(r)) = \begin{cases} \left[p\,\epsilon_{\alpha_j(r)-r}\, x_k^r\right], & \text{if } k=j, \\ 0, & \text{otherwise.} \end{cases}$$

Moreover,

$$q_k(\hat{\kappa}_j(r)) = \begin{cases} p\epsilon_{\alpha_j(r)-r} x_k^r, & \text{if } k=j, \\ 0, & \text{otherwise.} \end{cases}$$

For convenience, we define $\hat{\kappa}_j(0)$ to be $\hat{\kappa}_j$. Observe that, for $r \geq 1$, the elements $\hat{\kappa}_j(r)$ can also be constructed from the elements $\kappa D_j C_j(r)$ in the same way that the elements $\hat{\kappa}_j$ are constructed from the κD_j.

We begin our list of relations with the relation between any two of the c_{ij} and the relation between any two of the $C_j(r)$.

PROPOSITION 5.9. (a) Let i, j, r, and s be integers with $0 \leq i, j, r, s \leq m$ and $i \neq j$, $r \neq s$. Then

$$c_{ij} = \sigma_{\beta_{ij}-\beta_{rs}} c_{rs} + d_{ij}^{sj} \epsilon_{\beta_{ij}} + \sum_{k \neq s} \frac{d_{ij}^{kj} - d_{ij}^{sj} - d_{ij}^{rs} d_{rs}^{ks}}{p} \epsilon_{\beta_{ij}} \hat{\kappa}_k .$$

(b) Let $r \geq 1$ and let i and j be integers such that ϕ_i and ϕ_j are in $\Phi(r+1) - \Phi(r)$. Then

$$C_i(r) = \sigma_{\alpha_i(r)-\alpha_j(r)} C_j(r) + \sum_{k \neq j} \frac{\tilde{d}_{ki}^r - \tilde{d}_{kj}^r \tilde{d}_{ji}^r}{p} \mu_{\alpha_i(r)-\alpha_k(r)} \hat{\kappa}_k(r).$$

An obvious initial response to this result is to assume that $\underline{H}_G^* P(V)^+$ can be generated as an algebra over $\underline{H}_G^* S^0$ by any one of the c_{ij} and, for each r with $\Phi(r+1) - \Phi(r)$ nonempty, any one of the $C_j(r)$. The $\hat{\kappa}_k$ and $\hat{\kappa}_k(r)$ in the formulas spoil this simplification, especially since they are defined in terms of precisely the generators one would hope to omit. Solving this by taking the elements $\hat{\kappa}_k$ and $\hat{\kappa}_k(r)$ as part of a generating set is hardly satisfactory since, from a Mackey functor point of view, these are torsion elements (because $\rho(\hat{\kappa}_k)$ and $\rho(\hat{\kappa}_k(r))$ are zero).

The remaining results in this section describe the products of pairs of elements from either of the generating sets in terms of the smaller generating set. All of the relations in $\underline{H}_G^* P(V)^+$ follow from the relations in Proposition 5.9 and the relations below. If V is finite, then some of the elements appearing on the right hand side of these relations may not appear in the list of generators of $\underline{H}_G^* P(V)^+$. Any such element is to be regarded as zero. We begin with the products which land in dimensions where there is no torsion. These are easily computed using the maps $\tilde{q}_k$ and ρ.

PROPOSITION 5.10. (a) Let i, j, r, and s be integers with $0 \leq i, j, r, s \leq m$ and $i \neq j$, $r \neq s$. If $m \geq 2$, then

$$c_{ij} c_{rs} = d_{ij}^{0j} d_{rs}^{0s} \epsilon_{\beta_{ij}+\beta_{rs}} + (d_{ij}^{1j} d_{rs}^{1s} - d_{ij}^{0j} d_{rs}^{0s}) \epsilon_{\beta_{ij}+\beta_{rs}-\beta_{10}} c_{10} + \sigma_\alpha D_2 +$$

$$\sum_{k=2}^{m} \frac{d_{ij}^{kj} d_{rs}^{ks} - d_{ij}^{0j} d_{rs}^{0s} - (d_{ij}^{1j} d_{rs}^{1s} - d_{ij}^{0j} d_{rs}^{0s}) d_{10}^{k0} - d_{20}^{k0} d_{21}^{k1} d_{-\alpha}}{p} \epsilon_{\beta_{ij}+\beta_{rs}} \hat{\kappa}_k ,$$

where $\alpha = \beta_{ij} + \beta_{rs} - \gamma_2$.

If m = 1, then

$$c_{ij} c_{rs} = d_{ij}^{0j} d_{rs}^{0s} \epsilon_{\beta_{ij}+\beta_{rs}} + (d_{ij}^{1j} d_{rs}^{1s} - d_{ij}^{0j} d_{rs}^{0s}) \epsilon_{\beta_{ij}+\beta_{rs}-\beta_{10}} c_{10} + \xi_{\beta_{ij}+\beta_{rs}-\alpha_0(1)} C_0(1).$$

(b) Let i, j, and r be integers with $0 \le i, j \le m$, $i \ne j$, and $1 \le r < m$. Then

$$c_{ij} D_r = d_{ij}^{rj} \epsilon_{\beta_{ij}} D_r + \sigma_\alpha D_{r+1} + \sum_{k=r+1}^{m} \frac{(d_{ij}^{kj} - d_{ij}^{rj}) \prod_{s=0}^{r-1} d_{rs}^{ks} - d_{-\alpha} \prod_{s=0}^{r} d_{r+1,s}^{ks}}{p} \epsilon_{\beta_{ij}+\gamma_r} \hat{\kappa}_k,$$

where $\alpha = \beta_{ij} + \gamma_r - \gamma_{r+1}$.

(c) Let i, j be integers with $0 \le i, j \le m$ and $i \ne j$. Then

$$c_{ij} D_m = d_{ij}^{mj} \epsilon_{\beta_{ij}} D_m + \xi_{\beta_{ij}+\gamma_m-\alpha_0(1)} C_0(1).$$

(d) Let i, j, r, and s be integers with $0 \le i, j, s \le m$, $i \ne j$, $r \ge 1$, and $\phi_s \in \Phi(r+1) - \Phi(r)$. If $\phi_1 \in \Phi(r+1) - \Phi(r)$, then

$$c_{ij} C_s(r) = d_{ij}^{0j} \tilde{d}_{0s}^{r} \epsilon_{\beta_{ij}+\alpha_s(r)-\alpha_0(r)} C_0(r) + \sigma_\alpha D_1 C_1(r) + \sum_{\substack{k \ge 1 \\ \phi_k \in \Phi(r+1)-\Phi(r)}} \frac{d_{ij}^{kj} \tilde{d}_{ks}^{r} - d_{ij}^{0j} \tilde{d}_{0s}^{r} \tilde{d}_{k0}^{r} - d_{10}^{k0} \tilde{d}_{k1}^{r} d_{-\alpha}}{p} \epsilon_{\beta_{ij}+\alpha_s(r)-\alpha_k(r)} \hat{\kappa}_k(r),$$

where $\alpha = \beta_{ij} + \alpha_s(r) - \gamma_1 - \alpha_1(r)$.

If $\phi_1 \notin \Phi(r+1) - \Phi(r)$, then

$$c_{ij} C_s(r) = d_{ij}^{0j} \tilde{d}_{0s}^{r} \epsilon_{\beta_{ij}} C_0(r) + \xi_{\beta_{ij}+\alpha_0(r)-\alpha_0(r+1)} C_0(r+1).$$

(e) Let i, j, r, and s be integers with $0 \le i, j, s \le m$, $i \ne j$, $r \ge 1$, and $\phi_s \in \Phi(r+1) - \Phi(r)$. If $\phi_{s+1} \in \Phi(r+1) - \Phi(r)$, then

$$c_{ij} D_s C_s(r) = d_{ij}^{sj} \epsilon_{\beta_{ij}} D_s C_s(r) + \sigma_\alpha D_{s+1} C_{s+1}(r) + \sum_{\substack{k \ge s+1 \\ \phi_k \in \Phi(r+1)-\Phi(r)}} \frac{\tilde{d}_{ks}^{r} (d_{ij}^{kj} - d_{ij}^{sj}) \prod_{t=0}^{s-1} d_{st}^{kt} - \tilde{d}_{k,s+1}^{r} d_{-\alpha} \prod_{t=0}^{s} d_{s+1,t}^{kt}}{p} \epsilon_{\delta_k} \hat{\kappa}_k(r),$$

where $\alpha = \beta_{ij} + \gamma_s + \alpha_s(r) - \gamma_{s+1} - \alpha_{s+1}(r)$ and $\delta_k = \beta_{ij} + \gamma_s + \alpha_s(r) - \alpha_k(r)$.

If $\phi_{s+1} \notin \Phi(r+1) - \Phi(r)$, then

$$c_{ij} D_s C_s(r) = d_{ij}^{sj} \epsilon_{\beta_{ij}} D_s C_s(r) + \xi_{\beta_{ij}+\gamma_s+\alpha_s(r)-\alpha_0(r+1)} C_0(r+1).$$

(f) Let r, $s \geq 1$ and assume that $1 \leq j \leq m$. If the irreducibles that appear in $\Phi(r+s)$ appear with equal multiplicities, then

$$C_j(r) C_j(s) = C_j(r+s) + \sum_{\substack{k \neq j \\ \phi_k \in \Phi(r+s+1)-\Phi(r+s)}} \frac{\tilde{d}_{kj}^r \tilde{d}_{kj}^s - \tilde{d}_{kj}^{r+s}}{p} \mu_{\alpha_j(r+s)-\alpha_k(r+s)} \hat{\kappa}_k(r+s).$$

Moreover, the integers $\tilde{d}_{kj}^{r+s}$ may be selected to be the products $\tilde{d}_{kj}^r \tilde{d}_{kj}^s$ so that the $\hat{\kappa}_k(r+s)$ correction terms are not needed.

Since the elements $\hat{\kappa}_k(r)$ appear in so many formulas, we include a description of products involving them.

LEMMA 5.11. Let i, j, k, r, and s be integers with $0 \leq i,j,k \leq m$, $r, s \geq 0$, and $\phi_k \in \Phi(s+1) - \Phi(s)$.

(a) If $i \neq j$, then

$$c_{ij} \hat{\kappa}_k(s) = d_{ij}^{kj} \epsilon_{\beta_{ij}} \hat{\kappa}_k(s).$$

(b) If $\phi_j \in \Phi(r+1) - \Phi(r)$ and $\phi_k \in \Phi(r+s+1) - \Phi(r+s)$, then

$$C_j(r) \hat{\kappa}_k(s) = \tilde{d}_{kj}^r \epsilon_{\alpha_j(r)+\alpha_k(r)-\alpha_k(r+s)} \hat{\kappa}_k(r+s)$$

and

$$D_j C_j(r) \hat{\kappa}_k(s) = \tilde{d}_{kj}^r \left[\prod_{t=0}^{j-1} d_{jt}^{kt} \right] \epsilon_{\gamma_j + \alpha_j(r) + \alpha_k(s) - \alpha_k(r+s)} \hat{\kappa}_k(r+s).$$

In the formula for $C_j(r) \hat{\kappa}_k(s)$, replace $\epsilon_{\alpha_j(r)+\alpha_k(s)-\alpha_k(r+s)}$ by $\mu_{\alpha_j(r)+\alpha_k(s)-\alpha_k(r+s)}$ if $|\alpha_j(r) + \alpha_k(s) - \alpha_k(r+s)|$ is zero.

(c) If $\phi_j \in \Phi(r+1) - \Phi(r)$ and $\phi_k \notin \Phi(r+s+1) - \Phi(r+s)$, then $C_j(r) \hat{\kappa}_k(s)$ and $D_j C_j(r) \hat{\kappa}_k(s)$ are zero.

To complete our description of the multiplicative structure of $\underline{H}_G^* P(V)^+$ we need to describe the products of various pairs made from elements of the types $C_i(r)$,

$D_jC_j(r)$, and D_k. If we use the convention that $D_0 = C_j(0) = 1$, then the products we must describe are all special cases of the general product $(D_{i'}C_i(r))(D_{j'}C_j(s))$, where $r, s \geq 0$, $\phi_i \in \Phi(r+1) - \Phi(r)$, $\phi_j \in \Phi(s+1) - \Phi(s)$, i' is 0 or i, and j' is 0 or j. We may assume that $i' \geq j'$. Recall the formula given in Theorem 5.1(c) for the product $C(j)C(k)$ when $p = 2$ and $j+k > n_1$. Observe that this formula may be obtained from the binomial expansion of $(\epsilon^2 + \xi x)^{\bar{j}+\bar{k}-n_1}$ by replacing the powers of x by various generators C(t). The formula for our general product is related in a similar way to the expansion of an expression of the form $\prod_{i=0}^{n}(a_i + b_i x)$. The summands in this expansion are indexed on the subsets of the set $\{0, 1, \dots, n\}$. The summand corresponding to the subset I is

$$\Big(\prod_{i \notin I} a_i\Big)\Big(\prod_{i \in I} b_i\Big)x^{|I|},$$

where $|I|$ denotes the number of elements in I. To describe the analogous part of our formula for $(D_{i'}C_i(r))(D_{j'}C_j(s))$, we must specify the indexing set which replaces $\{0, 1, \dots, n\}$, the factors which replace $\prod a_i$ and $\prod b_i$, and the procedure for replacing the powers of x by the appropriate $D_k C_k(t)$.

In the $p = 2$ case, describing how the powers of x are to be replaced by the generators C(j) is very simple because, if $j \geq n_1$, then the next generator after C(j) is always $C(j+1)$. However, when p is odd, the generator after $D_k C_k(r)$ may be either $D_{k+1} C_{k+1}(r)$ or $C_0(r+1)$. To handle this complication, we introduce two functions f and g from the nonnegative integers to the nonnegative integers. These functions are to be chosen so that, for any $i \geq 0$, $C_{f(i+1)}(g(i+1))$ is the generator immediately following $C_{f(i)}(g(i))$ in our stairstep ordering. If $C_{f(n)}(g(n))$ is the last generator in $\underline{H}_G^* P(\Phi)^+$, then we define $f(i) = 0$ and $g(i) = g(n) + i - n$ for $i > n$ and use the convention that $D_j C_j(r)$ is to be regarded as zero if it does not appear in the list of generators of $\underline{H}_G^* P(\Phi)^+$. Each time we use this notation, the initial values, f(0) and g(0), of the functions will be specified to suit the particular application.

The indexing set which replaces the set $\{0, 1, \dots, n\}$ is related to the difference in dimension between the product $(D_{i'}C_i(r))(D_{j'}C_j(s))$ and the lowest dimensional generator $D_{i'}C_i(r+s)$ which should appear in its description. If $r \geq 0$

and $0 \leq j \leq m$, then define the subset $\Phi_j(r)$ of $\Phi(r+1)$ by

$$\Phi_j(r) = \Phi(r) \cup \{\phi_i : i < j \text{ and } \phi_i \in \Phi(r+1) - \Phi(r)\}.$$

Let $\Phi_{i'}(r) \sqcup \Phi_{j'}(s)$ denote the disjoint union of the sets $\Phi_{i'}(r)$ and $\Phi_{j'}(s)$. Our replacement for the set $\{0, 1, \dots, n\}$ is the set Ψ obtained by deleting from $\Phi_{i'}(r) \sqcup \Phi_{j'}(s)$ a subset equivalent to the set $\Phi_{i'}(r+s)$. We abuse notation by writing Ψ as $\Phi_{i'}(r) \sqcup \Phi_{j'}(s) - \Phi_{i'}(r+s)$. Observe that $\Phi_{j'}(s)$ is equivalent to the disjoint union of Ψ and $\Phi_{i'}(r+s) - \Phi_{i'}(r)$. Let u be $|\Psi| - 1$ and number the elements of Ψ from 0 to u. Let h be a function from the set $\{0, 1, \dots, u\}$ to the set $\{0, 1, \dots, m\}$ such that the ith element of Ψ is isomorphic to the irreducible representation $\phi_{h(i)}$.

One of the coefficients appearing in our formula is determined by a certain element α of RSO(G) with $|\alpha| = 0$ and $|\alpha^G| \leq 0$. This coefficient will be ξ_α if $|\alpha^G| < 0$ or σ_α if $|\alpha^G| = 0$. To simplify our notation, we write χ_α for either of these, relying on $|\alpha^G|$ to indicate whether ξ_α or σ_α is intended. Another coefficient will depend on a certain element β of RSO(G) with $|\beta^G| = 0$ and $|\beta| \geq 0$. This coefficient will be ϵ_β if $|\beta| > 0$ and μ_β if $|\beta| = 0$. We write θ_β for either of these, relying on $|\beta|$ to indicate which is intended.

PROPOSITION 5.12. Let i, i′, j, j′, r, and s be integers with $r, s \geq 0$, $\phi_i \in \Phi(r+1) - \Phi(r)$, $\phi_j \in \Phi(s+1) - \Phi(s)$, $i' = 0$ or i, $j' = 0$ or j, and $i' \geq j'$. Let $\Psi = \Phi_{i'}(r) \sqcup \Phi_{j'}(s) - \Phi_{i'}(r+s)$. Initialize the functions f and g by

$$f(0) = \begin{cases} i', & \text{if } \phi_{i'} \in \Phi(r+s+1), \\ 0, & \text{otherwise,} \end{cases}$$

and

$$g(0) = \begin{cases} r+s, & \text{if } \phi_{i'} \in \Phi(r+s+1), \\ r+s+1, & \text{otherwise.} \end{cases}$$

Let $u = |\Psi| - 1$ and number the elements of Ψ from 0 to u. Let $\Delta \subset \Psi$ and let s′ and s'' be the number of elements isomorphic to ϕ_j in Δ and $\Phi_{i'}(r+s)-\Phi_{i'}(r)$, respectively. If the subset Δ of Ψ contains the elements numbered $j_0, j_1, \ldots, j_w$, with $j_0 < j_1 < \ldots < j_w$, then let

$$d_\Delta^k = \left[\prod_{\substack{t=0\\ h(j_t)\neq j}}^{w} d_{j,h(j_t)}^{f(j_t-t),h(j_t)}\right]\left[\prod_{\substack{t=0\\ h(j_t)=j}}^{w} d_{jk}^{f(j_t-t),j}\right],$$

$$\epsilon_\Delta^k = \left[\prod_{\substack{t=0\\ h(j_t)\neq j}}^{w} \epsilon_{\beta_{j,h(j_t)}}\right]\left[\prod_{\substack{t=0\\ h(j_t)=j}}^{w} \epsilon_{\beta_{jk}}\right],$$

and

$$\chi_\Delta = \chi_\alpha,$$

where

$$\alpha = \phi_j^{-1}\left[\sum_{\substack{t=0\\ h(j_t)\neq j}}^{w} \phi_{h(j_t)} + \sum_{\substack{\phi_t \in \Phi_{i'}(r+s)-\Phi_{i'}(r)\\ t\neq 0,j}} \phi_t\right] +$$

$$\phi_i^{-1}\left[\sum_{\phi_t \in \Phi_{i'}(r)} \phi_t\right] + \left[(s'+s'')\,\phi_j^{-1}\phi_0\right]_{j\neq 0} +$$

$$2s - \phi_{f(|\Delta|)}^{-1}\left[\sum_{\phi_t \in \Phi_{f(|\Delta|)}(g(|\Delta|))} \phi_t\right].$$

The tag $j\neq 0$ on the bracket about the $(s'+s'')\,\phi_j^{-1}\phi_0$ indicates that this term is present only if $j\neq 0$. The 2s term in α indicates 2s copies of the real one-dimensional trivial representation. If $\alpha \in RSO_0(G)$, then let

$$\hat{d}_\Delta = d_\alpha.$$

If $\Delta = \emptyset$, then let d_Δ^k, ϵ_Δ^k, $\hat{d}_\Delta$, and χ_Δ be 1.

If $i' \leq k \leq m$ and $\phi_k \in \Phi(r+s+1) - \Phi(r+s)$, let

$$\theta_k = \theta_\beta,$$

where

$$\beta = \alpha_i(r) + \alpha_j(s) + \gamma_{i'} + \gamma_{j'} - \alpha_k(r+s),$$

and let A_k be

$$\frac{1}{p}\left[\tilde{d}^r_{ki}\tilde{d}^s_{kj}\left(\prod_{t=0}^{i'-1} d^{kt}_{i't}\right)\left(\prod_{t=0}^{j'-1} d^{kt}_{j't}\right) - \sum_{v=i'}^{k}\left[\tilde{d}^{r+s}_{kv}\left(\prod_{t=0}^{v-1} d^{kt}_{vt}\right)\sum_{\substack{\Delta\subset\Psi\\ |\Delta|=v-i'}} d^0_{\Psi-\Delta}\hat{d}_\Delta\right]\right].$$

Then

$$(D_{i'}C_i(r))(D_{j'}C_j(s)) = \sum_{\Delta\subset\Psi} d^0_{\Psi-\Delta}\,\epsilon^0_{\Psi-\Delta}\,\chi_\Delta\, D_{f(|\Delta|)}C_{f(|\Delta|)}(g(|\Delta|)) +$$

$$\sum_{\substack{k=i'\\ \phi_k\in\Phi(r+s+1)-\Phi(r+s)}}^{m} A_k\,\theta_k\,\hat{\kappa}_k(r+s).$$

REMARKS 5.13. (a) Let $r \geq 1$. If $\Phi(r)$ contains r copies of every irreducible complex G-representation, then $\alpha_i(r)$ is independent of i and it is easy to see that $C_i(r) = C_j(r)$ for every i and j such that $\phi_i, \phi_j \in \Phi(r+1) - \Phi(r)$. Moreover, $C_j(r) = C_j(1)^r$. Thus, if Φ contains every irreducible complex G-representation and these representations appear with equal multiplicities in Φ, then $C_i(r)$ generates a polynomial, or truncated polynomial, subalgebra of $\underline{H}^*_G P(\Phi)^+$. In this case, the elements D_j, for $1 \leq j \leq m$, and $C_i(1)$, for any i, generate $\underline{H}^*_G P(\Phi)^+$ as an algebra over $\underline{H}^*_G S^0$.

(b) If $p = 3$, then we may choose the integers d_α so that $d_\alpha = \pm 1$ for every α in $RSO_0(G)$. When this is done, the assignment of d_α to α is a homomorphism from the additive group of $RSO_0(G)$ to the multiplicative group $\{\pm 1\}$. With this choice of the integers d_α, all the relations among the d^{ij}_{rs} and the $\tilde{d}^r_{ij}$ given in Definitions 5.4, except the one involving a sum, hold in $\mathbb{Z}$ as well as in $\mathbb{Z}/3$. If $r \geq 1$ and $\phi_i, \phi_j \in \Phi(r+1)$, then

$$C_i(r) = \sigma_{\alpha_i(r)-\alpha_j(r)} C_j(r).$$

Thus, the only elements of the form $C_j(r)$ needed to generate $\underline{H}^*_G P(\Phi)^+$ as an algebra over $\underline{H}^*_G S^0$ are the elements $C_0(r)$ for $r \geq 1$. Also, a pair of elements c_{ij} and c_{rs} will generate D_1 and D_2 if $\tilde{q}_k(c_{ij}c_{rs})$ is nonzero for only one value of k. In particular, c_{01} and c_{10} generate D_1 and D_2. When all three irreducible complex G-representations of $\mathbb{Z}/3$ appear in Φ with equal multiplicities, c_{01}, c_{10}, and $C_0(1)$ generate $\underline{H}^*_G P(\Phi)^+$ as an algebra over $\underline{H}^*_G S^0$.

6. PROOFS. The results stated in section 5 are proved here. As indicated in Remark 5.7, our results for $p = 2$ are a special case of the results asserted for odd

primes. They have been presented separately only because they can be stated so simply. The proofs given here are independent of whether p is 2 or odd. We begin by construct the elements c_{ij} and $C_j(r)$. We then show that they generate $\underline{H}_G^* P(V)^+$ as an algebra over $\underline{H}_G^* S^0$. Finally, the relations stated at the end of section 5 are verified. Throughout this section, Φ is a set of irreducible complex representations of $\mathbb{Z}/p$ and $\Phi(0), \Phi(1), \ldots$ is a proper filtration of Φ. We order the elements of Φ in the standard proper ordering introduced in section 5. Recall the maps q_i and $\tilde{q}_i$ and the cohomology classes x and x_i from the introductory remarks in section 5 and the representations $\alpha_i(r)$, β_{ij}, and γ_j from Definitions 5.4 and Theorem 5.5(d). If $\Delta \subset \Psi$, then x also denotes the image of $x \in \underline{H}_G^2(P(\Phi)^+)(e)$ in $\underline{H}_G^2(P(\Delta)^+)(e)$; thus, the powers of x are thought of as the standard additive generators for the nonequivariant cohomology of all the sub-projective spaces of $P(\Psi)$. For each integer j with $0 \leq j \leq m$, let $P_j(\Phi)$ be the component of the fixed point space of $P(\Phi)$ associated to the irreducible representation ϕ_j.

The classes c_{ij} and $C_j(r)$ are constructed by defining them on the smallest possible projective space and then inductively lifting them to larger projective spaces.

CONSTRUCTION 6.1. (a) Let i and j be distinct integers with $0 \leq i, j \leq m$. The space $P(\{\phi_j\})$ is just a point and the space $P(\{\phi_i, \phi_j\})$ is G-homeomorphic to $S^{\beta_{ij}}$. The inclusion of $P(\{\phi_j\})$ into $P(\{\phi_i, \phi_j\})$ induces the cofibre sequence

$$P(\{\phi_j\})^+ \xrightarrow{q_j} P(\{\phi_i, \phi_j\})^+ \xrightarrow{\pi} S^{\beta_{ij}}.$$

Let $c_{ij} \in \underline{H}_G^{\beta_{ij}}(P(\{\phi_i, \phi_j\})^+)(1)$ be the image of $1 \in A(1) \cong \underline{H}_G^{\beta_{ij}}(S^{\beta_{ij}})(1)$ under π^*. Then $q_j(c_{ij}) = 0$ by exactness and $q_i(c_{ij}) = \epsilon_{\beta_{ij}}$ by the commutativity of the diagram

$$\begin{array}{ccc} P(\{\phi_i\})^+ & \xrightarrow{q_i} & P(\{\phi_i, \phi_j\})^+ \\ \downarrow & & \downarrow \\ S^0 & \xrightarrow{\epsilon_{\beta_{ij}}} & S^{\beta_{ij}}. \end{array}$$

These are the correct values for $q_i(c_{ij})$ and $q_j(c_{ij})$ because x_i and x_j are zero. Since the map $\pi^* : \underline{H}_G^{\beta_{ij}}(S^{\beta_{ij}})(e) \to \underline{H}_G^{\beta_{ij}}(P(\{\phi_i, \phi_j\})^+)(e)$ is an isomorphism in dimension β_{ij}, $\rho(c_{ij}) = x$.

Let Ψ be a subset of Φ which properly contains the set $\{\phi_i, \phi_j\}$ and assume that, for every proper subset Δ of Ψ containing $\{\phi_i, \phi_j\}$, c_{ij} has been defined in $\underline{H}_G^{\beta_{ij}}(P(\Delta)^+)(1)$ and has the proper images under the maps q_k and ρ. Pick an irreducible representation ϕ_t which appears in Ψ at least as often as any other irreducible. If no irreducible appears more than once in Ψ, then we may also insist that $t \neq i, j$. Let $\Delta = \Psi - \{\phi_t\}$, and let V be the representation $\phi_t^{-1} \sum_{\phi \in \Delta} \phi$. The inclusion of Δ into Ψ induces the cofibre sequences

$$P(\Delta)^+ \xrightarrow{\theta} P(\Psi)^+ \to S^V$$

and

$$P_t(\Delta)^+ \xrightarrow{\theta_t} P_t(\Psi)^+ \to S^{V^G}.$$

We will lift the class $c_{ij} \in \underline{H}_G^{\beta_{ij}}(P(\Delta)^+)(1)$ along the map

$$\theta^*(1): \underline{H}_G^{\beta_{ij}}(P(\Psi)^+)(1) \to \underline{H}_G^{\beta_{ij}}(P(\Delta)^+)(1)$$

induced by θ. To distinguish the class c_{ij} and its lifting, we will denote the class in $\underline{H}_G^{\beta_{ij}}(P(\Delta)^+)(1)$ by $\hat{c}_{ij}$. The maps q_k, for $k \neq t$, factor through $\theta^*(1)$, so any lifting of $\hat{c}_{ij}$ along $\theta^*(1)$ will have the right image under q_k, for $k \neq t$. Moreover, since $\theta^*(e)$ is an isomorphism in dimension β_{ij}, any lifting of $\hat{c}_{ij}$ will also have the right image under ρ.

It remains to show that we can choose a lifting of c_{ij} with the correct image under q_t. We have chosen t so that the long exact cohomology sequences associated to our cofibre sequences have zero boundary maps. If $|V^G| \geq 2$, then $\underline{H}_G^{\beta_{ij}}(S^V)(1) = 0$ and we take c_{ij} to be the unique lifting of $\hat{c}_{ij}$. If $|V^G| > 2$, then θ_t induces a cohomology isomorphism in dimension β_{ij} and this lifting of $\hat{c}_{ij}$ along $\theta^*(1)$ must have the correct image under q_t. If $|V^G| = 2$, then the short exact sequence

$$0 \to \underline{H}_G^{\beta_{ij}} S^2 \to \underline{H}_G^{\beta_{ij}} P_t(\Psi)^+ \xrightarrow{\theta_t^*} \underline{H}_G^{\beta_{ij}} P_t(\Delta)^+ \to 0$$

splits. The end terms are

$$\underline{H}_G^{\beta_{ij}} S^2 \cong R \quad \text{and} \quad \underline{H}_G^{\beta_{ij}} P_t(\Delta)^+ \cong \langle \mathbb{Z} \rangle.$$

The image of $1 \in \mathbb{Z} = R(1)$ in $\underline{H}_G^{\beta_{ij}} P_t(\Psi)^+$ is $\xi_{\beta_{ij}-2} x_t$. By our induction hypothesis,

$$\theta_t^*(1) q_t(c_{ij}) = q_t(\hat{c}_{ij}) = d_{ij}^{tj} \epsilon_{\beta_{ij}}.$$

Since $\rho(c_{ij}) = x$, $\rho q_t(c_{ij})$ is the generator of $\underline{H}_G^{\beta_{ij}}(P_t(\Psi)^+)(e)$. It follows that $q_t(c_{ij}) = d_{ij}^{tj} \epsilon_{\beta_{ij}} + \xi_{\beta_{ij}-2} x_t$.

If $|V^G| = 0$, then no irreducible appears more than once in Ψ and we have selected ϕ_t so that $t \neq i, j$. In the diagram

$$\begin{array}{ccccccccc}
0 & \to & \underline{H}_G^{\beta_{ij}} S^V & \to & \underline{H}_G^{\beta_{ij}} P(\Psi)^+ & \to & \underline{H}_G^{\beta_{ij}} P(\Delta)^+ & \to & 0 \\
 & & \downarrow \epsilon_V & & \downarrow q_t & & \downarrow q_t & & \\
0 & \to & \underline{H}_G^{\beta_{ij}} S^0 & \to & \underline{H}_G^{\beta_{ij}} P_t(\Psi)^+ & \to & 0 & &
\end{array}$$

comparing the cohomology sequences of our two cofibre sequences, we have that $\underline{H}_G^{\beta_{ij}}S^V$ and $\underline{H}_G^{\beta_{ij}}S^0$ are $\langle\mathbb{Z}\rangle$ and the map ϵ_V is multiplication by p. Thus, if z is a lifting of $\hat{c}_{ij}$, then by adding elements from the image of $\underline{H}_G^{\beta_{ij}}S^V$ to z, we can adjust $q_t(z)$ by any multiple of p. It now suffices to show that there is a lifting z with $q_t(z) \equiv d_{ij}^{tj}\epsilon_{\beta_{ij}}$ mod p. The lifting problems for $P(\Psi)$ and $P(\{\phi_i, \phi_j, \phi_t\})$ can be compared via the cohomology maps induced by the inclusion of $\{\phi_i, \phi_j, \phi_t\}$ into Ψ. This comparison indicates that it suffices to show that the lifting problem can be solved when $\Psi = \{\phi_i, \phi_j, \phi_t\}$. In this case, consider the diagram

$$\begin{array}{ccccccccc}
0 & \to & \underline{H}_G^{\beta_{ij}}S^V & \to & \underline{H}_G^{\beta_{ij}}P(\Psi)^+ & \overset{\theta^*}{\to} & \underline{H}_G^{\beta_{ij}}P(\Delta)^+ & \to & 0 \\
 & & \downarrow \epsilon & & \downarrow q & & \downarrow q_j & & \\
0 & \to & \underline{H}_G^{\beta_{ij}}S^{\beta_{tj}} & \overset{\gamma}{\to} & \underline{H}_G^{\beta_{ij}}P(\{\phi_j, \phi_t\})^+ & \overset{q_j}{\to} & \underline{H}_G^{\beta_{ij}}P(\{\phi_j\})^+ & \to & 0
\end{array}$$

comparing the cohomology exact sequences for the pairs $(P(\Psi), P(\Delta))$ and $(P(\{\phi_j, \phi_t\}), P(\{\phi_j\}))$. Let $\alpha = \beta_{ij} - \beta_{tj}$. If z is a lifting of $\hat{c}_{ij}$ along $\theta^*(1)$, then $q_j(z) = q_j q(z) = 0$. Thus, $q(z) = \gamma(y)$ for some $y \in \underline{H}_G^{\beta_{ij}}(S^{\beta_{tj}})(1)$. Since $\rho q(z)$ is the generator x of $\underline{H}_G^{\beta_{ij}}(P(\{\phi_j, \phi_t\})^+)(e)$, $\rho(y)$ must generate $\underline{H}_G^{\beta_{ij}}S^{\beta_{tj}}(e)$, and y must be $\sigma_\alpha + n\kappa_\alpha$ for some integer n. The diagram

$$\begin{array}{ccc}
\underline{H}_G^{\beta_{ij}}S^{\beta_{tj}} & \overset{\gamma}{\to} & \underline{H}_G^{\beta_{ij}}P(\{\phi_j, \phi_t\})^+ \\
\downarrow \epsilon & & \downarrow q_t \\
\underline{H}_G^{\beta_{ij}}S^0 & \overset{\cong}{\to} & \underline{H}_G^{\beta_{ij}}P(\{\phi_t\})^+
\end{array}$$

commutes and gives that $q_t(z) = q q_t(z) = \epsilon(y) \equiv \epsilon(\sigma_\alpha)$ mod p. By the definition of σ_α, $\epsilon(\sigma_\alpha) = d_{ij}^{tj}\epsilon_{\beta_{ij}}$.

(b) Let $r \geq 1$ and let $\phi_j \in \Phi(r+1)$. The cofibre sequence associated to the inclusion of $P(\Phi(r))$ into $P(\Phi(r) \cup \{\phi_j\})$ is

$$P(\Phi(r))^+ \to P(\Phi(r) \cup \{\phi_j\})^+ \overset{\pi}{\to} S^{\alpha_j(r)}.$$

Define $C_j(r) \in \underline{H}_G^{\alpha_j(r)}(P(\Phi(r) \cup \{\phi_j\})^+)(1)$ to be the image under $\pi^*(1)$ of $1 \in A(1) = \underline{H}_G^{\alpha_j(r)}(S^{\alpha_j(r)})(1)$. Since π^* is an isomorphism in dimension $\alpha_j(r)$, $\rho(C_j(r)) = x^{|\alpha_j(r)|/2}$. The cohomology diagram in dimension $\alpha_j(r)$ induced by the diagram

$$\begin{array}{ccc} P_j(\Phi(r) \cup \{\phi_j\})^+ & \xrightarrow{q_j} & P(\Phi(r) \cup \{\phi_j\})^+ \\ \Big\downarrow \pi_j & & \Big\downarrow \pi \\ S^r & \xrightarrow{\epsilon} & S^{\alpha_j(r)} \end{array}$$

indicates that $q_j(C_j(r)) = \epsilon_{\alpha_j(r)-r} x_j^r$. If $k \neq j$, $q_k(C_j(r)) = 0$ for dimensional reasons. As we did with the definition of c_{ij} in part (a), we extend the definition of $C_j(r)$ to $\underline{H}_G^* P(\Phi)^+$ by working inductively along a sequence of subsets of Φ between $\Phi(r) \cup \{\phi_j\}$ and Φ. The only difference between the argument given for c_{ij} and the one which should be used for $C_j(r)$ is that the liftings of $C_j(r)$ should be chosen to behave properly with respect to ρ and $\tilde{q}_k$ instead of ρ and q_k. This change is necessary because $q_k(C_j(r))$ is more complicated than $q_k(c_{ij})$. The behavior of the $C_j(r)$ with respect to the maps q_k is established in the lemma below.

LEMMA 6.2. Let $r \geq 1$ and $\phi_k \in \Phi(r+1) - \Phi(r)$. Then

$$q_k(C_k(r)) = x_k^r \left[\prod_{\substack{\phi_i \in \Phi(r) \\ i \neq k}} (\epsilon_{\beta_{ki}} + \xi_{\beta_{ki}-2} x_k) \right].$$

If $\phi_j \in \Phi(r+1) - \Phi(r)$ and $j \neq k$, then

$$q_k(C_j(r)) = x_k^r \left(d_{jk}^{kj} \epsilon_{\beta_{jk}} + \xi_{\beta_{jk}-2} x_k \right)^r \left[\prod_{\substack{\phi_i \in \Phi(r) \\ i \neq j,k}} (d_{ji}^{ki} \epsilon_{\beta_{ji}} + \xi_{\beta_{ji}-2} x_k) \right] +$$

$$\left[\tilde{d}_{kj}^r - (d_{jk}^{kj})^r \prod_{\substack{\phi_i \in \Phi(r) \\ i \neq j,k}} (d_{ji}^{ki}) \right] \epsilon_{\alpha_j(r)-r} x_k^r.$$

If $\phi_k \notin \Phi(r+1) - \Phi(r)$, then $q_k(C_j(r))$ is zero.

PROOF. If $\phi_k \notin \Phi(r+1) - \Phi(r)$, then $q_k(C_j(r))$ vanishes for dimensional reasons. Therefore, assume that $\phi_j, \phi_k \in \Phi(r+1) - \Phi(r)$. Let

$$\Psi = \Phi(r) \cup \{\phi : \phi \in \Phi - \Phi(r) \text{ and } \phi \cong \phi_k\}.$$

The image of the class $C_j(r)$ in $\underline{H}_G^* P(\Phi)^+$ under the map

$$\underline{H}_G^* P(\Phi)^+ \to \underline{H}_G^* P(\Psi \cup \{\phi_j\})^+$$

may be computed using the maps ρ and $\tilde{q}_i$. It is the class $C_j(r)$ in $\underline{H}_G^* P(\Psi \cup \{\phi_j\})^+$. The image of this class under the map

$$\underline{H}_G^* P(\Psi \cup \{\phi_j\})^+ \to \underline{H}_G^* P(\Psi)^+$$

is the class $\sigma_{\alpha_j(r)-\alpha_k(r)} C_k(r)$. Thus,

$$q_k(C_j(r)) = q_k(\sigma_{\alpha_j(r)-\alpha_k(r)} C_k(r)) = \sigma_{\alpha_j(r)-\alpha_k(r)} q_k(C_k(r)),$$

since $P_k(\Phi) = P_k(\Psi)$ and the map q_k for $P(\Phi)$ factors as the composite of the map $\underline{H}_G^* P(\Phi)^+ \to \underline{H}_G^* P(\Psi)^+$ and the map q_k for Ψ. Observe that

$$\sigma_{\alpha_j(r)-\alpha_k(r)} = (\sigma_{\beta_{jk}-\beta_{kj}})^r \left[\prod_{\substack{\phi_i \in \Phi(r) \\ i \neq j,k}} \sigma_{\beta_{ji}-\beta_{ki}} \right] + a\kappa_{\alpha_j(r)-\alpha_k(r)}$$

for some integer a. With this description of $\sigma_{\alpha_j(r)-\alpha_k(r)}$, it is easy to derive the formula for $q_k(C_j(r))$ from the formula for $q_k(C_k(r))$. The formula for $q_k(C_k(r)$ is derived using an iterative procedure. Let $s \geq r$ and pick $\phi_t \in \Psi$ with $t \neq k$. The image of $C_k(s) \in \underline{H}_G^*(P(\Psi)^+)(1)$ under the map $\underline{H}_G^* P(\Psi)^+ \to \underline{H}_G^* P(\Psi - \{\phi_t\})^+$ is

$$\epsilon_{\beta_{kt}} C_k(s) + \xi_{\beta_{kt}-2} C_k(s+1).$$

Iterating this process to eliminate from Ψ all the irreducible representations not isomorphic to ϕ_k, we move from $\underline{H}_G^* P(\Psi)^+$ to $\underline{H}_G^* P(n_k \phi_k)^+ \cong \underline{H}_G^* P_k(\Psi)^+$ and from $C_k(r)$ to the expansion of

$$x_k^r \left[\prod_{\substack{\phi_i \in \Phi(r) \\ i \neq k}} (\epsilon_{\beta_{ki}} + \xi_{\beta_{ki}-2} x_k) \right].$$

On the other hand, the image of $C_k(r)$ under this sequence of transformations must be $q_k(C_k(r))$.

Now that we have defined the classes c_{ij} and $C_j(r)$, we must show that they generate $\underline{H}_G^* P(\Phi)^+$ as an algebra over $\underline{H}_G^* S^0$.

PROPOSITION 6.3. The classes c_{ij}, for $\phi_i, \phi_j \in \Phi(1)$, and the classes $C_j(r)$, for $r \geq 1$ and $\phi_j \in \Phi(r+1) - \Phi(r)$, generate $\underline{H}_G^* P(\Phi)^+$ as an algebra over $\underline{H}_G^* S^0$.

PROOF. If Φ is infinite, then, by the proof of Theorem 2.6, $\underline{H}_G^* P(\Phi)^+$ is the limit of the $\underline{H}_G^* P(\Delta)^+$ where Δ runs over the finite subsets of Φ. Thus, it suffices to prove the result for Φ finite. Recall the functions f and g and the subsets $\Phi_j(r)$ of Φ defined in the remarks preceding Proposition 5.12. For this proof, initialize f and g by $f(0) = 0$ and $g(0) = 0$. We will show, by induction on n, that the classes c_{ij} and $C_j(r)$ which are defined in $\underline{H}_G^* P(\Phi_{f(n)}(g(n)))^+$ generate that Mackey functor as an algebra over $\underline{H}_G^* S^0$. The result is obvious for $n = 1$, since $\Phi_{f(1)}(g(1)) = \{\phi_0\}$ and

$P(\{\phi_0\})$ is a point. Assume the result for n. Denote $\alpha_{f(n+1)}(g(n+1)) + \gamma_{f(n+1)}$ by α. The boundary map is zero in the cohomology long exact sequence associated to the cofibre sequence

$$P(\Phi_{f(n)}(g(n)))^+ \xrightarrow{\theta} P(\Phi_{f(n+1)}(g(n+1)))^+ \to S^\alpha.$$

Thus, we have a split short exact sequence

$$0 \to \underline{H}_G^* S^\alpha \to \underline{H}_G^* P(\Phi_{f(n+1)}(g(n+1)))^+ \xrightarrow{\theta^*} \underline{H}_G^* P(\Phi_{f(n)}(g(n)))^+ \to 0.$$

All of the classes c_{ij} and $C_j(r)$ which are defined in $\underline{H}_G^* P(\Phi_{f(n)}(g(n)))^+$ are also defined in $\underline{H}_G^* P(\Phi_{f(n+1)}(g(n+1)))^+$. Moreover, θ^* takes these classes in $\underline{H}_G^* P(\Phi_{f(n+1)}(g(n+1)))^+$ to the corresponding classes in $\underline{H}_G^* P(\Phi_{f(n)}(g(n)))^+$. Thus, to generate $\underline{H}_G^* P(\Phi_{f(n+1)}(g(n+1)))^+$ as an algebra over $\underline{H}_G^* S^0$, it suffices to add to these classes the image z of the canonical generator of $A(1) = \underline{H}_G^\alpha(S^\alpha)(1)$. Clearly, $\rho(z)$ is the generator of $\underline{H}_G^\alpha(P(\Phi_{f(n+1)}(g(n+1)))^+)(e)$. Moreover, for $k \neq f(n+1)$, $\tilde{q}_k(z) = 0$ since $\tilde{q}_k$ factors through $\underline{H}_G^* P(\Phi_{f(n)}(g(n)))^+$. Finally,

$$\tilde{q}_{f(n+1)}(z) = \left[\epsilon_{\alpha - g(n+1)}\left(x_{f(n+1)}\right)^{g(n+1)}\right]$$

since the diagram

$$\begin{array}{ccc}
P_{f(n+1)}(\Phi_{f(n+1)}(g(n+1)))^+ & \xrightarrow{q_{f(n+1)}} & P(\Phi_{f(n+1)}(g(n+1)))^+ \\
\downarrow{\scriptstyle \pi_j} & & \downarrow{\scriptstyle \pi} \\
S^{g(n+1)} & \xrightarrow{\epsilon} & S^\alpha
\end{array}$$

commutes. The elements z and $D_{f(n+1)} C_{f(n+1)}(g(n+1))$ must be equal since they have the same image under the maps $\tilde{q}_k$ and ρ.

The equations in Propositions 5.9 and 5.10 describe elements in dimensions where there is no torsion. As a result, these equations can be checked easily by applying the maps ρ and $\tilde{q}_k$ to both sides. The equations in Lemma 5.11 are easily checked using the maps ρ and q_k because the images of the classes $\hat{\kappa}_j(r)$ under the maps q_k are so simple. However, the formula in Proposition 5.12 is more difficult to verify.

PROOF OF PROPOSITION 5.12. We may assume that $|\Phi| \geq |\Phi_{i'}(r)| + |\Phi_{j'}(s)|$ so that all of the $D_{f(|\Delta|)}C_{f(|\Delta|)}(g(|\Delta|))$ on the right hand side of the equation are nonzero. If $|\Phi|$ is too small, then form a sufficiently large set Φ' by adding enough copies of ϕ_0 to Φ. The proof below applies to Φ'; the result for Φ is obtained using the cohomology map induced by the inclusion of Φ into Φ'. We show the equality of the images of the two sides of the equation under the maps ρ and q_k. Since the map ρ preserves products, $\rho(D_{i'}C_i(r)D_{j'}C_j(s))$ is the generator of $\underline{H}^*_G(P(\Phi)^+)(e)$ in the appropriate dimension. The only term on the right hand side of the equation in Proposition 5.12 which is not in the kernel of ρ is the summand corresponding to Ψ regarded as a subset of itself. This term is $\chi_\Psi D_{f(u)}C_{f(u)}(g(u))$ and its image under ρ is the generator of $\underline{H}^*_G(P(\Phi)^+)(e)$ in the same dimension. Thus, the expressions on the two sides of the equation have the same image under ρ.

Let k be an integer with $0 \le k \le m$. If $\phi_k \notin \Phi(r+s+1) - \Phi(r+s)$, then both sides of the equation vanish under q_k. If $\phi_k \in \Phi(r+s+1) - \Phi(r+s)$, then expand the polynomial obtained by applying q_k to $D_{i'}C_i(r)D_{j'}C_j(s)$. Each term in the expansion consists of the product of an integer, a power of x_k, and an element of the form ϵ_β, ξ_α, or $\epsilon_\beta\xi_\alpha$ from $\underline{H}^*_GS^0$. We classify these terms according to the factor from $\underline{H}^*_GS^0$. There is exactly one term with a ξ_α; its integer coefficient is one. There is exactly one term with an ϵ_β; its integer coefficient may be zero. This term is exactly the part of q_k which is detected by $\tilde{q}_k$. There may be any number, including zero, of terms containing a product $\epsilon_\beta\xi_\alpha$. These terms are all torsion elements of order p.

Expand the polynomial obtained by applying q_k to the right hand side of the equation and observe that the same three types of terms appear. The summand indexed on Ψ regarded as a subset of itself is the only source of a ξ_α. It is easy to see that this ξ_α term exactly matches the corresponding term from the left hand side of the equation. If $i' > k$, then the expansion of the image of the right hand side under q_k will contain no ϵ_β term. In this case, $\tilde{q}_k(D_{i'})$ is zero and the image of the left hand side under q_k also lacks an ϵ_β term. If $i' \le k$, then numerous summands contribute to the ϵ_β term of the left hand side, but the coefficient of the $\hat{\kappa}_k(r+s)$ term is explicitly designed to ensure that the ϵ_β terms of the expansions of both sides match. The only problem here is that it is not obvious that the coefficient A_k of $\hat{\kappa}_k(r+s)$ is an integer. To show that A_k is an integer, it suffices to show that, modulo p, the image under q_k of the left hand side is equal to the image of the part of the right hand side indexed on the subsets of Ψ. Since the $\epsilon_\beta\xi_\alpha$ terms are all torsion of order p and the $\hat{\kappa}_t(r+s)$ summands on the right hand side contribute nothing to them, proving the equation

$$q_k(D_{i'}C_i(r)D_{j'}C_j(s)) \equiv q_k\Big(\sum_{\Delta\subset\Psi} d_{\Psi-\Delta}\,\epsilon_{\Psi-\Delta}\,\chi_\Delta\, D_{f(|\Delta|)}C_{f(|\Delta|)}(g(|\Delta|))\Big) \mod p$$

also shows that the $\epsilon_\beta\xi_\alpha$ terms of the two sides agree and so completes the proof of the proposition.

We prove this equation modulo p by transforming the right hand side into the left. In Theorem 5.5(c), $q_k(C_j(r))$ is described as a sum of two terms when $j \neq k$. The second term can be ignored in this transformation process because it vanishes modulo p. Recall that each χ_Δ is a χ_α for some virtual representation α. We accomplish our transformation by writing α as a sum of differences $\eta - \phi$ of irreducible complex representations. We then rewrite $\chi_\Delta = \chi_\alpha$ as the product of the elements $\chi_{\eta-\phi}$. To see that such a rewriting is justified, recall that if β and γ in RSO(G) are chosen so that the elements below are defined, then in $\underline{H}^*_G(S^0)(1)$

$$\xi_\beta \xi_\gamma = \xi_{\beta+\gamma} \qquad \xi_\beta \kappa_\gamma = 0 \qquad \epsilon_\beta \kappa_\gamma = p\,\epsilon_{\beta+\gamma}$$

and

$$\sigma_\beta \sigma_\gamma = \sigma_{\beta+\gamma} + A\kappa_{\beta+\gamma},$$

where A is some integer depending on β and γ. Now observe that every summand in the expansion of $q_k(D_{f(|\Delta|)} C_{f(|\Delta|)}(g(|\Delta|)))$ contains either an ϵ_β or a ξ_β. Thus, the $\kappa_{\beta+\gamma}$ error terms that might arise in the rewriting of χ_Δ as the product of the $\chi_{\eta-\phi}$ are killed by the ϵ_β and ξ_β from $q_k(D_{f(|\Delta|)} C_{f(|\Delta|)}(g(|\Delta|)))$.

We perform our transformation of the left hand side in four stages. During the first three stages, we think of the left hand side as a sum indexed on the subsets of Ψ and work on each summand separately. Therefore, fix a subset Δ of Ψ and let α be the virtual representation such that $\chi_\Delta = \chi_\alpha$. Recall that s′ and s″ are the number of elements isomorphic to ϕ_j in Δ and $\Phi_{i'}(r+s) - \Phi_{i'}(r)$, respectively. Recall that $u = |\Psi| - 1$, that the elements of Ψ are numbered from 0 to u, and that h is a function from the set $\{0, 1, \ldots, u\}$ to the set $\{0, 1, \ldots, m\}$ such that the ith element in Ψ is isomorphic to $\phi_{h(i)}$. Assume that the elements of Ψ numbered $j_0, j_1, \ldots, j_w$, with $j_0 < j_1 < \ldots < j_w$, are in Δ and that the elements numbered $i_0, i_1, \ldots, i_v$, with $i_0 < i_1 < \ldots < i_v$, are in $\Psi - \Delta$. For any integers q and t, with $0 \leq q, t \leq m$, abbreviate $\epsilon_{\beta_{qt}}$ and $\xi_{\beta_{qt}-2}$ by ϵ_{qt} and ξ_{qt}. Define the elements α_1, α_2, and α_3 of RSO(G) by

$$\alpha_1 = \left(\phi_i^{-1} - \phi_{f(|\Delta|)}^{-1}\right)\left[\sum_{\substack{\phi_t \in \Phi_{i'}(r) \\ t \neq f(|\Delta|), i, k}} \phi_t\right] + \left[r(\phi_i^{-1}\phi_k - \phi_{f(|\Delta|)}^{-1}\phi_i)\right]_{i \neq f(|\Delta|), k} +$$

$$\left[(r+\delta)(\phi_i^{-1}\phi_{f(|\Delta|)} - \phi_{f(|\Delta|)}^{-1}\phi_k)\right]_{f(|\Delta|) \neq i,k} +$$

$$\left[\phi_i^{-1}\phi_k - \phi_{f(|\Delta|)}^{-1}\phi_k\right]_{f(|\Delta|)>i>k} + \left[\phi_i^{-1}\phi_k \cdot 2\right]_{i>f(|\Delta|),\,k}$$

$$\alpha_2 = \left(\phi_j^{-1} - \phi_{f(|\Delta|)}^{-1}\right)\left[\sum_{\substack{\phi_t \in \Phi_{i'}(r+s) - \Phi_{i'}(r) \\ t \neq f(|\Delta|),j,k}} \phi_t\right] +$$

$$\left[(s - s' - s'')(\phi_j^{-1}\phi_k - \phi_j^{-1}\phi_0)\right]_{0 \neq j,k} +$$

$$\left[s''(\phi_j^{-1}\phi_k - \phi_{f(|\Delta|)}^{-1}\phi_j)\right]_{j \neq f(|\Delta|),\,k} +$$

$$\left[s(\phi_j^{-1}\phi_{f(|\Delta|)} - \phi_{f(|\Delta|)}^{-1}\phi_k)\right]_{f(|\Delta|) \neq j,k}$$

and

$$\alpha_3 = \alpha - \alpha_1 - \alpha_2,$$

where

$$\delta = \begin{cases} 1, & \text{if } i' > f(|\Delta|), \\ 0, & \text{otherwise.} \end{cases}$$

In the first stage of our transformation, χ_{α_1} is used to convert

$$d^0_{\Psi-\Delta}\,\epsilon^0_{\Psi-\Delta}\,\chi_\Delta\, q_k\Big(D_{f(|\Delta|)}\,C_{f(|\Delta|)}(g(|\Delta|))\Big)$$

into the product of

$$d^0_{\Psi-\Delta}\,\epsilon^0_{\Psi-\Delta}\,\chi_{\alpha_2+\alpha_3}\, q_k\Big(D_{i'}\,C_i(r)\Big)$$

and

$$x_k^{g(|\Delta|)-r-\delta'}\left[\prod_{\substack{\phi_t\in\Phi_{f(|\Delta|)}(g(|\Delta|))-\Phi_{i'}(r)\\ t\neq f(|\Delta|),k}}\left(d^{kt}_{f(|\Delta|),t}\,\epsilon_{f(|\Delta|),t}+\xi_{f(|\Delta|),t}\,x_k\right)\right]\cdot$$

$$\left[\left(d^{k,f(|\Delta|)}_{f(|\Delta|),k}\,\epsilon_{f(|\Delta|),k}+\xi_{f(|\Delta|),k}\,x_k\right)^{g(|\Delta|)-r-\delta}\right]_{f(|\Delta|)\neq k}\left[\xi_{f(|\Delta|),k}\,x_k\right]_{\substack{i'>f(|\Delta|)>k \text{ or }\\ f(|\Delta|)>k\geq i'}}$$

Here, δ is as in the definition of α_1 and

$$\delta'=\begin{cases}1, & \text{if } i'>f(|\Delta|),\ k;\\ 0, & \text{otherwise.}\end{cases}$$

In the second stage of the transformation, χ_{α_2} is used to convert this product into the product of $d^k_{\Psi-\Delta}\,\epsilon^k_{\Psi-\Delta}\,\chi_{\alpha_3}\,q_k\left(D_{i'}, C_i(r)\right)$ with the three factors

$$x_k^{s}\left[\prod_{\substack{\phi_t\in\Phi_{i'}(r+s)-\Phi_{i'}(r)\\ t\neq j,k}}\left(d^{kt}_{jt}\,\epsilon_{jt}+\xi_{jt}\,x_k\right)\right]\left[\left(d^{kj}_{jk}\,\epsilon_{jk}+\xi_{jk}\,x_k\right)^{s''}\right]_{j\neq k},$$

$$x_k^{g(|\Delta|)-r-s-\delta'}\left[\prod_{\substack{t=0\\ f(t)\neq f(|\Delta|),k}}^{w}\left(d^{k,f(t)}_{f(|\Delta|),f(t)}\,\epsilon_{f(|\Delta|),f(t)}+\xi_{f(|\Delta|),f(t)}\,x_k\right)\right],$$

and

$$\left[\left(d^{k,f(|\Delta|)}_{f(|\Delta|),k}\,\epsilon_{f(|\Delta|),k}+\xi_{f(|\Delta|),k}\,x_k\right)^{g(|\Delta|)-r-s-\delta}\right]_{f(|\Delta|)\neq k}\left[\xi_{f(|\Delta|),k}\,x_k\right]_{\substack{i'>f(|\Delta|)>k \text{ or }\\ f(|\Delta|)>k\geq i'}}$$

Observe that the $d^0_{\Psi-\Delta}\,\epsilon^0_{\Psi-\Delta}$ factor has been transformed into a $d^k_{\Psi-\Delta}\,\epsilon^k_{\Psi-\Delta}$ factor. This is accomplished by the $\left[(s-s'-s'')(\phi_j^{-1}\phi_k-\varphi_j^{-1}\phi_0)\right]_{0\neq j,k}$ summand in α_2. If $k=0$, then obviously no such transformation is needed. If $j=0$, then there will not be any elements of Ψ isomorphic to ϕ_j, and the value of $d^k_{\Psi-\Delta}\,\epsilon^k_{\Psi-\Delta}$ will not depend on k. In the description of the factor above indexed on t, for $0\leq t\leq w$, and throughout the third stage of the transformation, the set $\Phi_{f(|\Delta|)}(g(|\Delta|))-\Phi_{i'}(r+s)$ is identified with the set $\{\phi_{f(t)} : 0 \leq t \leq w\}$. By this identification, constructions that would naturally be indexed on $\Phi_{f(|\Delta|)}(g(|\Delta|)) - \Phi_{i'}(r+s)$ may be indexed on t. The description of the set $\{\phi_{f(t)} : 0 \leq t \leq w\}$ involves our usual abuse of notation in that, whenever $q \neq t$ and $f(q) = f(t)$, the representations $\phi_{f(q)}$ and $\phi_{f(t)}$ are intended to be distinct, but isomorphic, elements of the set.

The factor

$$q_k\Big(D_{i'}, C_i(r)\Big)x_k^s\left[\prod_{\substack{\phi_t \in \Phi_{i'}(r+s)-\Phi_{i'}(r) \\ t\neq j,k}}\left(d_{jt}^{kt}\epsilon_{jt} + \xi_{jt}x_k\right)\right]\left[\left(d_{jk}^{kj}\epsilon_{jk} + \xi_{jk}x_k\right)^{s''}\right]_{j\neq k}$$

appears in every summand of the transformation of the right hand side of the equation. We therefore factor it out of the sum and ignore it for the rest of the transformation. Observe that this factor consists of $q_k\Big(D_{i'}, C_i(r)\Big)$ and that part of $q_k\Big(D_{j'}, C_j(s)\Big)$ which is associated with the set $\Phi_{i'}(r+s) - \Phi_{i'}(r)$ when $\Phi_{i'}(s)$ is regarded as the disjoint union of Ψ and $\Phi_{i'}(r+s) - \Phi_{i'}(r)$. Thus, we must transform what remains of the sum after this factor is removed into the part of $q_k\Big(D_{j'}, C_j(s)\Big)$ coming from Ψ.

In the third stage of the transformation, χ_{α_3} is used to transform the remaining part of the Δ summand into

$$d_{\Psi-\Delta}^{k}\epsilon_{\Psi-\Delta}^{k}\left[\prod_{\substack{t=0 \\ h(j_t)\neq j}}^{w}\left(d_{j,h(j_t)}^{k,f(t)}\epsilon_{j,h(j_t)} + \xi_{j,h(j_t)}x_k\right)\right]\left[\prod_{\substack{t=0 \\ h(j_t)=j}}^{w}\left(d_{jk}^{k,f(t)}\epsilon_{jk} + \xi_{jk}x_k\right)\right].$$

For the fourth stage of the transformation, consider the subsets Δ of Ψ that contain the last element $\phi_{h(u)}$ of Ψ. The summands indexed on Δ and $\Delta - \{\phi_{h(u)}\}$ contain the common factor

$$\left[\prod_{\substack{t=0 \\ h(i_t)\neq j}}^{v-1} d_{j,h(i_t)}^{f(i_t-t),h(i_t)}\right]\left[\prod_{\substack{t=0 \\ h(i_t)=j}}^{v-1} d_{jk}^{f(i_t-t),j}\right]\left[\prod_{\substack{t=0 \\ h(i_t)\neq j}}^{v-1}\epsilon_{j,h(i_t)}\right]\left[\prod_{\substack{t=0 \\ h(i_t)=j}}^{v-1}\epsilon_{jk}\right].$$

$$\left[\prod_{\substack{t=0\\ h(j_t)\neq j}}^{w-1}\left(d^{k,f(t)}_{j,h(j_t)}\epsilon_{j,h(j_t)}+\xi_{j,h(j_t)}x_k\right)\right]\left[\prod_{\substack{t=0\\ h(j_t)=j}}^{w-1}\left(d^{k,f(t)}_{jk}\epsilon_{jk}+\xi_{jk}x_k\right)\right],$$

which we have written down using the i_t and j_t numbering of the elements in $\Psi-\Delta$ and Δ. Each of the two summands contains exactly one term not in this common factor. If $h(u)\neq j$, then these terms are

$$d^{f(w),h(u)}_{j,h(u)}\epsilon_{j,h(u)}+d^{k,f(w)}_{j,h(u)}\epsilon_{j,h(u)}+\xi_{j,h(u)}x_k = d^{k,h(u)}_{j,h(u)}\epsilon_{j,h(u)}+\xi_{j,h(u)}x_k.$$

If $h(u)=j$, then these terms are

$$d^{f(w),j}_{j,k}\epsilon_{j,k}+d^{k,f(w)}_{j,k}\epsilon_{j,k}+\xi_{j,k}x_k = d^{k,j}_{j,k}\epsilon_{j,k}+\xi_{j,k}x_k.$$

In either case, the result is independent of Δ and may be factored out of the sum. Moreover, this factor is exactly the contribution that $\phi_{h(u)}$ should make to $q_k\left(D_{j'},C_j(s)\right)$ when $\phi_{h(u)}$ is regarded as an element of $\Phi_{j'}(s)$ under the identification of $\Phi_{j'}(s)$ with the disjoint union of Ψ and $\Phi_{i'}(r+s)-\Phi_{i'}(r)$.

The sum that remains after the factor associated to $\phi_{h(u)}$ is removed may be regarded as one indexed on the subsets Δ of $\Psi-\{\phi_{h(u)}\}$. We now pair the summand indexed on a subset Δ containing the last element $\phi_{h(u-1)}$ of $\Psi-\{\phi_{h(u)}\}$ with the summand indexed on $\Delta-\{\phi_{h(u-1)}\}$ to obtain the factor of $q_k\left(D_{j'},C_j(s)\right)$ associated to $\phi_{h(u-1)}$. Repeating this process until the elements of Ψ are exhausted, we recover the part of $q_k\left(D_{j'},C_j(s)\right)$ associated with Ψ.

APPENDIX. Computing $\underline{H}^*_G S^0$. Here, we outline the calculation of $\underline{H}^*_G S^0$. The computation of the additive structure and, for $G=\mathbb{Z}/2$ or $\mathbb{Z}/3$, the computation of the multiplicative structure are unpublished work of Stong.

Three cofibre sequences suffice for the computation of the additive structure of $\underline{H}^*_G(S^0)$. Recall that ζ is the real 1-dimensional sign representation of $\mathbb{Z}/2$. Let η be a nontrivial irreducible complex representation of $G=\mathbb{Z}/p$, for any prime p. Let $G^+\to S\eta^+$ be the inclusion of an orbit and let $S\eta^+\to S^0$ and $S\zeta^+\to S^0$ be the maps collapsing the unit spheres $S\eta$ and $S\zeta$ to the non-basepoint in S^0. The cofibre sequences associated to these maps are

$$G^+\to S\eta^+\to \Sigma G^+$$

$$S\eta^+\to S^0\xrightarrow{\epsilon} S^\eta$$

and

$$G^+ \cong S\zeta^+ \to S^0 \xrightarrow{\epsilon} S^\zeta.$$

The first step in the computation is obtaining the values of $\underline{H}^G_* S\eta^+$ and $\underline{H}^*_G S\eta^+$ from the first cofibre sequence.

LEMMA A.1. For any nontrivial irreducible complex representation η of G,

$$\underline{H}^G_\alpha S\eta^+ = \begin{cases} L, & \text{if } |\alpha| = 0 \text{ and } |\alpha^G| \text{ is even,} \\ L_-, & \text{if } |\alpha| = 0 \text{ and } |\alpha^G| \text{ is odd,} \\ R, & \text{if } |\alpha| = 1 \text{ and } |\alpha^G| \text{ is odd,} \\ R_-, & \text{if } |\alpha| = 1 \text{ and } |\alpha^G| \text{ is even,} \\ 0, & \text{otherwise,} \end{cases}$$

$$\underline{H}_G^\alpha S\eta^+ = \begin{cases} R, & \text{if } |\alpha| = 0 \text{ and } |\alpha^G| \text{ is even,} \\ R_-, & \text{if } |\alpha| = 0 \text{ and } |\alpha^G| \text{ is odd,} \\ L, & \text{if } |\alpha| = 1 \text{ and } |\alpha^G| \text{ is odd,} \\ L_-, & \text{if } |\alpha| = 1 \text{ and } |\alpha^G| \text{ is even,} \\ 0, & \text{otherwise.} \end{cases}$$

PROOF. The next map $\Sigma G^+ \to \Sigma G^+$ in the first cofibre sequence is $1-g$, the difference of the identity map and the multiplication by g map, for some element g of G which depends on η. The homology and cohomology long exact sequences associated to the first cofibre sequence have the form

$$\ldots \to \underline{H}^G_\alpha G^+ \to \underline{H}^G_\alpha G^+ \to \underline{H}^G_\alpha S\eta^+ \to \underline{H}^G_{\alpha-1} G^+ \to \underline{H}^G_{\alpha-1} G^+ \to \ldots$$

and

$$\ldots \to \underline{H}_G^{\alpha-1} G^+ \to \underline{H}_G^{\alpha-1} G^+ \to \underline{H}_G^\alpha S\eta^+ \to \underline{H}_G^\alpha G^+ \to \underline{H}_G^\alpha G^+ \to \ldots .$$

The Mackey functor $\underline{H}^G_\alpha G^+$ may be identified with the Mackey functor $(\underline{H}^G_\alpha S^0)_G$ defined in Examples 1.1(f). The difference $1-g$ may be regarded as a map in B(G). Under the identification of $\underline{H}^G_\alpha G^+$ with $(\underline{H}^G_\alpha S^0)_G$, the first map in the part of the homology long exact sequence displayed above becomes the map from $(\underline{H}^G_\alpha S^0)_G$ to $(\underline{H}^G_\alpha S^0)_G$ induced by the map $1-g$ in B(G). It follows that the cokernel of the map $(1-g)_*: \underline{H}^G_\alpha G^+ \to \underline{H}^G_\alpha G^+$ is the Mackey functor $L(\underline{H}^G_\alpha(S^0)(e))$ defined in Examples 1.1(e). Similar observations reduce the homology and cohomology long exact sequences of the first cofibre sequence to the short exact sequences

$$0 \to L(\underline{H}^G_\alpha(S^0)(e)) \to \underline{H}^G_\alpha S\eta^+ \to R(\underline{H}^G_{\alpha-1}(S^0)(e)) \to 0$$

and

$$0 \to L(\underline{H}_G^{\alpha-1}(S^0)(e)) \to \underline{H}_G^\alpha S\eta^+ \to R(\underline{H}_G^\alpha(S^0)(e)) \to 0.$$

Since $\underline{H}_\alpha^G(S^0)(e) \cong H_{|\alpha|}(S^0; \mathbb{Z})$, $L(\underline{H}_\alpha^G(S^0)(e))$ is zero if $|\alpha| \neq 0$. If $|\alpha| = 0$, then $L(\underline{H}_\alpha^G(S^0)(e))$ is $L(\mathbb{Z})$ for some action of G on $\mathbb{Z}$. This action is the sign action of $\mathbb{Z}/2$ on $\mathbb{Z}$ when $p = 2$ and α contains an odd number of copies of ζ; otherwise, the action is trivial. Similar remarks apply to $L(\underline{H}_G^{\alpha-1}(S^0)(e))$, $R(\underline{H}^G_{\alpha-1}(S^0)(e))$, and $R(\underline{H}_G^\alpha(S^0)(e))$.

Notice the frequency with which $\underline{H}_G^\alpha S\eta^+$ and $\underline{H}_\alpha^G S\eta^+$ vanish. From the dimension axiom, we also obtain that $\underline{H}_G^\alpha G^+ = \underline{H}_\alpha^G G^+ = 0$ if $|\alpha| \neq 0$. These vanishing results determine most of the homological and cohomological behavior of the maps ϵ in our second and the third cofibre sequences.

LEMMA A.2. Let $\alpha \in RSO(G)$.

(a) The map $\epsilon^*: \underline{H}_G^{\alpha-\eta}S^0 \cong \underline{H}_G^\alpha(S^\eta) \to \underline{H}_G^\alpha(S^0)$

$$\text{is} \begin{cases} \text{mono} & \text{for } |\alpha| \neq 1, 2, \\ \text{epi} & \text{for } |\alpha| \neq 0, 1, \\ \text{iso} & \text{for } |\alpha| \neq 0, 1, 2. \end{cases}$$

(b) If $p = 2$, then the map $\epsilon^*: \underline{H}_G^{\alpha-\zeta}S^0 \cong \underline{H}_G^\alpha(S^\zeta) \to \underline{H}_G^\alpha(S^0)$

$$\text{is} \begin{cases} \text{mono} & \text{for } |\alpha| \neq 1, \\ \text{epi} & \text{for } |\alpha| \neq 0, \\ \text{iso} & \text{for } |\alpha| \neq 0, 1. \end{cases}$$

The divisibility results involving Euler classes in Lemmas 4.2, 4.6, and 4.8 follow from this lemma. Moreover, from this lemma and the vanishing of $\underline{H}_G^n S^0$, for $n \in \mathbb{Z}$ and $n \neq 0$, one can derive all of the zeroes in the first and third quadrants of our standard plot of $\underline{H}_G^* S^0$.

LEMMA A.3. Let $\alpha \in RSO(G)$. Then $\underline{H}_G^\alpha S^0 = 0$ if $|\alpha|$ and $|\alpha^G|$ are both positive or both negative.

Lemma A.2 indicates that all of $\underline{H}_G^* S^0$ can be determined from the values of $\underline{H}_G^\alpha S^0$ for the α in RSO(G) with $-2 \leq |\alpha| \leq 2$. If $p = 2$, it suffices to know $\underline{H}_G^\alpha S^0$ for the α in RSO(G) with $-1 \leq |\alpha| \leq 1$. The next lemma describes $\underline{H}_G^* S^0$ on the edges of these two ranges of values for $|\alpha|$.

LEMMA A.4. Let $\alpha \in RSO(G)$ and let η be any nontrivial irreducible complex representation of G.

(a) If $|\alpha| = 2$, then

$$\underline{H}_G^\alpha S^0 \cong \text{coker}\,(\tau: \underline{H}_G^{\alpha-\eta} G^+ \to \underline{H}_G^{\alpha-\eta} S^0).$$

(b) If $|\alpha| = -2$, then

$$\underline{H}_G^\alpha S^0 \;\cong\; \ker\,(\rho : \underline{H}_G^{\alpha+\eta} S^0 \to \underline{H}_G^{\alpha+\eta} G^+).$$

(c) If $p = 2$ and $|\alpha| = 1$, then

$$\underline{H}_G^\alpha S^0 \;\cong\; \operatorname{coker}\,(\tau : \underline{H}_G^{\alpha-\zeta} G^+ \to \underline{H}_G^{\alpha-\zeta} S^0).$$

(d) If $p = 2$ and $|\alpha| = -1$, then

$$\underline{H}_G^\alpha S^0 \;\cong\; \ker\,(\rho : \underline{H}_G^{\alpha+\zeta} S^0 \to \underline{H}_G^{\alpha+\zeta} G^+).$$

Moreover, in all four cases, $\underline{H}_G^\alpha(S^0)(e) = 0$.

PROOF. Part (d) follows immediately from the cohomology long exact sequence associated to the third cofibre sequence. Part (c) follows via duality from the homology long exact sequence associated to the third cofibre sequence. For part (b), consider the diagram

$$\begin{array}{ccccccc} 0 & \to & \underline{H}_G^\alpha S^0 & \to & \underline{H}_G^{\alpha+\eta} S^0 & \xrightarrow{f} & \underline{H}_G^{\alpha+\eta} S\eta^+ \\ & & & & & & \downarrow h \\ & & & & & & \underline{H}_G^{\alpha+\eta} G^+ \end{array}$$

in which the row is from the cohomology exact sequence of the second cofibre sequence and the vertical arrow comes from the inclusion of an orbit G into $S\eta$. Clearly, $\underline{H}_G^\alpha S^0 \cong \ker f$. By our computation of $\underline{H}_G^* S\eta^+$, the map h is mono, so $\ker f \cong \ker hf$. The composite hf is just ρ. The proof for part (a) is similar, but uses the homology long exact sequence to describe $\underline{H}^G_{-\alpha} S^0$ as the cokernel of the map $\underline{H}^G_{\eta-\alpha} G^+ \to \underline{H}^G_{\eta-\alpha} S^0$ induced by the collapse map $G^+ \to S^0$. Dualizing the homology Mackey functors to cohomology Mackey functors gives the result since the transfer is the dual of the collapse map. In all four cases, the group $\underline{H}_G^\alpha(S^0)(e)$ is zero either because $\tau(e)$ is surjective or because $\rho(e)$ is injective.

Most of the values of $\underline{H}_G^\alpha S^0$ for $|\alpha| = 0$ and $|\alpha^G| \neq 0$ follow immediately from the cohomology long exact sequence of the second cofibre sequence and Lemmas A.1 and A.3.

LEMMA A.5. Let $\alpha \in RSO(G)$ with $|\alpha| = 0$. Then

$$\underline{H}_G^\alpha S^0 = \begin{cases} R, & \text{if } |\alpha^G| \le -2 \text{ and } |\alpha^G| \text{ is even,} \\ R_-, & \text{if } |\alpha^G| \le -1 \text{ and } |\alpha^G| \text{ is odd,} \\ L, & \text{if } |\alpha^G| \ge 2 \text{ and } |\alpha^G| \text{ is even,} \\ L_-, & \text{if } |\alpha^G| \ge 3 \text{ and } |\alpha^G| \text{ is odd.} \end{cases}$$

PROOF. Let η be any nontrivial irreducible complex representation. If $|\alpha^G| < 0$, then consider the portion

$$\underline{H}_G^{\alpha-\eta}S^0 \cong \underline{H}_G^{\alpha}S^{\eta} \to \underline{H}_G^{\alpha}S^0 \to \underline{H}_G^{\alpha}S\eta^+ \to \underline{H}_G^{\alpha+1}S^{\eta} \cong \underline{H}_G^{\alpha+1-\eta}S^0$$

of the cohomology long exact sequence of the second cofibre sequence. The left hand term is zero by Lemma A.3 and the right hand term is zero by the same lemma unless $|\alpha^G|$ is -1. If $|\alpha^G| = -1$, then $p = 2$, $\alpha = \zeta - 1$, $\underline{H}_G^{\alpha}S\eta^+$ is R_- by Lemma A.1, and $\underline{H}_G^{\alpha+1-\eta}S^0$ is $\langle \mathbb{Z} \rangle$ by Lemma A.4. The last identification is based on the observations that η must be 2ζ and $\underline{H}_G^0 S^0$ is A. By inspection, there are no nontrivial maps from R_- to $\langle \mathbb{Z} \rangle$. Thus, if $|\alpha^G| < 0$, the middle arrow must be an isomorphism.

If $|\alpha^G| \geq 2$, then consider the portion

$$\underline{H}_G^{\alpha+\eta-1}S^0 \to \underline{H}_G^{\alpha+\eta-1}S\eta^+ \to \underline{H}_G^{\alpha+\eta}S^{\eta} \cong \underline{H}_G^{\alpha}S^0 \to \underline{H}_G^{\alpha+\eta}S^0$$

of the cohomology long exact sequence for the second cofibre sequence. The left and right hand terms in this portion of the sequence must be zero by Lemma A.3. Therefore, the middle arrow is an isomorphism.

If $p = 2$, then the results above reduce the computation of $\underline{H}_G^{*}S^0$ to the determination of $\underline{H}_G^0 S^0$, which is A by the dimension axiom, and $\underline{H}_G^{1-\zeta}S^0$, which is given by the following lemma.

LEMMA A.6. If $p = 2$, then $\underline{H}_G^{1-\zeta}S^0 \cong R_-$.

PROOF. Consider the portion

$$\underline{H}_G^0 S^0 \to \underline{H}_G^0 G^+ \to \underline{H}_G^1 S^{\zeta} \cong \underline{H}_G^{1-\zeta}S^0 \to \underline{H}_G^1 S^0$$

of the cohomology long exact sequence of the third cofibre sequence. By the dimension axiom, the right hand term is zero and the first two terms from the left are A and A_G, respectively. The value of $\underline{H}_G^{1-\zeta}S^0$ follows by computation.

If $p \neq 2$, then we must still determine the value of $\underline{H}_G^{\alpha}S^0$ when $|\alpha| = \pm 1$ or $\alpha \in RSO_0(G)$. The next three lemmas dispose of the α with $|\alpha| = \pm 1$ which are not already covered by Lemma A.3.

LEMMA A.7. Let M be a Mackey functor and $f: L \to M$ be a map. If $f(e)$ is a monomorphism, then so is f.

PROOF. The composite $f(e)\,\rho$ is a monomorphism and $\rho\, f(1) = f(e)\,\rho$.

LEMMA A.8. If $p \neq 2$, $\alpha \in RSO(G)$, $|\alpha| = 1$, and $|\alpha^G| < 0$, then $\underline{H}_G^\alpha S^0 = 0$.

PROOF. Consider the portion

$$\underline{H}_G^{\alpha-\eta} S^0 \cong \underline{H}_G^\alpha S^\eta \to \underline{H}_G^\alpha S^0 \to \underline{H}_G^\alpha S\eta^+ \xrightarrow{f} \underline{H}_G^{\alpha+1} S^\eta \cong \underline{H}_G^{\alpha+1-\eta} S^0$$

of the cohomology long exact sequence associated to the second cofibre sequence. The left hand term must be zero by Lemma A.3. By Lemma A.1, $\underline{H}_G^\alpha S\eta^+ \cong L$. Since $|\alpha+1-\eta| = 0$, $\underline{H}_G^{\alpha+1-\eta}(S^0)(e)$ is $\mathbb{Z}$. The map $f: \underline{H}_G^\alpha S\eta^+ \to \underline{H}_G^{\alpha+1-\eta} S^0$ is induced by the geometric map $S^\eta \to \Sigma S\eta^+$ which identifies the points 0 and ∞ in S^η. From this description, it follows that f(e) is an isomorphism. By the lemma above, f is a monomorphism. Therefore, $\underline{H}_G^\alpha S^0$ must be zero.

LEMMA A.9. Assume that $p \neq 2$, $\alpha \in RSO(G)$, $|\alpha| = -1$, and $|\alpha^G| > 0$. Then for any nontrivial irreducible complex representation η,

$$\underline{H}_G^\alpha S^0 \cong \text{coker}\,(\underline{H}_G^{\alpha+\eta-1} S^0 \to \underline{H}_G^{\alpha+\eta-1} S\eta^+).$$

Moreover, if $|\alpha^G| > 1$,

$$\underline{H}_G^\alpha S^0 \cong \langle \mathbb{Z}/p \rangle.$$

PROOF. Consider the portion

$$\underline{H}_G^{\alpha+\eta-1} S^0 \xrightarrow{h} \underline{H}_G^{\alpha+\eta-1} S\eta^+ \to \underline{H}_G^{\alpha+\eta} S^\eta \cong \underline{H}_G^\alpha S^0 \to \underline{H}_G^{\alpha+\eta} S^0$$

of the cohomology long exact sequence for the second cofibre sequence. The right hand term must be zero by Lemma A.3. The first part of the lemma follows immediately. By Lemma A.1, $\underline{H}_G^{\alpha+\eta-1} S\eta^+ \cong R$. The map h is induced by the collapse map $S\eta^+ \to S^0$. Since $|\alpha+\eta-1| = 0$,

$$\underline{H}_G^{\alpha+\eta-1}(S^0)(e) = \underline{H}_G^{\alpha+\eta-1}(S\eta^+)(e) = \mathbb{Z}.$$

The map h(e) is an isomorphism by an obvious computation in nonequivariant cohomology. If $|\alpha^G| > 1$, then by Lemma A.5, $\underline{H}_G^{\alpha+\eta-1} S^0 \cong L$. The only two maps h from L to R with h(e) an isomorphism have cokernel $\langle \mathbb{Z}/p \rangle$.

If $d \not\equiv 0 \bmod p$, then the only maps $h: A[d] \to R$ with h(e) an isomorphism are surjective. Therefore, once we have shown that $\underline{H}_G^\beta S^0$ is $A[d_\beta]$ when $\beta \in RSO_0(G)$, it will follow from the lemma above that $\underline{H}_G^\alpha S^0 = 0$ when $|\alpha| = -1$ and $|\alpha^G| = 1$.

Lemma 4.6 follows from Lemma A.9.

PROOF OF LEMMA 4.6. Let α and β be elements of RSO(G) with $|\alpha| = -1$, $|\alpha^G| > 0$, $|\beta| = 0$, and $|\beta^G| \leq 0$. Let η be a nontrivial irreducible complex representation. Consider the diagram

$$\begin{array}{ccccc} R \cong \underline{H}_G^{\alpha+\eta-1}S^0 & \to & \underline{H}_G^{\alpha}S^0 & \to & 0 \\ \downarrow & & \downarrow & & \\ R \cong \underline{H}_G^{\alpha+\beta+\eta-1}S^0 & \to & \underline{H}_G^{\alpha+\beta}S^0 & \to & 0 \end{array}$$

in which the vertical arrows are given by multiplication by ξ_β or μ_β. The rows of this diagram are exact by the proof of Lemma A.9. Let $y \in \underline{H}_G^{\alpha+\eta-1}(S^0)(1)$ be a generator and let $x \in \underline{H}_G^{\alpha}(S^0)(1)$ be its image. Since ρ preserves products, $\rho(\xi_\beta y)$ must be a generator. Thus, $\xi_\beta y$ must be a generator and so must $\xi_\beta x$. Similarly, $\rho(\mu_\beta y)$ is d_β times a generator, so $\mu_\beta y$ is d_β times a generator. It follows that $\mu_\beta x$ is a generator. This proves Lemma 4.6 in the special case where $|\alpha| = -1$ and $|\alpha^G| > 0$. The general case follows from the special case and Lemma A.2.

Let α be an element of $RSO_0(G)$. The main difficulty in identifying $\underline{H}_G^{\alpha}S^0$ with $A[d_\alpha]$ is that we must select a representative for α in $\tilde{R}_0(G)$ in order to define μ_α and d_α. To circumvent this difficulty, we work primarily with elements of $\tilde{R}_0(G)$ instead of elements of $RSO_0(G)$ in the remainder of our discussion of the additive structure of $\underline{H}_G^{*}S^0$. If α is in $\tilde{R}_0(G)$, we write $\underline{H}_G^{\alpha}S^0$ for the cohomology Mackey functor associated to the image of α in $RSO(G)$. To work with elements of $\tilde{R}_0(G)$, we must introduce variants of Definitions 4.5(a) and 4.5(d).

DEFINITION A.10. Observe that the procedure used to produce the element μ_α in Definitions 4.5(a) actually associates a map $\mu\colon S^{\Sigma\eta_i} \to S^{\Sigma\phi_i}$ to any element $\sum\phi_i - \eta_i$ of $\tilde{R}_0(G)$. If α is a nonzero element of $\tilde{R}_0(G)$, denote this map, and its image in $\underline{H}_G^{\alpha}(S^0)(1)$, by $\tilde{\mu}_\alpha$. Let $\tilde{\mu}_0$ denote the identity map of S^0 and $1 \in \underline{H}_G^{0}(S^0)(1)$. If ϕ is a nontrivial irreducible complex representation, then let $\epsilon_{\alpha,\phi}\colon S^{\Sigma\eta_i} \to S^{\phi+\Sigma\phi_i}$ denote the smash product of the map $\epsilon\colon S^0 \to S^\phi$ and the map $\tilde{\mu}_\alpha$. We also use $\epsilon_{\alpha,\phi}$ to denote the corresponding element in $\underline{H}_G^{\alpha+\phi}(S^0)(1)$.

If α and β are elements in $\tilde{R}_0(G)$ which represent the same element in $RSO_0(G)$, then $\tilde{\mu}_\alpha$ and $\tilde{\mu}_\beta$ need not be the same class in $\underline{H}_G^{\alpha}(S^0)(1)$. However, the class $\epsilon_{\alpha,\phi}$ in $\underline{H}_G^{\alpha+\phi}(S^0)(1)$ is uniquely determined by the sum $\alpha+\phi$ in $RSO(G)$. This uniqueness can be exploited to resolve the problems caused by dependence of $\tilde{\mu}_\alpha$ on α.

LEMMA A.11. Let α and β be in $\tilde{R}_0(G)$ and let ϕ and η be nontrivial irreducible complex representations such that $\alpha+\phi$ and $\beta+\eta$ represent the same element in $RSO(G)$. Then the cohomology classes $\epsilon_{\alpha,\phi}$ and $\epsilon_{\beta,\eta}$ in $\underline{H}_G^{\alpha+\phi}(S^0)(1)$ are equal.

PROOF. We establish the result for three special cases and then argue that the general case follows from them. Let η, η_1, η_2, ϕ, ϕ_1, and ϕ_2 be nontrivial irreducible complex representations and let $c: S^{\phi_1+\phi_2} \to S^{\phi_2+\phi_1}$ be the switch map. Regard $\alpha_1 = \phi_1 - \eta$, $\alpha_2 = \phi_2 - \eta$, and $\alpha = \phi_1 + \phi_2 - 2\eta$ as elements of $\tilde{R}_0(G)$. Let $\epsilon: S^0 \to S^\eta$ be the usual Euler class. The two maps $1 \wedge \epsilon$ and $\epsilon \wedge 1$ from S^η to $S^{\eta+\eta}$ are obviously equivariantly homotopic. On the level of maps,

$$\epsilon_{\alpha_2,\phi_1} = \tilde{\mu}_\alpha(\epsilon \wedge 1) \quad \text{and} \quad \epsilon_{\alpha_1,\phi_2} = c\,\tilde{\mu}_\alpha(1 \wedge \epsilon).$$

Therefore, $\epsilon_{\alpha_2,\phi_1}$ and $c\,\epsilon_{\alpha_1,\phi_2}$ are equivariantly homotopic. Thus, $\epsilon_{\alpha_2,\phi_1}$ and $\epsilon_{\alpha_1,\phi_2}$, regarded as cohomology classes, are equal. Here, the map c is, of course, absorbed in the passage to an RSO(G)-grading for $\underline{H}_G^* S^0$.

If η and ϕ_1 are equal and $\epsilon': S^0 \to S^{\phi_2}$ is the inclusion, then the trick used above can also be used to show that $1 \wedge \epsilon': S^\eta \to S^{\phi_1+\phi_2}$ is equivariantly homotopic to $\epsilon_{\alpha_2,\phi_1}$. Thus, if $\alpha_3 = \phi_1 - \phi_1 \in \tilde{R}_0(G)$, then ϵ' and $\epsilon_{\alpha_3,\phi_2}$ are equal in $\underline{H}_G^{\phi_2}(S^0)(1)$.

Regard $\beta_1 = (\phi_1 - \eta_1) + (\phi_2 - \eta_2)$ and $\beta_2 = (\phi_1 - \eta_2) + (\phi_2 - \eta_1)$ as elements of $\tilde{R}_0(G)$. By three applications of the result just proved for $\epsilon_{\alpha_2,\phi_1}$ and $\epsilon_{\alpha_1,\phi_2}$, it is possible to show that $\epsilon_{\beta_1,\phi}$ and $\epsilon_{\beta_2,\phi}$ are equal in $\underline{H}_G^{\beta_1+\phi}(S^0)(1)$.

If α and β are in $\tilde{R}_0(G)$ and ϕ and η are nontrivial irreducible complex representations such that $\alpha + \phi$ and $\beta + \eta$ represent the same element in RSO(G), then we can convert the pair (α, ϕ) into the pair (β, η) by some combination of the three basic transformations for which the lemma has already been proved. Thus, $\epsilon_{\alpha,\phi}$ and $\epsilon_{\beta,\eta}$ must be equal in $\underline{H}_G^{\alpha+\phi}(S^0)(1)$.

This lemma establishes that the element ϵ_β of Definition 4.5(d) does not depend on the choice of α and V used in its definition.

LEMMA A.12. If $\alpha \in RSO_0(G)$, then $\underline{H}_G^\alpha S^0 \cong A[d_\alpha]$. Moreover, if η is any nontrivial irreducible complex representation, then μ_α is the unique element of $\underline{H}_G^\alpha(S^0)(1)$ such that $\epsilon_\eta \mu_\alpha = \epsilon_{\alpha+\eta}$ and $\rho(\mu_\alpha) = d_\alpha \iota_\alpha$.

PROOF. Recall the map $s: RSO_0(G) \to \tilde{R}_0(G)$ introduced in section 2. Let $\alpha \in RSO_0(G)$ and assume that $s(\alpha) = \sum_{i=1}^{n} \phi_i - \eta_i$. Let α_0 be $0 \in \tilde{R}_0(G)$ and, for

$1 \leq k \leq n$, let α_k be the element $\sum_{i=1}^{k} \phi_i - \eta_i$ of $\tilde{R}_0(G)$. Denote by $d(\alpha_k)$ the integer associated to α_k by our homomorphism from $\tilde{R}_0(G)$ to $\mathbb{Z}$. For $0 \leq k \leq n$, let β_k be the element $\alpha_k + \phi_{k+1}$ of RSO(G). We will show by induction on k that

i) $\underline{H}_G^{\alpha_k} S^0$ is isomorphic to $A[d(\alpha_k)]$,
ii) $\tilde{\mu}_{\alpha_k}$ and $\tau(\iota_{\alpha_k})$ generate $\underline{H}_G^{\alpha_k}(S^0)(1)$,
iii) $\underline{H}_G^{\beta_k} S^0$ is isomorphic to $\langle \mathbb{Z} \rangle$, and
iv) ϵ_{β_k} generates $\underline{H}_G^{\beta_k} S^0$.

By the dimension axiom and Lemma A.4, these statements are true for $k = 0$. Consider the portion

$$\underline{H}_G^{\beta_k - 1}(S\eta_{k+1})^+ \to \underline{H}_G^{\beta_k} S^{\eta_{k+1}} \cong \underline{H}_G^{\alpha_{k+1}} S^0 \to \underline{H}_G^{\beta_k} S^0 \to \underline{H}_G^{\beta_k}(S\eta_{k+1})^+$$

of the cohomology long exact sequence of the second cofibre sequence. By Lemma A.1, The left hand term is isomorphic to L and the right hand term is zero. By Lemma A.7, the left hand arrow is a monomorphism. Thus, we have a short exact sequence

$$0 \to L \xrightarrow{f} \underline{H}_G^{\alpha_{k+1}} S^0 \to \underline{H}_G^{\beta_k} S^0 \to 0.$$

Assume that the assertions above hold for some integer k. The element μ_{k+1} in $\underline{H}_G^{\alpha_{k+1}}(S^0)(1)$ hits the generator ϵ_{β_k} in $\underline{H}_G^{\beta_k}(S^0)(1)$ by Lemma A.11. Since f(e) is an isomorphism, we may assume that f(e) takes the generator $1 \in \mathbb{Z} = L(e)$ to the generator $\iota_{\alpha_{k+1}}$ of $\underline{H}_G^{\alpha_{k+1}}(S^0)(e)$. It follows that $\tilde{\mu}_{\alpha_{k+1}}$ and $\tau(\iota_{\alpha_{k+1}})$ generate $\underline{H}_G^{\alpha_{k+1}}(S^0)(1)$. Since

$$\rho(\mu_{\alpha_{k+1}}) = d(\alpha_{k+1})\,\iota_{\alpha_{k+1}} \quad \text{and} \quad \rho\,\tau(\iota_{\alpha_{k+1}}) = p\iota_{\alpha_{k+1}},$$

$\underline{H}_G^{\alpha_{k+1}} S^0$ is isomorphic to $A[d(\alpha_{k+1})]$. By Lemma A.4, $\underline{H}_G^{\beta_{k+1}} S^0$ is isomorphic to $\langle \mathbb{Z} \rangle$ and is generated by $\epsilon_{\beta_{k+1}}$. Since $\tilde{\mu}_{\alpha_n} = \mu_\alpha$ and $d(\alpha_n) = d_\alpha$, $\underline{H}_G^{\alpha} S^0$ is isomorphic to $A[d_\alpha]$.

Replacing α_{k+1} by α, η_{k+1} by η, and β_k by $\alpha + \eta$ in the cohomology long exact sequence above, we obtain the short exact sequence

$$0 \to L \to \underline{H}_G^{\alpha} S^0 \xrightarrow{h} \underline{H}_G^{\alpha+\eta} S^0 \to 0.$$

Our characterization of μ_α in terms of $\epsilon_\eta\, \mu_\alpha = h(\mu_\alpha)$ and $\rho(\mu_\alpha)$ follows directly from this sequence.

Two general observations suffice for the proofs of many of the multiplicative relations. Any product involving at least one element in the image of the transfer map τ is easily computed using the Frobenius property

$$x\,\tau(y) \;=\; \tau(\rho(x)\,y).$$

Any relation involving an element, like $\epsilon^{-m}\kappa$, obtained by divided some other element by an Euler class may be checked by eliminating the division by the Euler class and checking the resulting relation. The original relation then follows by Lemma A.2.

PROOF OF THEOREM 4.1. We will describe the individual Mackey functors $\underline{H}_G^\alpha S^0$ of $\underline{H}_G^* S^0$ by their positions in our standard plot of $\underline{H}_G^* S^0$. Since $\underline{H}_G^\alpha(S^0)(e) \cong H^{|\alpha|}(S^0;\mathbb{Z})$, it is easy to check that the elements $\iota_{1-\zeta}$ and $\iota_{\zeta-1}$ generate $\underline{H}_G^*(S^0)(e)$ and satisfy no relations in $\underline{H}_G^*(S^0)(e)$ other than the obvious relation $\iota_{1-\zeta}\,\iota_{\zeta-1} = \rho(1)$. It follows immediately from the structure of the Mackey functors R_-, L, and L_- that the elements $\tau(\iota_{1-\zeta}^n)$, for $n \geq 1$, generate the part of $\underline{H}_G^*(S^0)(1)$ on the positive horizontal axis. For any positive integer n, $\rho(\xi^n) = \iota_{\zeta-1}^{2n}$. Therefore, ξ^n must generate $\underline{H}_G^{2n(\zeta-1)}(S^0)(1)$. The relation $\tau(\iota_{\zeta-1}^m) = 2\,\xi^m$ follows from the additive structure. No other relations involving only ξ and $\iota_{\zeta-1}$ are permitted by the additive structure. Lemmas A.2 and A.4 ensure that the powers of ϵ generate the part of $\underline{H}_G^*(S^0)(1)$ on the positive vertical axis. These two lemmas also indicate that the elements $\epsilon^m\,\xi^n$, for $m, n \geq 1$, generate the part of $\underline{H}_G^*(S^0)(1)$ in the second quadrant. The same two lemmas indicate that the elements $\epsilon^{-m}\kappa$ and the elements $\epsilon^{-m}\,\tau(\iota_{1-\zeta}^{2n+1})$ generate the parts of $\underline{H}_G^*(S^0)(1)$ on the negative vertical axis and in the fourth quadrant, respectively. The relations not already verifed follow easily from the additive structure of $\underline{H}_G^* S^0$ or from our general observations. The additive structure of $\underline{H}_G^* S^0$ eliminates the possibility of any unlisted relations involving a single element. Since we have described every possible nonzero product of a pair of generators in terms of the generators, no further relations involving products are possible.

PROOF OF THEOREM 4.9. Again, we describe the individual Mackey functors $\underline{H}_G^\alpha S^0$ in terms of their positions in our plot of $\underline{H}_G^* S^0$. Since $\underline{H}_G^\alpha(S^0)(e) \cong H^{|\alpha|}(S^0;\mathbb{Z})$, it is easy to check that the relation $\iota_\alpha\,\iota_\beta = \iota_{\alpha+\beta}$ holds for any $\alpha, \beta \in RSO(G)$ with $|\alpha| = |\beta| = 0$ and that no other relations in $\underline{H}_G^*(S^0)(e)$ hold among the ι_α. Therefore, for any $\beta \in RSO(G)$ with $|\beta| = 0$, ι_β can be written as a product of the ι_α included in the proposed list of generators of $\underline{H}_G^* S^0$. The elements ι_β, for $\beta \in RSO(G)$ with $|\beta| = 0$, generate $\underline{H}_G^*(S^0)(e)$ and the elements $\tau(\iota_\beta)$, for $\beta \in RSO(G)$ with $|\beta| = 0$ and $|\beta^G| > 0$, generate the part of $\underline{H}_G^*(S^0)(1)$ on the positive horizontal axis.

Let α and β be in $RSO_0(G)$ and let γ be an element of $RSO(G)$ such that $|\gamma| > 0$ and $|\gamma^G| = 0$. The relation $\mu_\alpha\,\epsilon_\gamma = \epsilon_{\alpha+\gamma}$ follows from Lemma A.11. The relation

$$\mu_\alpha \mu_\beta = \mu_{\alpha+\beta} + [(d_\alpha d_\beta - d_{\alpha+\beta})/p]\,\tau(\iota_{\alpha+\beta})$$

follows from our characterization in Lemma A.12 of $\mu_{\alpha+\beta}$ as an element of $\underline{H}_G^{\alpha+\beta}(S^0)(1)$. From this relation, it follows that all of the elements μ_α can be constructed from the μ_β and ι_β in our proposed list of generators. By Lemma A.12, the elements μ_α and ι_α generate all of the $\underline{H}_G^\alpha S^0$ which are plotted at the origin. The relation $\mu_\alpha \epsilon_\gamma = \epsilon_{\alpha+\gamma}$ indicates that we can construct all the elements ϵ_γ from our proposed list of generators. By Lemmas A.2 and A.4, these elements generate all of the $\underline{H}_G^\alpha S^0$ on the positive vertical axis.

Let $\alpha \in RSO_0(G)$ and $\beta,\gamma \in RSO(G)$ with $|\beta| = |\gamma| = 0$ and $|\beta^G|, |\gamma^G| < 0$. The element σ_α can be obtained from μ_α and ι_α. The relations

$$\rho(\mu_\alpha \xi_\beta) = d_\alpha \iota_{\alpha+\beta} = \rho(d_\alpha \xi_{\alpha+\beta}),$$

$$\rho(\sigma_\alpha \xi_\beta) = \iota_{\alpha+\beta} = \rho(\xi_{\alpha+\beta}),$$

and

$$\rho(\xi_\beta \xi_\gamma) = \iota_{\beta+\gamma} = \rho(\xi_{\beta+\gamma})$$

follow from the fact that ρ is a ring homomorphism. They imply the relations $\mu_\alpha \xi_\beta = d_\alpha \xi_{\alpha+\beta}$, $\sigma_\alpha \xi_\beta = \xi_{\alpha+\beta}$, and $\xi_\beta \xi_\gamma = \xi_{\beta+\gamma}$ since ρ is a monomorphism in dimensions $\alpha+\beta$ and $\beta+\gamma$. These relations indicate that all of the elements ξ_β can be produced from our proposed list of generators. These elements generate the part of $\underline{H}_G^* S^0$ on the negative horizontal axis. By Lemmas A.2 and A.4, the elements $\epsilon_\delta \xi_\beta$ generate the part of $\underline{H}_G^* S^0$ in the second quadrant.

The relations $\mu_\gamma(\epsilon_\beta^{-1}\kappa_\alpha) = \epsilon_\beta^{-1}\kappa_{\alpha+\beta}$ and $\epsilon_\beta^{-1}\kappa_\alpha = \epsilon_\gamma^{-1}\kappa_\delta$, for $\alpha+\gamma = \beta+\delta$, may be checked by our general procedure for relations involving division by an Euler class. Together, these relations indicate that our proposed set of generators suffices to construct all of the elements $\epsilon_\beta^{-1}\kappa_\alpha$ and therefore to generate the part of $\underline{H}_G^* S^0$ on the negative vertical axis.

Let $\beta \in RSO_0(G)$ and let $\alpha \in RSO(G)$ with $|\alpha| < 0$ and $|\alpha^G| > 0$. Recall the class ν_α and the virtual representation $<\alpha>$ from Definitions 4.7. By definition, $<\alpha+\beta> = <\alpha>$, and by the Frobenius relation, $\nu_{<\alpha>}\tau(\iota_{\alpha+\beta}) = 0$. Therefore,

$$\begin{aligned}\mu_\beta \nu_\alpha &= \mu_\beta \mu_{\alpha-<\alpha>} \nu_{<\alpha>} \\ &= \mu_{\alpha+\beta-<\alpha>} \nu_{<\alpha>} \\ &= \nu_{\alpha+\beta}.\end{aligned}$$

This relation indicates that our proposed set of generators suffices to produce all of the elements ν_α and therefore the part of $\underline{H}_G^* S^0$ in the fourth quadrant.

We have now shown that our proposed set of generators does generate $\underline{H}_G^* S^0$. Seven of the relations we have not already established deserve comments. The relation $\epsilon_\alpha \epsilon_\beta = \epsilon_{\alpha+\beta}$ follows easily from the definition of the Euler classes, the Frobenius relation and the product relation for the classes μ_γ. The relation $\epsilon_\beta \xi_\alpha = d_{\delta-\alpha} \epsilon_\gamma \xi_\delta$, for $\alpha+\beta = \gamma+\delta$, follows from the sequence of equations

$$\epsilon_\beta \xi_\alpha = \mu_{\beta-\gamma} \epsilon_\gamma \xi_\alpha$$

$$= \epsilon_\gamma \mu_{\delta-\alpha} \xi_\alpha$$
$$= d_{\delta-\alpha} \epsilon_\gamma \xi_\delta.$$

The relations $\kappa_\alpha \kappa_\delta = p\kappa_{\alpha+\delta}$ and $\kappa_\gamma \nu_\alpha = 0$ can be confirmed from the definitions, the Frobenius property, and the relations which have already been established. Given these equations, the relations

$$\epsilon_\gamma (\epsilon_\beta^{-1} \kappa_\alpha) = \epsilon_{\beta-\gamma}^{-1} \kappa_\alpha,$$
$$(\epsilon_\beta^{-1} \kappa_\alpha)(\epsilon_\gamma^{-1} \kappa_\delta) = p \epsilon_{\beta+\gamma}^{-1} \kappa_{\alpha+\delta},$$

and

$$(\epsilon_\beta^{-1} \kappa_\gamma) \nu_\alpha = 0$$

follow from our general procedure for checking relations involving classes divided by Euler classes. For the relations $\epsilon_\beta \nu_\alpha = \nu_{\alpha+\beta}$ and $\xi_\beta \nu_\alpha = d_{<\beta>-\beta} \nu_{\alpha+\beta}$, observe that ξ_β can be written as $\sigma_{\beta-<\beta>} \xi_{<\beta>}$ and that ϵ_β can be written as $\mu_\gamma \epsilon_{n\lambda}$, for some $\gamma \in RSO_0(G)$ and some positive integer n. The relations now follow by straightforward computations using the definitions, the Frobenius property, and the previously established relations. All of the remaining relations in the theorem follow directly from the definitions or the additive structure of $\underline{H}_G^* S^0$. The additive structure of $\underline{H}_G^* S^0$ eliminates the possibility of any unlisted relations involving a single element. Since we have described every possible nonzero product of a pair of generators in terms of the generators, no further relations involving products are possible.

REFERENCES

[tDP] T. tom Dieck and T. Petrie, Geometric modules over the Burnside ring. Inventiones Math. 47 (1978), 273-287.

[DRE] A. Dress, Contributions to the theory of induced representations. Springer Lecture Notes in Mathematics, vol. 342, 1973, 183-240.

[ILL] S. Illman, Equivariant singular homology and cohomology I. Memoirs Amer. Math. Soc. vol. 156, 1975.

[LE1] L. G. Lewis, Jr., The equivariant Hurewicz map. Preprint.

[LE2] L. G. Lewis, Jr. An introduction to Mackey functors (in preparation).

[LMM] L. G. Lewis, Jr., J. P. May, and J. E. McClure, Ordinary RO(G)-graded cohomology. Bull. Amer. Math. Soc. 4 (1981), 208-212.

[LMSM] L. G. Lewis, Jr., J. P. May, and M. Steinberger (with contributions by J. E. McClure). Equivariant stable homotopy theory. Springer Lecture Notes in Mathematics, vol. 1213, 1986.

[LIN] H. Lindner, A remark on Mackey functors. Manuscripta Math. 18 (1976), 273-278.

[LIU] A. Liulevicius, Characters do not lie. Transformation Groups. London Math. Soc. Lecture Notes Series, vol. 26, 1976, 139-146.

[MAT] T. Matumoto, On G-CW complexes and a theorem of J. H. C. Whitehead. J. Fac. Sci. Univ. Tokyo 18 (1971/72), 363-374.

[WIR] K. Wirthmüller, Equivariant homology and duality. Manuscripta Math. 11 (1974), 373-390.

THE EQUIVARIANT DEGREE

by

Wolfgang Lück

0. Introduction

Abstract. In this paper we study the possible values $\deg f^H$, $H \subset G$ for a G-map $f : M \to N$ if M and N are compact smooth G-manifolds and G a compact Lie group. We generalize results about maps between spheres of G-representations. We give applications to one-fixed point actions and G-surgery. We prove that the unstable H-homotopy type of the sphere of the H-normal slice $S\nu(M^H,M)_x$ for $x \in M^H$ is a G-homotopy invariant of M.

Survey. As an illustration we state a consequence of our main result in a very special situation where it is easy to formulate.

Let G be finite. Consider a compact smooth G-manifold M such that M^H is non-empty, connected and orientable for all $H \subset G$. Assume either that G is nilpotent or that $\dim M^H \geq \dim M^K - 2$ holds for $H \underset{\neq}{\subset} K$, $H, K \in \mathrm{Iso}(M) = \{G_x \mid x \in M\}$. Here and elsewhere G_x denotes the isotropy group $\{g \in G \mid gx = x\}$ of $x \in M$. The set of finite G-sets S with $\mathrm{Iso}(S) \subset \mathrm{Iso}(M)$ is an abelian semi-group under disjoint union. Let $A(G,\mathrm{Iso}(M))$ be its Grothendieck group. The cartesian product induces the structure of a commutative ring with unit on it. Let $\mathrm{Con}(G)$ be the set of conjugacy classes of subgroups of G and $C(G)$ be the ring $\prod_{\mathrm{Con}(G)} \mathbb{Z}$. Then $A(G,\mathrm{Iso}(M))$ is a subring of $C(G)$ by identifying S with $(\mathrm{card}\, S^H \mid (H) \in \mathrm{Con}(G))$. For a G-selfmap $f : M \to M$ define $\mathrm{DEG}(f) \in C(G)$ by $(\deg f^H \mid (H) \in \mathrm{Con}(G))$.

Theorem A.

a) $\mathrm{DEG}(f) \in A(G,\mathrm{Iso}(M)) \subset C(G)$.

b) If $H \subset G$ is a p-group then:

$\deg f \equiv \deg f^H \mod p$.

c) If G has odd order and $\deg f^H \in \{\pm 1\}$ for each $H \subset G$, then we have for all $H \subset G$:
$\deg f = \deg f^H$. □

This theorem is well known for M as the one-point compactification V^c of a G-representation V. The proof for V^c uses the equivariant Lefschetz index and Smith theory. These methods do not suffice for M a G-manifold. Our main tool is quasi-transversality and the notion of a local degree.

The notion of the degree is used to classify G-homotopy classes of G-maps $f : V^c \to W^c$ (see tom Dieck [6'], p. 213, Laitinen [14], Tornehave [21]) and G-homotopy types of G-homotopy representations (see tom Dieck-Petrie [8]). It plays also a role in equivariant surgery theory (see for example Dovermann-Petrie [11], Lück-Madsen [17]). We give a survey over the various sections.

In section one we define the fibre transport tp_M of the tangent bundle of a G-manifold and the notion of an O(G)-transformation $\varphi : f^* tp_N \to tp_M$ for a G-map $f : M \to N$. Roughly speaking, φ assigns to each point x in M a G-map (not necessarily a G_x-homotopy equivalence) $TN^c_{fx} \to TM^c_x$ such that certain compatibility conditions hold. Using φ we get a one-to-one-correspondence between local orientations of M^H at x and N^H at fx for each $H \subset G$ and $x \in M^H$. This enables us to define the equivariant degree $DEG(f,\varphi) \in C(N) = \prod_{(H) \in Con(G)} \prod_{\pi_0(N^H)/WH} \mathbb{Z}$ in section two.

In section three the Burnsidering $A(G,\mathfrak{F})$ of a compact Lie group with respect to a family $\mathfrak{F}$ is treated. We identify $[V^c,V^c]^G$ and

$A(G,Iso(V))$ for a G-representation V. We introduce in section four a multiplicative submonoid $End_{tp_N} \subset C(N)$ and prove $DEG(f,\varphi) \in End_{tp_N}$ for any f and φ in section five. We will see that End_{tp_N} does not involve f and φ but depends only on the component structure of N. The main idea of the proof is best explained in the special case where G is finite and all N^H are non-empty and connected. Then $C(N) = C(G) = \prod_{Con(G)} \mathbb{Z}$ and End_{tp_N} is $A(G,Iso(N))$.

Choose y in N^G and make f quasi-transverse to y. Then $f^{-1}(y)$ is finite and for each $x \in f^{-1}(y)$ f looks like a (not necessarily linear) norm-preserving G_x-map $T_xM \to T_yN$ in a G_x-neighbourhood of x. Consider a G-orbit c of $f^{-1}(y)$. For each x in c we obtain G_x-maps $TM_x^c \to TN_y^c$ by f and $TN_y^c \to TM_x^c$ by φ. Their composition $TN_y^c \to TN_y^c$ defines an element in $A(G_x,Iso(TN_y))$. Its image in $A(G,Iso(N))$ under the induction homomorphism for $G_x \subset G$ is independent of the choice of x and denoted by $d(c)$. Let d be the sum $\Sigma d(c)$ running over $c \in f^{-1}(y)/G$. Since the global degree can be computed by local degrees, $DEG(f,\varphi) \in C(G)$ is just $d \in A(G,Iso(N))$. Roughly speaking, we have counted the local degrees orbitwise in the Burnside ring to get the global degree.

Section six contains some examples to illustrate our results. We give an elementary proof of the following known statement (see Atiyah-Bott [1], Browder [4], Ewing-Stong [12]).

Corollary B. There is no closed G-manifold M with $\dim M \geq 1$ such that each M^H is connected and orientable and M^G a single point if G is the product of a p-group and a torus. □

It is of special interest to choose $\varphi : f^*tp_N \to tp_M$ as an $O(G)$-equivalence i. e. all $TN^c_{fx} \to TM^c_x$ are G_x-homotopy equivalences. Then another choice of φ would change the equivariant degree only by a unit. Moreover, we have:

Theorem C. *A normal* G-*map* $f : M \to N$ *can be changed into a* G-*homotopy equivalence by equivariant surgery only if there is an* $O(G)$-*equivalence* $\varphi : tp_N \to f^*tp_M$ *with* $DEG(f,\varphi) = 1$. □

The existence of an $O(G)$-equivalence φ is related to the notion of the equivariant first Stiefel Whitney class w_M of a G-manifold. In section seven we relate tp_M und w_M and show that the existence of an $O(G)$-equivalence $\varphi : f^*tp_N \to tp_M$ is equivalent to $f^*w_N = w_M$. We prove:

Theorem D. *If* $f : M \to N$ *is a* G-*homotopy equivalence we have* $f^*w_N = w_M$. □

This implies the unstable version of the stable result in Kawakubo [13].

Corollary E. *If* $f : M \to N$ *is a* G-*homotopy equivalence, we get for* $x \in M$:

$$TM^c_x \simeq_{G_x} TN^c_{fx} \ . \ \square$$

Our setting and proofs would be much simpler if we supposed that all fixed point sets are non-empty, connected and orientable. Unfortunately, such conditions are unrealistic in the study of G-manifolds. Hence we make no assumptions about the existence of

G-fixed points or about the connectivity or orientability of the fixed point sets and do not demand $\pi_0(f^H)$ being bijective.

Our notion of the equivariant degree using O(G)-transformations has some advantages compared with the one using fundamental classes. It is in this generality much easier to state elementary properties like bordism invariance or the computation by local degrees in our language. We have the global choice of φ instead of the various choices of fundamental classes $[M^H]$ and $[N^H]$. Notice that the choice of $[M^H]$ is independent of the one of $[M^K]$ for $(K) \neq (H)$ and $[N^K]$. Hence in the case of fundamental classes the interaction between the various fixed point sets are not taken into account, what is done in our setting. It seems to be difficult, or even impossible, to state some of our results by means of fundamental classes. For example, the statement of example 6.5 makes no sense if it is formulated with fundamental classes and in example 6.3 there must appear signs because we can substitute $[M^H]$ by $-[M^H]$ and thus change the corresponding degree by a sign. The advantages of our approach for the notion of an equivariant normal map is worked out in Lück-Madsen [17]. (see also theorem C above and example 2.8).

<u>Conventions:</u> We denote by G a compact Lie group unless it explicitly is stated differently. Subgroups are assumed to be closed. A G-representation is always real. A G-manifold M is a compact smooth G-manifold with smooth G-action and possibly non-empty boundary. We call a component C of M^H an isotropy component if there is a x in C with isotropy group $G_x = H$. We say that M fullfills condition (*) if it satisfies the conditions

i) and ii) or the conditions i) and iii) below.

i) $C \neq \{point\}$ for all $C \in \pi_o(M^H)$, $H \subset G$.

ii) If $C \in \pi_o(M^H)$ is an isotropy component, $C^{>H}$ is $\{x \in C \mid G_x \neq H\}$ and $H \subset G$ we have $\dim C^{>H} + 2 \leq \dim C^H$.

iii) G is finite and nilpotent.

A G-map $f : M \to N$ respects always the boundary and we assume $\dim C = \dim D$ for all $C \in M_o(M^H)$, $D \in M_o(N^H)$, $H \subset G$ with $f^H(C) \subset D$.

Acknowledgement. The author wishes to thank the topologists at Århus for their hospitality and support during 1985 - 1986 when the main part of this paper was written. The author is indebted to Ib Madsen and Erkki Laitinen for their useful comments.

1. The fibre transport.

We organize the book-keeping of the components of the various fixed point sets and their fundamental groups for a G-space as follows. We recall that an object of the fundamental groupoid $\Pi(Y)$ of a space Y is a point in y and a morphism $y_o \to y_1$ is a homotopy class of paths from y_1 to y_o. The orbit category $O(G)$ has the homogenous spaces G/H as objects and G-maps as morphisms.

Definition 1.1. The fundamental $O(G)$-groupoid $\Pi^G X$ of a G-space X is the contravariant functor $\Pi^G X : O(G) \to \{groupoids\}$ sending G/H to $\Pi(X^H) = \Pi(map(G/H,X)^G)$. □

In general an $O(G)$-category resp. $O(G)$-groupoid is a contravariant functor from O(G) into the category of small categories resp. groupoids. We recall that a groupoid is a category whose morphisms are all isomorphisms. An O(G)-functor $F : C \to D$ between O(G)-categories is a natural transformation. Let I be the category of two objects 0 and 1 and three morphisms $ID : 0 \to 0$, $ID : 1 \to 1$ and $u : 0 \to 1$. We define an O(G)-transformation $\varphi : F_0 \to F_1$ between O(G)-functors F_0 and $F_1 : C \to D$ as an O(G)-functor $\varphi : C \times I \to D$ with $C \mid i = F_i$. Given a second O(G)-transformation $\psi: F_1 \to F_2$, let the composition $\psi \circ \varphi : F_0 \to F_2$ be determined by $\psi \circ \varphi(id,u) = \psi(id,u) \circ \varphi(id,u) : (x,0) \to (x,1)$ for all $x \in C$. One should think of an O(G)-functor $F : C \to D$ as a collection of functors $F(G/H) : C(G/H) \to D(G/H)$ and of an O(G)-transformation $\varphi : F_0 \to F_1$ as a collection of natural transformations $\varphi(G/H) : F_0(G/H) \to F_1(G/H)$ fitting nicely together. An O(G)-transformation $\varphi : F_0 \to F_1$ is called an O(G)-equivalence if there is an O(G)-transformation $\psi : F_1 \to F_0$ with both compositions the identity.

A G-map $f : X \to Y$ induces an O(G)-functor $\Pi^G f : \Pi^G X \to \Pi^G Y$ whereas a G-homotopy $h : X \times I \to Y$ between f and g determines an O(G)-equivalence $\Pi^G f \to \Pi^G g$.

A G-S^n-Hurewicz-fibration $\eta \downarrow X$ is called locally linear if there exists a G_x-neighbourhood U_x for each x in X such that U_x is G_x-fibre homotopy equivalent to $U_x \times SV_x$ for some G_x-representation V_x. We call a locally linear G-S^n-Hurewicz-fibration briefly a a G-S^n-fibration. An example is the fibrewise one-point compactification ξ^c of a G-$\mathbb{R}^n$-bundle ξ. Denote by $bf_{G,n}(X)$ the category of G-S^n-fibrations over X with G-fibre homotopy classes of fibrewise G-maps as morphisms. We obtain an O(G)-category $bf_{G,n}$ by letting X vary over all homogenous spaces. One should notice that $bf_{G,n}(G/H)$ is equivalent to the category with spheres of H-representations and H-homotopy classes of H-maps as morphisms. We prefer $bf_{G,n}(G/H)$ because of its better transformation behaviour in view of O(G).

The fibre transport of a G-S^n-fibration $\eta \downarrow X$ defines an O(G)-functor $tp_\eta : \Pi^G X \rightarrow bf_{G,n}$ analogously to the non-equivariant case (see [19], p. 343). The functor $tp(G/H) : \Pi(X^H) \rightarrow bf_{G,n}(G/H)$ sends a point in X^H given by $x : G/H \rightarrow X$ to $x^*\eta$. Let $h : G/H \times I \rightarrow X$ be a G-homotopy from y to x representing a morphism $x \rightarrow y$. Choose a solution $\bar{h}$ of the G-homotopy lifting problem

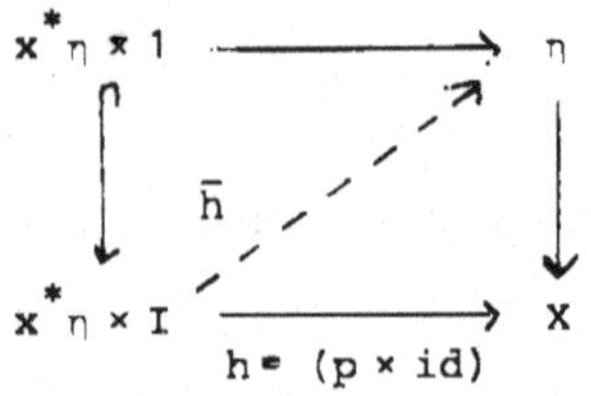

Define $x^*\eta \rightarrow y^*\eta$ by the pull-back property and $\bar{h}_o$.

<u>Definition 1.2. We call</u> $tp_\eta : \pi^G M \rightarrow bf_{G,n}$ <u>the fibre transport of</u> $\eta \downarrow X$. <u>The fibre transport</u> tp_M <u>of a</u> G-<u>manifold</u> M <u>is</u> $tp_{TM}c$.

2. The equivariant degree.

We consider a G-map $f : M \rightarrow N$ between G-manifolds an an O(G)-

transformation $\varphi : f^*tp_N \to tp_M$ with $f^*tp_N := tp_N \circ \pi^G f$. We want to define its equivariant degree $DEG(f,\varphi)$ lying in a certain ring $C(N)$.

We consider the case $G = 1$ and both M and N connected first. Recall that we always assume $\dim M = \dim N$. Suppose that $\varphi(x) : TN^c_{fx} \to TM^c_x$ is not nullhomotopic for one (and hence all) $x \in M$. Otherwise define $DEG(f,\varphi) \in \mathbb{Z}$ to be zero. Let u be any loop in M at x. By functoriality of φ we get $\varphi(x) \circ tp_N(f \circ u) \simeq tp_M(u) \circ \varphi(x)$. Since the first Stiefel-Whitney class $w_1(M) \in H^1(M,\mathbb{Z}/2) = HOM(\pi_1(M),\mathbb{Z}/2)$ sends u to $\deg tp_M(u)$ we have $f^*w_1(N) = w_1(M)$. Let $p : \hat{M} \to M$ be the orientation covering if $w_1(M)$ is non-trivial and the identity $\hat{M} = M \to M$ otherwise and define $p : \hat{N} \to N$ analogously. Then $\hat{M}$ and $\hat{N}$ are orientable connected manifolds and we can choose a lift $\hat{f} : \hat{M} \to \hat{N}$. If $f(\hat{M}) \subset \partial\hat{N}$ let $DEG(f,\varphi)$ be zero. Otherwise choose a point $\hat{x} \in \hat{M} \setminus \partial\hat{M}$ with $\hat{f}\hat{x} \in \hat{N} \setminus \partial\hat{N}$. Write $x = p\hat{x}$. Let $c : \hat{M} \to T\hat{M}^c_x$ and $c : \hat{N} \to T\hat{N}^c_{fx}$ be the collaps maps uniquely determined up to homotopy by the property that the differentials at x and fx are the identity. Let d be the degree of the following endomorphism of $\mathbb{Z}$.

$$
\begin{array}{ccc}
\mathbb{Z} = H_n(\hat{M},\partial\hat{M}) & \xrightarrow{\hat{f}_*} & H_n(\hat{N},\partial\hat{N}) \\
\cong \downarrow c_* & & \cong \downarrow c_* \\
H_n(T\hat{M}^c_{\hat{x}}) & & H_n(T\hat{N}^c_{\hat{f}\hat{x}}) \\
\cong \downarrow (Tp_{\hat{x}})_* & & \cong \downarrow (Tp_{\hat{f}\hat{x}})_* \\
H_n(TM^c_x) & \xleftarrow[\varphi(x)_*]{} & H_n(TN^c_{fx})
\end{array}
$$

A straightforward calculation shows that d is independent of the choices of $\hat{f}$ and $\hat{x}$. Now define $DEG(f,\varphi)$ as $2d$ if $w_1(M) = 0$ and $w_1(N) \neq 0$, and as d otherwise. The factor 2 in the case $w_1(M) = 0$ and $w_1(N) \neq 0$ is due to the fact that then $\hat{M}$ is only one of the two components of the pullback of the orientation covering of N.

The global degree has an easy description by local degrees. Let y be a point in $N \setminus \partial N$. Assume that $f^{-1}(y)$ is finite and f looks in a neighbourhood of x like a proper map $k(x) : (TM_x, 0) \to (TN_y, 0)$ with $k(x)^{-1}(0) = 0$ if we identify the tangent space with neighbourhoods by an exponential map. Then:

Proposition 2.1

$$DEG(f,\varphi) = \sum_{x \in f^{-1}(y)} \deg(k(x)^c \circ \varphi(x)^c : TN_y^c \to TN_y^c)$$

Proof. Use [9], p. 267. □

As an illustration consider the example of a n-fold covering $p : M \to N$ between connected manifolds. Its differential induces an $O(1)$-transformation $\varphi_p : p^* tp_N \to tp_M$. By proposition 2.1 $DEG(p,\varphi_p)$ is n. This applies in particular to $p : S^{2m} \to \mathbb{R}P^{2m}$. Notice that S^{2m} is orientable but $\mathbb{R}P^{2m}$ not.

Now we treat the general case. Let Con(G) be the set of conjugacy classes of subgroups of G. The set of isomorphism classes $\bar{x}$ of objects x in a category C is denoted by $\bar{C}$. Given an $O(G)$-groupoid $\mathcal{G}$, we write $CON(\mathcal{G})$ for $\prod_{(H) \in Con(G)} \overline{\mathcal{G}(G/H)}/WH$ and $C(\mathcal{G})$ for the ring of functions $CON(\mathcal{G}) \to \mathbb{Z}$. Let $CON(X)$ and $C(X)$ be $CON(\Pi^G X)$ and $C(\Pi^G X)$ for a G-space X.

We will define $DEG(f,\varphi)$ in $C(N)$ by specifying integers $DEG(f,\varphi)(D,H)$ for all $H \subset G$ and $D \subset \pi_0(N^H)$. Let $C_1,\dots,C_r$ be the components of M^H with $f^H(C_i) \subset D$ and $f_i : C_i \to D$ be the map induced by f^H. Because of $(TM \mid M^H)^H = T(M^H)$ we obtain from φ non-equivariant transformations $\varphi_i : f_i^* tp_D \to tp_{C_i}$ by restriction and taking the H-fixed point sets. We have introduced $DEG(f_i,\varphi_i)$ above. Define:

$$DEG(f,\varphi)(D,H) = \sum_{i=1}^{r} DEG(f_i,\varphi_i).$$

The sum shall be zero for $r = 0$.

Definition 2.2. We call $DEG(f,\varphi)$ in $C(N)$ the equivariant degree of f with respect to φ. □

Finally we state the elementary properties. Consider a G-map of triads $(F,f,f_+) : (P,M,M_+) \to (Q,N,N_+)$ and $O(G)$-transformations $\varphi : f^* tp_N \to tp_M$ and $\Phi : F^* tp_Q \to tp_P$. Identifying $TP_x = TM_x \oplus \mathbb{R}$ using the inward normal we get $tp_P \mid M = tp_{(TM \oplus \mathbb{R})^c}$ and analogously $tp_Q \mid N = t_{p(TN \oplus \mathbb{R})^c}$. We assume that $\Phi \mid N$ and φ fit together under these identifications

Let $j^* : C(Q) \to C(N)$ be the ring homomorphism given by composition with the obvious map $CON(N) \to CON(Q)$. Then the equivariant degree turns out to be a bordism invariant.

2.3 $$DEG(f,\varphi) = j^* DEG(F,\Phi)$$

The equivariant degree is a homotopy invariant in the following sense. Given a G-homotopy $h : M \times I \to N$ between f and g we get an O(G)-equivalence $\psi_h : g^* tp_N \to f^* tp_N$ by the fibre transport. Then:

2.4 $$DEG(f,\varphi) = DEG(g,\varphi \circ \psi_h)$$

Consider G-maps $f : L \to M$ and $g : M \to N$ and O(G)-equivalences $\varphi : f^* tp_M \to tp_L$ and $\psi : g^* tp_N \to tp_M$. Provided that $\pi_o(g^H) : \pi_o(M^H) \to \pi_o(N^H)$ is bijective for all $H \subset G$, we obtain the <u>composition formula:</u>

2.5 $$DEG(g \circ f, \varphi \circ f^* \psi) = DEG(g,\psi) \cdot (g^*)^{-1}(DEG(f,\varphi))$$

The following examples illustrate our definitions.

<u>Example 2.6.</u> Let $f : V^c \to W^c$ be a G-map for two G-representations V and W with $\dim V^G$, $\dim W^G \geq 1$. Any G-map $\Phi : W^c \to V^c$ can be interpreted as an O(G)-transformation $\varphi : f^* tp_{V^c} \to tp_{W^c}$ using the facts that $TV^c \oplus (V^c \times \mathbb{R}) = V^c \times (V \oplus \mathbb{R})$ holds and the suspension $[V^c, W^c]^G \to [(V \oplus \mathbb{R})^c, (W \oplus \mathbb{R})^c]^G$ is bijective. Then $DEG(f,\varphi)$ lies in $C(W^c) = C(G)$ and $DEG(f,\varphi)(H)$ is just $\deg(\Phi^H \circ f^H)$ for $(H) \in Con(G)$. □

<u>Example 2.7.</u> Let M be a G-manifold such that the components of M^H are orientable for all $H \subset G$. If $f : M \to M$ is a G-map with $\pi_o(f^H) : \pi_o(M^H) \to \pi_o(M^H)$ the identity for all $H \subset G$ we can define its degree $DEG(f) \in C(M)$ by the collection $(\deg(f^H|C : C \to C)$ $C \in \pi_o(M^H)$, $H \subset G)$. The orientability condition ensures that we get a well-defined O(G)-equivalence $\varphi : f^* tp_M \to tp_M$ uniquely determined by the property that $\varphi(G/H)(x)_{eH} : tp_M(G/H)(fx)_{eH} \to tp_M(G/H)(x)_{eH}$ is given by the fibre

transport of the H-bundle $TM|M^H$ along any path in M^H from x to fx. One easily checks $DEG(f) = DEG(f,\varphi)$. □

The following non-equivariant example indicates the advantage of our notion of the degree with the one using fundamental classes for surgery.

Example 2.8. Let M and N be closed orientable connected manifolds with fundamental classes [M] and [N] and let M^- be M with $-[M]$. Consider a normal map $f:M \longrightarrow N$, $\hat{f}: TM \oplus \mathbb{R}^k \longrightarrow \xi$ of degree one taken with respect to the fundamental classes. If M_o is $M + M^- + M$ disjoint union gives a normal map of degree one $g = f + f + f$: $M_o \longrightarrow N$, $\hat{g} = \hat{f} + \hat{f} + \hat{f}$. The reader should figure out by himself that it is impossible with these bundle data and orientations to convert f by surgery into a normal map $f_+ : M_+ \longrightarrow N$ of degree one with connected M_+. We can see this using our degree as follows. Fix an $O(1)$ - equivalence $\varphi : tp_\xi$ —— $tp_{TN} \oplus \mathbb{R}^k$. Let $\psi_{\hat{g}} : g^* tp_N \longrightarrow tp_M$ be the $O(1)$ - equivalence uniquely determined by the property that its suspension is $(\varphi \circ tp_{\hat{g}})^{-1} : f^* tp_{TN} \oplus \mathbb{R}^k \longrightarrow tp_{TM} \oplus \mathbb{R}^k$. Since "normally bordant" includes the bundle data, $DEG(g,\psi_{\hat{g}})$ is a normal bordism invariant. But $DEG(g\ \psi_{\hat{g}})$ is ± 1 by Proposition 3.1.

3. The Burnside ring of a compact Lie group.

The Burnside ring of a compact Lie group G was introduced and examined by tom Dieck [5] and [6], p. 103 ff. Since we need some modifications of this material and want to keep the paper self-contained we make some remarks about it in this section.

A prefamily $\mathfrak{F}$ is a subset of $S(G) = \{H | H \subset G\}$ closed under conjugation. We call $\mathfrak{F}$ a family if it is also closed under intersection and finite if $\{(H) \in \mathrm{Con}(G) \mid H \in \mathfrak{F}\}$ is finite. The set of isotropy groups $\mathrm{Iso}(X) = \{G_x \mid x \in X\}$ of a finite G-CW-complex X is a finite prefamily. If X is a G-manifold with connected fixed point sets, Iso(X) is a finite family for finite G, but not in general. A counterexample is the sphere in the SO(3)-representation $\mathbb{R}^3 \oplus \mathbb{R}^3$ if SO(3) acts in the obvious way on both summands. If $\mathfrak{F}$ is a prefamily and χ denotes the Euler characteristic let $A(G,\mathfrak{F})$ be the set of equivalence classes of finite G-CW-complexes X with $\mathrm{Iso}(X) \subset \mathfrak{F}$ under the equivalence relation $X \sim Y \Leftrightarrow \chi(X^H) = \chi(Y^H)$ for all $H \subset G$. The disjoint union defines an abelian group structure. Moreover, the cartesian product induces the structure of an associative commutative ring with unit if $\mathfrak{F}$ is a family containing G. We can identify $A(G) := A(G,S(G))$ with the Burnside ring in [6] p. 103. Let $C(G,\mathfrak{F})$ be the ring of functions $\{(H) \in \mathrm{Con}(G) \mid H \in \mathfrak{F}\} \to \mathbb{Z}$ and $C(G) = C(G,S(G))$. For each $K \subset G$ we obtain a ring homomorphism

$$ch_K : A(G,\mathfrak{F}) \longrightarrow \mathbb{Z} \qquad [X] \longmapsto \chi(X^H).$$

Since WH acts freely on G/H^K and WH contains a circle for infinite

WH we get $ch_K(G/H) = 0$ for all K if WH is infinite. For any prefamily $\mathfrak{F}$ let $\mathfrak{F}_f$ be $\{H \in \mathfrak{F} \mid WH \text{ finite}\}$. Using the ideas in [6] p. 3, 4, 104, 119 one proves that ch is given by the product of the ch_K-s:

<u>Proposition 3.1.</u> Let $\mathfrak{F}$ <u>be a finite prefamily. Then</u> $\{[G/H] \mid H \in \mathfrak{F}_f\}$ <u>is a</u> $\mathbb{Z}$<u>-base of</u> $A(G,\mathfrak{F})$. <u>The homomorphism</u>

$$ch : A(G,\mathfrak{F}) \to C(G,\mathfrak{F}_f)$$

<u>is injective with a finite cokernel of order</u> $\prod_{\{(H) \mid H \in \mathfrak{F}_f\}} |WH|$. <u>Moreover, each</u> $ch(G/H)$ <u>is divisible by</u> $|WH|$ <u>and</u> $\{\frac{1}{|WH|} ch(G/H) \mid H \in \mathfrak{F}_f\}$ <u>is a</u> $\mathbb{Z}$<u>-base for</u> $C(G,\mathfrak{F}_f)$. □

Now we introduce the equivariant Lefschetz index following [14], chapter 1 to produce a bijection $[V^c,V^c]^G \to A(G,Iso(V))$ for an appropriate G-representation V.

Consider a G-self map $f : X \to X$ of a finite G-CW-complex X. Let $L(f^H,f^{>H})$ be the Lefschetz index of the self map $(f^H,f^{>H})$ of the pair of CW-complexes $(X^H,X^{>H})$.

<u>Definition 3.2. The equivariant Lefschetz index</u> $L^G(f)$ <u>in</u> $A(G,Iso(X))$ <u>is defined as</u>

$$L^G(f) = \sum_{\{(H) \mid H \in Iso(X)_f\}} \frac{1}{|WH|} \cdot L(f^H,f^{>H}) \cdot [G/H] \quad .\square$$

Since $(X^H,X^{>H})$ is WH-free, $L(f^H,f^{>H})$ is divisible by $|WH|$. Proposition 1.8 in [14] extends to compact Lie groups:

Lemma 3.3. $ch_K(L^G(f)) = L(f^K)$ for $K \subset G$.

Proof. Since the Lefschetz index is additive ([9], p. 213) one can reduce the problem by induction over the orbit bundles and dimensions to the case $X = \coprod_r G/H \times D^n / \coprod_r G/H \times S^{n-1}$ where one has to show with $*$ the obvious base-point:

$$L(f^K,*) = \begin{cases} \frac{1}{|WH|} \cdot L(f^H,*) \cdot \chi(G/H^K) & \text{if WH is finite} \\ 0 & \text{otherwise} \end{cases}$$

The second case follows from the fact that WH acts freely relative $*$ on X and X^K and contains a circle. The canonical inclusions and projections of the wedge X yield a pair of inverse isomorphisms between $H_*(X,*)$ and $\bigoplus_r H_*((G/H \times S^n)/(G/H \times *),*)$ where $*$ denotes the various base points. Now an easy homological computation reduces the proof of the first case to $X = (G/H \times S^n)/(G/H \times *)$ with WH finite. Then f^H is a self-map of $(WH \times S^n)/(WH \times *)$. The Künneth formula and the obvious map $G/H \times (WH \times S^n)/(WH \times *) \to X$ induce a chain homotopy equivalence such that the following diagram commutes up to homotopy

$$\begin{array}{ccc} C(G/H^K) \otimes_{\mathbb{Z}WH} C(WH \times S^n/WH \times *,*) & \xrightarrow{\cong} & C(X^K,*) \\ \Big\downarrow \scriptstyle id \otimes_{\mathbb{Z}WH} C(f^H,*) & & \Big\downarrow \scriptstyle C(f^K,*) \\ C(G/H^K) \otimes_{\mathbb{Z}WH} C(WH \times S^n/WH \times *,*) & \xrightarrow{\cong} & C(X^K,*) \end{array}$$

Notice that $C(WH \times S^n/WH \times *,*)$ is concentrated in dimension n and is $\mathbb{Z}WH$ there. Let $\Sigma a_w \cdot w \in \mathbb{Z}WH$ be the element determined by $C(f^H,*)$.

Then $L(f^H,*)$ is $|WH|\cdot a_1$ and $L(f^K,*)$ is $\chi(G/H^K)\cdot a_1$ since $C(G/H^K)$ is $\mathbb{Z}WH$-free. This finishes the proof. □

A G-<u>homotopy</u> <u>representation</u> X of G is a finite-dimensional G-complex of finite orbit type such that for each subgroup H of G the fixed point set X^H is an n(H)-dimensional CW-complex homotopy equivalent to $S^{n(H)}$. If dim $X^G \geq 1$ and Iso(X) is a family, we equip $[X,X]^G$ and A(G,Iso(X)) with the monoid structure given by composition and multiplication. If 1 denotes [G/G] and $\chi^G(X) := L^G(id_X)$ we have the unit $\chi^G(X) - 1$ in A(G,Iso(X)) and maps

$$\lambda : [X,X]^G \longrightarrow A(G,Iso(X)) \qquad [f] \longrightarrow (L^G(f) - 1)\cdot(\chi^G(X) - 1)$$

$$DEG : [X,X]^G \longrightarrow C(G) \qquad [f] \longrightarrow \{\deg f^H \mid (H) \in Con(G)\}$$

The main result of this section is:

<u>Theorem 3.4.</u> <u>Let</u> X <u>be</u> <u>a</u> G-<u>homotopy</u> <u>representation</u> <u>with</u> dim $X^G \geq 1$ <u>satisfying</u> <u>condition</u> (*) <u>defined</u> <u>in</u> <u>the</u> <u>introduction</u>.

a) $L^G - 1 : [X,X]^G \rightarrow A(G,Iso(X))$ <u>is</u> <u>bijective</u>.

b) <u>If</u> Iso(X) <u>is</u> <u>a</u> <u>family</u> <u>the</u> <u>monoid</u> <u>map</u> $\lambda : [X,X]^G \rightarrow A(G,Iso(X))$ <u>is</u> <u>bijective</u> <u>and</u> $ch \circ \lambda = DEG$. □

Theorem 3.4 follows from proposition 3.1, lemma 3.3 and the equivariant Hopf theorem 3.5 below. For its proof and further explanations we refer to [6] p. 213, [7] II.4., [14], [18] and [21].

<u>Theorem 3.5.</u> <u>Let</u> X <u>and</u> Y <u>be</u> G-<u>homotopy</u> <u>representations</u> <u>with</u> dim X^H = dim Y^H <u>for</u> <u>all</u> $H \subset G$ <u>satisfying</u> <u>condition</u> (*). <u>Choose</u>

<u>fundamental classes for</u> X^H <u>and</u> Y^H <u>such that</u> deg f^H <u>for a G-map</u> $f : X \to Y$ <u>is defined</u>.

<u>Then</u> $[X,Y]^G$ <u>is non-empty. Elements</u> [f] <u>are determined by the set</u> $\{\deg f^H \mid H \in Iso(Y)_f\}$. <u>The degree</u> deg f^H <u>is</u> modulo $|WH|$ <u>determined by the</u> deg f^K, $K \underset{\neq}{\supset} H$, <u>and fixing these degrees</u> deg f^K <u>the possible</u> deg f^H <u>fill the whole residue class</u> mod $|WH|$. □

We end with some remarks about induction and restriction for an inclusion $j : H \longrightarrow G$ of compact Lie groups.

Let $\mathcal{F}$ be a prefamily for H. Then $j_*\mathcal{F} = \{g^{-1}j(K)g \mid g \in G, K \in \mathcal{F}\}$ is a prefamily for G. We want to define an abelian group homomorphism

$$\mathrm{ind}_j : A(H,\mathcal{F}) \to A(G,j_*\mathcal{F})$$

by sending [X] to $[G \times_j X]$. The following formula and proposition 3.1 show that this is well-defined.

3.6 $\chi((G\times_j X)^K) = \sum_{gH \in G/H^K} \chi(X^{gKg^{-1}})$ for $K \subset G$, WK finite.

Notice that G/H^K has only finitely many WK-orbits ([2], p. 87) and is therefore finite if WK is finite.

Given a prefamily $\mathcal{F}$ for G, we have the prefamily $j^*\mathcal{F} = \{j^{-1}(K) \mid K \in \mathcal{F}\}$ for H. We obtain an abelian group homomorphism

$$res_j : A(G,\mathcal{F}) \to A(H,j^*\mathcal{F})$$

by restriction: $[X] \to [res_j X]$. If $\mathcal{F}$ is a family containing H and G then $j^*\mathcal{F}$ is a family with $H \in j^*\mathcal{F}$ and res_j is a ring homomorphism.

4. The monoid of endomorphisms of the fibre transport.

If we want to examine the dependency of $DEG(f,\varphi)$ on φ we have to compute in view of the composition formula 2.5 the $O(G)$-transformations $\varphi : tp_N \to tp_N$ and the possible values $DEG(ID,\varphi)$ in $C(N)$.

More generally we consider the monoid End(tp) of $O(G)$-transformations $\varphi : tp \to tp$ of any $O(G)$-functor $tp : \mathcal{G} \to bf_{G,n}$. The group of invertible elements $End(tp)^*$ consists of the $O(G)$-equivalences $\varphi : tp \to tp$.

Consider $C(\mathcal{G})$ as monoid by its multiplicative structure. The monoid map

$$DEG : End(tp) \to C(\mathcal{G})$$

maps φ to $DEG(\varphi)$ specified by the following function $CON(\mathcal{G}) \to \mathbb{Z}$. For $H \subset G$ and x in $\mathcal{G}(G/H)$ we get a G-fibre map $\varphi(G/H)(x)$. Let $DEG(\varphi)(x,H)$ be the degree of the induced self map on the H-fixed point set $tp(G/H)(x)^H_{eH}$ of the fibre over eH. Recall that $tp(G/H)(x)_{eH}$ is H-homotopic to SV for some H-representation V. We want to show that $DEG : End(tp) \to C(\mathcal{G})$ is an embedding of monoids and describe its image.

We say that an $O(G)$-transformation $tp : \mathcal{G} \to bf_{G,n}$ <u>satisfies condition</u> (*) if for any $H \subset G$ and $x \in \mathcal{G}(G/H)$ $tp(G/H)(x)_{eH}$ does and has an H-fixed point. If furthermore $Iso(tp(G/H)(x)_{eH})$ is a family we call tp <u>admissible</u>. Consider a G-manifold N satisfying condition (*). Then tp_N satisfies condition (*) and is even admissible if G is finite. If G is finite nilpotent and N a G-manifold such that no component of N^H is a point for $H \subset G$ then tp_N is admissible.

We recall the notion of the homotopy colimit $\Gamma(\mathcal{G})$ (see [20] p. 1625). Objects are pairs (x,H) with $x \in \mathcal{G}(G/H)$ and $H \subset G$. A morphism $(\sigma,u) : (x,H) \to (y,K)$ consists of a G-map $\sigma : G/H \to G/K$ and a morphism $u : x \to \sigma^* y$ with $\sigma^* = \mathcal{G}(\sigma) : \mathcal{G}(G/K) \to \mathcal{G}(G/H)$. Composition is defined by the "semi-direct product formula" $(\tau,v) \circ (\sigma,u) = (\tau \circ \sigma, \sigma^* v \circ u)$. Notice that $\overline{\Gamma(\mathcal{G})}$ is $Con(\mathcal{G})$ (see section 2). The fundamental group category of a G-space X appearing in [7] p. 57 and [15] is $\Gamma(\pi^G X)$. We now introduce contravariant functors A_{tp}, $C_{\mathcal{G}}$ and E_{tp} and relate their inverse limits to $End(tp)$ and $C(\mathcal{G})$. The contravariant functor into the category of monoids

$$E_{tp} : \Gamma(\mathcal{G}) \to MONO$$

maps (x,H) to $[tp(G/H)(x)_{eH}, tp(G/H)(x)_{eH}]^H$. Given a morphism $(\sigma,u) : (x,H) \to (y,K)$ choose g in G with $\sigma(eH) = gK$ so that we obtain a group homomorphism $c(g) : H \to K \quad h \to g^{-1}hg$. If $l(g^{-1})$ is multiplication with g^{-1} we get a H-homotopy equivalence $\alpha : tp(G/H)(x)_{eH} \to res_{c(g)} tp(G/K)(y)_{eK}$ by $l(g^{-1}) \circ tp(G/H)(u)_{eH}$.

Define $E_{tp}(\sigma,u)$: $[tp(G/K)(y)_{eK}, tp(G/K)(y)_{eK}]^K \to [tp(G/H)(x)_{eH}, tp(G/H)(x)_{eH}]^H$ by restriction with $c(g)$ and conjugation with α. This is well defined since conjugation with an H-self-equivalence induces the identity on $[X,X]^H$ for a G-homotopy representation X (theorem 3.4).

The contravariant functors

$$A_{tp} : \Gamma(\mathcal{G}) \to \text{MONO}$$

$$C_{\mathcal{G}} : \Gamma(\mathcal{G}) \to \text{MONO}$$

send (x,H) to $A(H,Iso(tp(G/H)(x)_{eH}))$ and $C(H)$. Given a morphism $(\sigma,u) : (x,H) \to (y,K)$ let $g \in G$ and $c(g) : H \to K$ be as above. Define $A_{tp}(\sigma,u)$ and $C_{\mathcal{G}}(\sigma,u)$ as the restriction with $c(g)$.

Let the transformation

$$D : E_{tp} \to C_{\mathcal{G}}$$

$$\Lambda : E_{tp} \to A_{tp}$$

$$CH : A_{tp} \to C_{\mathcal{G}}$$

be induced by the degree and the maps of section three

$\lambda : [tp(G/H)(x)_{eH}, tp(G/H)(x)_{eH}]^H \to A(H,Iso(tp(G/H)(x)_{eH}))$

$ch : A(H,Iso(tp(G/H)(x)_{eH})) \to C(H)$

The <u>inverse limit</u> of a contravariant functor $F : C \to \text{MONO}$ is the submonoid inv F of $\prod_{x \in C} F(x)$ consisting of those elements

$(a_x \mid x \in C)$ such that $F(f)(a_x) = a_y$ holds for any morphism $f : y \to x$.

We define a monoid map

$$\beta : \text{inv lim } C_{\mathcal{G}} \to C(\mathcal{G}) = \prod_{\overline{\Gamma(\mathcal{G})}} \mathbb{Z}$$

as follows. Let $pr_H : C(H) \to \mathbb{Z}$ be the projection onto the factor belonging to $(H) \in \mathrm{Con}(H)$. An element in the inverse limit given by $\{u(x,H) \in C(H) \mid (x,H) \in \Gamma(\mathcal{G})\}_*$ is sent to $\{pr_H(u(x,H)) \in \mathbb{Z} \mid \overline{(x,H)} \in \overline{\Gamma(\mathcal{G})}\}$.

Let $\alpha(x,H) : \mathrm{End}(tp) \to E_{tp}(x,H)$ be the monoid map sending φ to $\varphi(G/H)(x)_{eH}$. We obtain a homomorphism of monoids

$$\alpha : \mathrm{End}(tp) \to \text{inv lim } E_{tp}$$

<u>Theorem 4.1.</u>

a) <u>If</u> $tp : \mathcal{G} \to bf_{G,n}$ <u>fullfills</u> <u>condition</u> (*), <u>the</u> <u>following</u> <u>diagram</u> <u>of</u> <u>monoids</u> <u>commutes</u>. <u>All</u> <u>maps</u> <u>are</u> <u>injective</u> <u>and</u> α <u>is</u> <u>bijective</u>.

$$\begin{array}{ccc} \mathrm{End}(tp) & \xrightarrow[\cong]{\alpha} & \text{inv lim } E_{tp} \\ {\scriptstyle DEG}\downarrow & & \downarrow{\scriptstyle \text{inv lim } D} \\ C(\mathcal{G}) & \xleftarrow[\beta]{} & \text{inv lim } C_{\mathcal{G}} \end{array}$$

b) <u>If</u> tp <u>is admissible the following diagram of monoids commutes.</u>

<u>All maps are injective and</u> inv lim Λ <u>is bijective.</u>

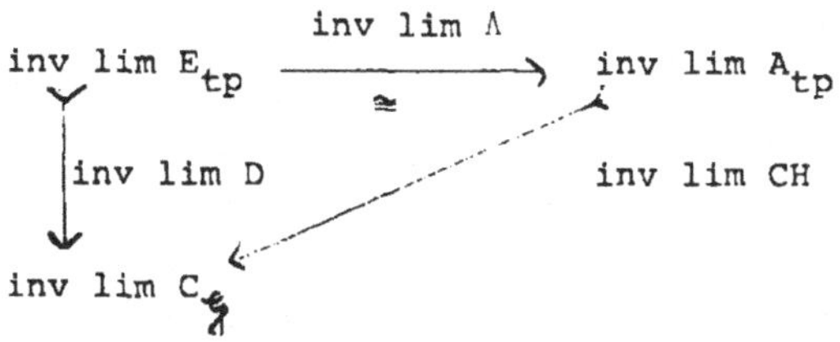

<u>Proof.</u> Everything follows directly from theorem 3.4 and the definitions. □

Let * : MONO → GROUPS be the functor "invertible elements". Since the inverse limit is compatible with * and $End(tp)^*$ is the group Aut(tp) of O(G)-equivalences tp → tp we conclude:

<u>Corollary 4.2.</u> <u>For admissible</u> tp <u>the following diagram of abelian groups commutes. The maps</u> α^* <u>and</u> inv lim Λ^* <u>are bijective the others injective.</u>

$$
\begin{array}{ccc}
\mathrm{Aut}(tp) & \xrightarrow[\alpha^*]{\cong} & \mathrm{inv\,lim}\ E^*_{tp} \\
 & & \Big\downarrow \cong \ \mathrm{inv\,lim}\ \Lambda^* \\
\mathrm{DEG}^* \Big\downarrow & & \mathrm{inv\,lim}\ A^*_{tp} \\
 & & \Big\downarrow \ \mathrm{inv\,lim}\ CH \\
C(\mathcal{G})^* & \xleftarrow[\beta^*]{} & \mathrm{inv\,lim}\ C^*_{\mathcal{G}}
\end{array}
\qquad \square
$$

Corollary 4.3. Let N be a connected G-manifold satisfying condition (*).

a) *If N^H is connected and non-empty for all $H \subset G$ and Iso(N) a family then:*

$$\mathrm{End}(tp_N) = A(G,\mathrm{Iso}(N)) \subset C(N) = C(G)$$

b) *Let G be a torus. Assume that any component of N^H contains a G-fixed point for $H \subset G$. Then we have for $y \in N$ the bijection*

$$\mathrm{End}(tp_N) \to \mathbb{Z} \quad \varphi \to \deg(\varphi(G/1)(y))$$

c) *If G is finite of odd order we get for $y \in N$ an isomorphism.*

$$\mathrm{Aut}(tp_N)^* \to \{\pm 1\} \quad \varphi \to \deg(\varphi(G/1)(y))$$

Proof:

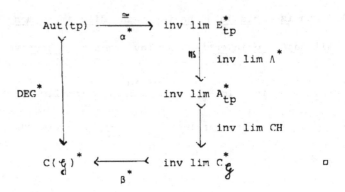

a) If x is a G-fixed point, we have for any object (y,H) in $\Gamma(\pi^G X)$ a morphism $(y,H) \to (x,G)$. Two such morphisms define the same map $E_{tp_N}(x,G) \to E_{tp_N}(y,H)$ and inv lim $E_{tp_N}(x,G)$ is $End(tp_N)$.

Hence $End(tp_N) = E_{tp_N}(x,G) = A(G,Iso(N))$.

b) If X is a G-homotopy representation of the torus G with $\dim X^G \geq 1$ then $[X,X]^G \to \mathbb{Z}\ [f] \to \deg f$ is bijective by proposition 3.1 and theorem 3.4.

c) $ch_1^* = A(G)^* \to \{\pm 1\}$ is a isomorphism by [6] p. 8 if G has odd order. □

If N is a G-manifold and tp_N is admissible, $End(tp_N) \cong$ inv lim A_{tp} depends only on the component structure of N and the sets $Iso(TN_x)_f$ for all $x \in N$ which can be read off from the dimension function.

5. The degree relations.

In this section we state the central result of this paper. In the following we identify $End(tp_N)$ with its image in C(N) under the embedding DEG.

Theorem 5.1. *Let* $f : M \to N$ *be a G-map of n-dimensional G-manifolds satisfying condition (*) and* $\varphi : f^*tp_N \to tp_M$ *be an O(G)-transformation. Then:*

$$DEG(f,\varphi) \in End(tp_N) \subset C(N) \quad \square$$

The rest of this section deals with its proof. Examples to illustrate its meaning are given in the next section. The most important ingredients are the concept of quasi-transversality which we will extend to compact Lie groups (see [10], chapter 3 for finite G) and local degrees.

We call the G-map $f : M \to N$ of G-manifolds *quasi-transverse* to y in N if the following is true.

i) The preimage $f^{-1}(y)$ consists of finitely many orbits G_y/H with all $W_{G_y}H$ finite.

ii) Equip the G-normal bundle $\nu(f^{-1}(G/G_y),M)$ and $\nu(G/G_y,N)$ with equivariant metrics. There is a norm preserving G-fibre map

$$\begin{array}{ccc} (f^{-1}(G/G_y),M) & \xrightarrow{k} & (G/G_y,N) \\ \downarrow & & \downarrow \\ f^{-1}(G/G_y) & \xrightarrow{f|} & G/G_y \end{array}$$

such that f looks like k in a tubular neighbourhood.

Lemma 5.2. We can change f up to G-homotopy such that f is quasi-transverse to y. (see also [10, ch. 3]).

Proof. Let $K_1, K_2, \ldots, K_r$ be a complete system of representatives of conjugacy classes (K) of subgroups of G with K occuring as isotropy group in M and $K \subset G_y$. We construct inductively an open G-set U_i containing $M^{K_1}, \ldots, M^{K_i}$ such that i) and ii) hold if one substitutes $f^{-1}(y)$ and $f^{-1}(G/G_y)$ by their intersections with U_i. We can assume $(K_i) \subset (K_j) \Rightarrow i \geq j$.

The induction begin $i = 0$ is trivial: $U = \emptyset$. In the induction step from $i-1$ to i write $U = U_{i-1}$, $K = K_i$. By possibly shrinking U we can suppose the existence of a closed G-set V with $\mathrm{int}(V) \supset \mathrm{clos}(U)$

and $f^{-1}(G/G_y) \cap V \setminus U = \emptyset$. Let M_o be $M^K \setminus (U \cap M^K)$. By induction hypothesis WK acts freely on M_o. If f_o is $f^K | M_o$ consider the non-equivariant map

$$(f_o \times id)/WK : M_o/WK \rightarrow (N^K \times M_o)/WK$$

We can change it homotopically relative $V \cap M_o/WK$ into f_1 such that f_1 is transverse to $(G/G_y^K \times M_o)/WK$. By a cofibration argument we can assume $(f_o \times id)/WK = f_1$. Now G/G_y^K is a finite disjoint union of WK-orbits $\coprod_{i=1}^{r} WK \cdot (g_i G_y)$ (see [2], p. 87). One easily checks $\dim (f_o/WK)^{-1}(WK \cdot g_i G_y) = \dim (f_o \times id/WK)^{-1}(WK \cdot g_i G_y \times M_o/WK) = -\dim NK \cap G_{g_i y}/K$. Hence $(f_o/WK)^{-1}(WK \cdot g_i G_y)$ consists of finitely many points if $NK \cap g_i G_y/K$ is finite and is empty otherwise. In other words $f^{-1}(y) \cap GM_o$ consists of finitely many orbits G_y/K_j such that $W_{G_y} K_j = NK_j \cap G_y/K_j$ is finite. We can treat any such orbit separately.

Consider any x in GM_o with $f(x) = y$ so that f maps $G/G_x \rightarrow G/G_y$ by the projection. We identify $\nu(G/G_x, M)$ with a tubular G-neighbourhood of G/G_x and analogously for y. We have $\dim \nu(G/G_x, M)_x^L \leq \dim \nu(G/G_y, N)_y^L$ for all $L \subset G_x$ so that we can extend any non-equivariant map $S\nu(G/G_x, M)_x^{G_x} \rightarrow S\nu(G/G_y, N)_y^{G_x}$ to a G_x-map $S\nu(G/G_x, M)_x \rightarrow S\nu(G/G_y, N)_y$. Since $\nu(G/G_x, M) = G \times_{G_x} \nu(G/G_x, M)_x$, $\nu(G/G_y, N) = G \times_{G_y} \nu(G/G_y, N)_y$ and $(K) = (G_x)$ holds we can construct a norm preserving fibre map

$$\begin{array}{ccc} \nu(G/G_x, M) & \xrightarrow{\;k\;} & \nu(G/G_y, N) \\ \downarrow & & \downarrow \\ G/G_x & \longrightarrow & G/G_y \end{array}$$

such that the restriction of k to the K-fixed point set agrees with f^K. By a cofibration argument we can change f in a small G-neighbourhood of G/G_x relative to M^K such that k coincides with f on all $\partial(G/G_x, M)$. Now one easily enlarges U to the desired U_i. This finishes the proof of lemma 5.2. □

<u>Proof of theorem 5.1.</u> We have to construct $\Delta \in \mathrm{End}(tp_N)$ such that $\mathrm{DEG} : \mathrm{End}(tp_N) \to C(\Pi^G N)$ sends Δ to $\mathrm{DEG}(f,\varphi)$. Let $\varepsilon(y,H) \in C_{\Pi^G N}(y,H) := C(H)$ for $(y,H) \in \Gamma(\Pi^G N)$ be defined by $\mathrm{Con}(H) \longrightarrow \mathbb{Z}$ $(K) \longrightarrow \mathrm{DEG}(f,\varphi)(y,K)$. One checks directly that $\{\varepsilon(y,H) \mid (y,H) \in \Gamma(\Pi^G N)\}$ determines an element in inv lim $C_{\Pi^G N}$ mapped by β : inv lim $C_{\Pi^G N} \to C(\Pi^G N)$ to $\mathrm{DEG}(f,\varphi)$. Suppose that we can construct for each $(y,H) \in \Gamma(\Pi^G N)$ a H-self-map $\delta(y,H)$ of $tp_N(G/H)(y)_{eH} = TN_y^c$ such that $D(y,H) : E_{tp_N}(y,H) \to C_{\Pi^G N}(y,H)$ sends $\delta(y,H)$ to $\varepsilon(y,H)$. Then $\{\delta(y,H) \mid (y,H) \in \Gamma(\Pi^G N)\}$ defines an element Δ' in inv lim E_{tp_N}. By theorem 4.1 there is $\Delta \in \mathrm{End}(tp_N)$ such that $\alpha : \mathrm{End}(tp_N) \longrightarrow$ inv lim E_{tp_N} maps Δ to Δ', and Δ has the desired property.

Let $ch' : A(H, \mathrm{Iso}(TN_y)) \to C(H)$ be the composition of the inclusion $A(H,\mathrm{Iso}(TN_y)) \to A(H)$, multiplication with the unit $\chi^H(TN_y^c)-1 : A(H) \to A(H)$ and $ch : A(H) \to C(H)$. The map $L^H-1 : [TN_y^c, TN_y^c]^H \to A(H,\mathrm{Iso}(TN_y))$ is a bijection (theorem 3.4.) and $ch' \circ (L^H-1) = \mathrm{DEG}$. Hence theorem 5.1. is true if we can construct for any $(y,H) \in \Gamma(\Pi^G N)$ an element $d \in A(H,\mathrm{Iso}(TN_y))$ satisfying:

5.3. $\quad ch'_K(d) = \mathrm{DEG}(f,\varphi)(y,K)$ for all $K \subset H$.

Now we construct d. We can assume in view of 2.4 and lemma 5.2 that f is quasi-transverse to y. Furthermore we can suppose $H = G_y$. We want to assign to each H-orbit c in $f^{-1}(y)$ an element d(c) in $A(H, Iso(TN_y))$. Choose x in c. Then $TM_x = (TG/G_x)_x \oplus \nu(G/G_x, M)_x$ and $TN_y = (TG/G_y)_y \oplus \nu(G/G_y, N)_y$. Split $Tp_x : (TG/G_x)_x \to (TG/G_y)_y$ induced from the projection p as $0 \oplus q_x : (TG_y/G_x)_x \oplus V \to (TG/G_y)_y$ with q_x a G_x-linear isomorphism. Checking the dimensions and using elementary obstruction theory we can extend the G_x-map $k_x : \nu(G/G_x, M)_x \to \nu(G/G_y, N)_y$ coming from k appearing in the definition of quasi-transverse to a norm preserving G_x-map $k'_x : \nu(G/G_x, M)_x \oplus (TG_y/G_x)_x \to \nu(G/G_y, N)_y$. Since $k'_x \oplus q_x : TM_x \to TN_y$ is norm-preserving we obtain a G_x-self map $l_x : TN_y^c \to TN_y^c$ by $(k'_x \oplus q_x)^c \circ \varphi(G/G_x)_{eG_x}$. As H/G_x^K contains only finitely many W_HK-orbits (see [2], p. 87) H/G_x^K is finite and $(TG_y/G_x)_x^K = \{0\}$ for $K \subset H$ with finite W_HK. Hence

5.4. $$l_x^K = (Tp_x^K \oplus k_x^K)^c \circ \varphi(G/G_x)(x)_{eG_x}^K \quad \text{if } W_HK \text{ is finite.}$$

Denote the image of l_x under $L^{G_x}-1 : [TN_y^c, TN_y^c]^{G_x} \to A(G_x, Iso(TN_y))$ by d(x) and the image of d(x) under $ind_{G_x}^H : A(G_x, Iso(TN_y)) \to A(H, Iso(TN_y))$ by d(c). For $u \in A(H)$ and $v \in A(G_x)$ one easily checks $ind_{G_x}^H(res_{G_x}^H(u) \cdot v) = u \cdot ind_{G_x}^H(v)$. We obtain from 3.6. and 5.4. and $res_{G_x}^H(\chi^H(TN_y^c) - 1) = \chi^{G_x}(TN_y^c) - 1$

5.5. $$ch'_K \circ ind_{G_x}^H(d(x)) =$$

$$\sum_{hG_x \in H/G_x^K} \deg((Tp_x^{hKh^{-1}} \oplus k_x^{hKh^{-1}})^c \circ \varphi(G/G_x)(x)_{eG_x}^{hKh^{-1}}) =$$

$$\sum_{z \in c^K} \deg((Tp_z^K \oplus k_z^K)^c \circ \varphi(G/G_z)(z)_{eG_z}^K) \quad \text{if } W_HK \text{ is finite.}$$

This shows in particular using proposition 3.1. that $d(c)$ does not depend on the choice of x. Define $d = \sum_{c \in f^{-1}(y)/H} d(c)$.

If $f^K : M^K \to N^K$ and $\varphi^K : f^{K*}tp_N K \to tp_M K$ are induced by f and $DEG(f^K,\varphi^K)$ is the non-equivariant degree we have by definition $DEG(f,\varphi)(y,K) = DEG(f^K,\varphi^K)(y)$ and by proposition 2.1.:

5.6. $$DEG(f,\varphi)(y,K) = \sum_{z \in (f^K)^{-1}(y)} \deg((Tp_z^K \oplus k_z^K) \circ \varphi(G/G_z)(z)^K_{eG_z}$$

if $W_H K$ is finite.

Combining 5.5. and 5.6. gives

5.7. $ch'_K(d) = DEG(f,\varphi)(y,K)$ if $W_H K$ is finite.

Let $K \subset H$ be any subgroup. We can find a bigger subgroup K' with $K \subset K' \subset H$ such that K'/K is a torus T, $W_H K'$ is finite and $ch_K = ch_{K'}$, and hence $ch'_K = ch'_{K'}$, holds (see [6] p. 113). If we can show $DEG(f,\varphi)(y,K) = DEG(f,\varphi)(y,K')$ the assertion 5.3 is a consequence of 5.7. Since T acts on N^K with fixed point set $N^{K'}$ this follows from:

<u>Lemma 5.9.</u> <u>Let</u> $g : P \to Q$ <u>be</u> <u>a</u> T-<u>map</u> <u>between</u> T-<u>manifolds</u> <u>and</u> $\psi : g^* tp_Q \to tp_P$ an $O(T)$-<u>transformation</u>. <u>Then</u> <u>we</u> <u>get</u> $DEG(g,\psi)(y,T) = DEG(g,\psi)(y,1)$ <u>for</u> <u>a</u> T-<u>fixed</u> <u>point</u> y.

<u>Proof.</u> We can assume that g is quasi-transverse to y. Since WL for $L \subset T$ is finite only for $L = T$ the preimage $f^{-1}(y)$ is a finite set of T-fixed points $x_1, .. x_r$. By proposition 2.1 we obtain

for certain T-self maps $l_i : TN_y \to TN_y$ that $DEG(f,\psi)(y,T) = \Sigma \deg(l_i^T)$ and $DEG(f,\psi)(y,1) = \Sigma \deg(l_i)$. Now apply proposition 3.1 and theorem 3.4. This finishes the proof of lemma 5.9 and of theorem 5.1. □

6. Some examples

The theorem 5.1 is very general so that it is necessary to give some examples to explain its meaning. The general problem is to calculate inv lim A_{tp} as a subring of $C(\mathcal{G})$. This can be done in special cases where $\Gamma(\mathcal{G})$ is rather simple or $A(H,\mathfrak{I}) \subset C(H)$ is well understood for all subgroups H of G. We recall that $DEG(f,\varphi)$ lies in $C(N) = \prod_{Con(G)} \pi_0(N^H)/WH$ for a G-map $f : M \to N$ and a O(G)-transformation $\varphi : f^* tp_N \to tp_M$ and $DEG(f,\varphi)(z,H)$ is the integer belonging to the component of N^H containing $z \in N^H$.

In the following we always assume that N fulfills condition (*) and is connected.

Example 6.1. Assume that N^H is non-empty and connected for all $H \subset G$. Suppose that Iso(N) is a family. This follows already from our assumption if G is finite. Then we have $C(N) = C(G)$ and by corollary 4.3 and theorem 5.1.

$$DEG(f,\varphi) \in A(G, Iso(N)) \subset C(G)$$

Hence we obtain the same relations as in the special case $M = N = V^c$ with V a G-representation (see theorem 3.4.). The assumption $N^G \neq \emptyset$ is essential. If N is connected and free we get $End(tp_N) = A(1) = C(N) = \mathbb{Z}$. Indeed, each integer d can be realized as the degree of a self-map of some connected free orientable G-manifold N. Take any connected free orientable G-manifold N_o and a map $f : S^1 \to S^1$ of degree d then $id \times f : N_o \times S^1 \to N_o \times S^1$ is an example. However, if N is the sphere of a free G-representation the degree of f is 1 modulo |G for finite G and 1 for infinite G. One explanation for this phenomenon is that the suspension of a manifold is not a manifold in general but the suspension of a homotopy representation is again a homotopy representation. □

<u>Example 6.2.</u> Let G be a torus T^n and assume that each component of N^H contains a G-fixed point for $H \subset G$.

We get from corollary 4.3 and theorem 5.1

$$DEG(f,\varphi)(z,H) = DEG(f,\varphi)(y,1)$$

for all $H \subset G$ and $z \in N^H$. □

<u>Example 6.3.</u> If H is a p-group the homomorphism ch_1 and $ch_H : A(H) \to \mathbb{Z}$ fulfill $ch_1 \equiv ch_H$ mod p. If H is a torus ch_1 and ch_H agree. Hence we get for each (z,H) by theorem 5.1 (see 5.3).

$$DEG(f,\varphi)(z,H) \equiv DEG(f,\varphi)(z,1) \bmod p, \text{ if } H \text{ is a p-group}$$

$$DEG(f,\varphi)(z,H) = DEG(f,\varphi)(z,1) \qquad , \text{ if } H \text{ is a torus}$$

If G is itself a p-group we obtain for all (z,H) (see also [3], [11] p. 10).

$$\mathrm{DEG}(f,\varphi)(z,H) \equiv \mathrm{DEG}(f,\varphi)(z,1) \bmod p \quad \square$$

Remark 5.4. Now we give the proof of corollary B stated in the introduction. Assume the existence of M. Since M has finite orbit type (see [6] p. 121) we can find a finite p-group $L \subset G$ with $M^G = M^L$. Hence we can suppose that G itself is a finite p-group. We use induction over $|G|$. The induction begin $G = \mathbb{Z}/p$ is done in [1] or by the following argument reflecting the results of example 6.3. Let $c : M \to TM_x^c$ be the collaps map. If ∞ is the point at infinity $c^{-1}(\infty) \cap M^G$ is empty. Since $c^{-1}(\infty)$ is contained in the free part of M we can use non-equivariant transversality to change c up to G-homotopy such that c is transverse to ∞ in the non-equivariant sense and still $c^{-1}(\infty) \cap M^G = \emptyset$ holds. We can assume that G acts orientation preserving, otherwise consider $M \times M$. Hence the local degree of c at x and gx for $x \in c^{-1}(\infty)$ and $g \in G$ agree. Each orbit in the finite set $c^{-1}(\infty)$ consists of p elements. Therefore the degree of c must be divisible by p. A contradiction, since computing deg c by its local degrees at $0 \in TM_y^c$ yields one. In the induction step choose a central subgroup C in G with $C = \mathbb{Z}/p$. If $M^C \neq M^G$ we get a contradiction to the induction hypothesis. Namely, consider the G/C-action on M^C. But $M^C = M^G$ is impossible by the induction begin applied to the C-action on M. This finishes the proof of corollary B.

If one drops the assumption in corollary B that all M^H are connected the result remains true for G an abelian p-group and is false for G a non-abelian p-group provided that p is odd (see [4], [12]). A complete classification of compact Lie groups with one-fixed point actions on (orientable) closed G-manifolds is given in [12]. □

Example 6.5. Let G be a finite group of odd order. If DEG(f,φ) lies in $C(N)^*$ we get from corollary 4.3. and theorem 5.1.

$$DEG(f,\varphi)(z,H) = DEG(f,\varphi)(y,1)$$

for all (z,H). □

Remark 6.6. All the relations we get for DEG(f,φ) also hold in the case of an endomorphism $f : M \rightarrow M$ of a G-manifold with $\pi_o(f^H) : \pi_o(M^H) \rightarrow \pi_o(M^H)$ the identity and all components of M^H orientable for $H \subset G$. Then we can define DEG(f) without specifying φ (see theorem A in the introduction and example 2.7.). □

Finally we mention a consequence of theorem 4.1. and theorem 5.1. and the composition formula 2.5.

Corollary 6.7. *If* DEG(f,φ) *lies in* $C(N)^*$ *for some* O(G)-*transformation* φ *then there is an* O(G)-*equivalence* ψ *with*

$$DEG(f,\psi) \equiv 1 \quad \square$$

7. The fibre transport and the first equivariant Stiefel-Whitney class.

In this section we analyze the fibre transport from a bundle theoretic point of view. We relate it and the question when an O(G)-equivalence $\varphi : f^*tp_N \rightarrow tp_M$ exists to the equivariant analogue of the first Stiefel-Whitney class.

We have introduced the notion of an $O(G)$-groupoid, $O(G)$-functor and $O(G)$-transformation in section one. We call two $O(G)$-functors F_o and $F_1 : \mathcal{G}_o \to \mathcal{G}_1$ homotopic if there exists an $O(G)$-equivalence $\varphi : F_o \to F_1$. Let $[\mathcal{G}_o, \mathcal{G}_1]^{O(G)}$ be the set of homotopy classes of $O(G)$-functors $\mathcal{G}_o \to \mathcal{G}_1$. A G-map $f : X \to Y$ induces an $O(G)$-functor $\pi^G f : \pi^G X \to \pi^G Y$. A G-homotopy $h : X \times I \to Y$ defines an $O(G)$-equivalence $\pi^G h_o \to \pi^G h_1$. Hence we get a well-defined map $[X,Y]^G \to [\pi^G X, \pi^G Y]^{O(G)}$ $[f] \to [\pi^G f]$.

Let $\eta = \eta(G,n) \downarrow BF(G,n)$ be the classifying G-S^n-fibration. It is characterized by the property that the map $[X,BF(G,n)]^G \to bf_{G,n}(X)$ sending $[f]$ to the G-fibre homotopy class of $f^*\eta$ is bijective.

Definition 7.1. Let $\xi \downarrow X$ be a G-S^n-fibration and $f_\xi : X \to BF(G,n)$ be a classifying map. We call $w_\xi = [\pi^G f] \in [\pi^G X, \pi^G BF(G,n)]^{O(G)}$ the first equivariant Stiefel-Whitney class of ξ. Let w_M be w_{TM^c} for a G-manifold M. □

This notion reduces for $G = 1$ to the ordinary definition of the first Stiefel-Whitney class $w_1(M) \in H^1(M,\mathbb{Z}/2) = \mathrm{Hom}(\pi_1(M), \pi_1(BF(n)))$. It is related to the fibre transport by:

Proposition 7.2. Let $tp : \pi^G B(G,n) \to bf_{G,n}$ be the fibre transport of the universal G-S^n-fibration.

a) For each $H \subset G$ $tp_\eta(G/H) : \pi(B(G,n)^H) \to bf_{G,n}(G/H)$ is an equivalence of categories.

b) For any G-complex X we get a bijection

$(tp_\eta)_* : [\pi^G X, \pi^G B(G,n)]^{O(G)} \to [\pi^G X, bf_{G,n}]^{O(G)}$.

c) If $\xi \downarrow X$ is a G-S^n-fibration $(tp_\eta)_*$ sends w_ξ to $[tp_\xi]$.

d) Let ξ_1 and ξ_2 be G-S^n-fibrations over the same one-dimensional G-complex X. Then ξ_1 and ξ_2 are G-fibre homotopy equivalent if and only if $w_{\xi_1} = w_{\xi_2}$ holds.

Proof:

a) We must show that $tp_\eta(G/H)$ induces a bijection between the sets of isomorphism classes of objects and for any object $x \in \pi(B(G,n)^H)$ an isomorphism $Aut(x) \to Aut(tp_\eta(G/H)(x))$ $u \to tp_\eta(G/H)(u)$. The first assertion follows directly from the universal property. The second follows from the observation that H-fibre homotopy classes of H-S^n-fibrations over S^1 equipped with the trivial H-action are in one to one correspondence to H-homotopy classes of self H-maps of the fibre by the fibre transport. b) and d) follow directly from [16] whereas c) is obvious. □

Given two G-S^n-fibrations ξ and η over X we want to analyze when $w_\xi = w_\eta$ holds. If $w_1(\xi^H)$ and $w_1(\eta^H)$ are the (non-equivariant) first Stiefel-Whitney classes in $H^1(X^H, Z/2)$ we have the following obvious conditions for $w_\xi = w_\eta$:

i) $\xi_x \simeq_{G_x} \eta_x$ for each $x \in X$.

ii) $w_1(\xi^H) = w_1(\eta^H)$ for $H \subset G$.

The following example shows that they are not sufficient in general. If $\mathbb{R}$ is the trivial and $\mathbb{R}_-$ the non-trivial $\mathbb{Z}/2$ representation and

$\mathbb{Z}/2$ acts freely on S^1, consider $\xi = S^1 \times \mathbb{R}^c$ and $\eta = S^1 \times \mathbb{R}^c_-$.

Under certain conditions, however, i) and ii) are sufficient.

Theorem 7.3. Let ξ *and* η *be* G-S^n-*fibrations over* X. *Then* $w_\xi = w_\eta$ *holds if one of the following conditions is satisfied.*

i) X^H *is connected and* $w_1(\xi^H) = w_1(\eta^H)$ *for all* $H \subset G$. *There is a* x *in* X^G *with* $\xi_x \simeq_{G_x} \eta_x$.

ii) *The group* G *is finite of odd order and* $\xi_x \simeq_{G_x} \eta_x$ *for all* $x \in M$. *We have after forgetting the group action* $w_1(\xi) = w_1(\eta)$.

Proof: We have to specify for each $H \subset G$ and $x \in X^H$ a G-fibre homotopy equivalence $\varphi(G/H)(x) : tp\,(G/H)(x) \to tp\,(G/H)(x)$. We do this by determining a H-homotopy equivalence $\varphi(G/H)(x)_{eH}$: $\xi_x \to \eta_x$ between the fibres over eH. The independence of the choice of the path u below follows from the assumptions about the first Stiefel-Whitney classes.

i) Fix y in X^G and a G-homotopy equivalence $\Phi : \xi_y \to \eta_y$. Define $\varphi(G/H)(x)_{eH}$ by requiring that the following diagram commutes up to H-homotopy for a path u from y to x in X^H.

$$\begin{array}{ccc} \xi_x & \xrightarrow{\varphi(G/H)(x)_{eH}} & \eta_x \\ {\scriptstyle tp\,(G/H)(u)_{eH}}\downarrow & & \downarrow{\scriptstyle tp\,(G/H)(u)_{eH}} \\ \xi_y & \xrightarrow{\Phi} & \eta_y \end{array}$$

ii) Without loss of generality we can suppose that X is connected. Fix a point y in X and a non-equivariant homotopy equivalence $\Phi : \xi_y \to \eta_y$. By assumption there is a H-homotopy equivalence

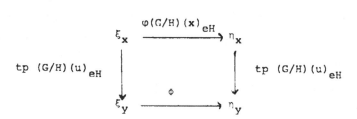

$\xi_x \to \eta_x$ and there are only two up to homotopy because of $A(H)^* = A(H, ISO(\eta_x))^* = \{\pm 1\}$ (see [6], p. 8) and theorem 3.4. Let $\varphi(G/H)(x)_{eH}$ be the one making the following non-equivariant diagram commutative up to homotopy for a path u between y and x.

$$\begin{array}{ccc} \xi_x & \xrightarrow{\varphi(G/H)(x)_{eH}} & \eta_x \\ {\scriptstyle tp(G/1)(u)_e}\downarrow & & \downarrow{\scriptstyle tp(G/1)(u)_e} \\ \xi_y & \xrightarrow{\Phi} & \eta_y \end{array}$$

□

Now we examine whether w_M is a homotopy invariant. If $f : M \to N$ is a G-homotopy equivalence $f^*(TN \oplus V)^c$ and $(TM \oplus V)^c$ are G-fibre homotopy equivalent for appropriate V by [13], theorem 2.3. If this could be destabilized to $f\, TN^c \simeq_G TM^c$ we would in particular obtain $f^* w_N = w_M$. Unfortunately, this is not possible in general, theorem 3.4. gives counterexamples over a point. Now we prove the unstable result $f^* w_N = w_M$.

<u>Theorem 7.4.</u> <u>Let</u> $f : M \to N$ <u>be a G-map between G-manifolds such that</u> $\pi_0(f^H) : \pi_0(M^H) \to \pi_0(N^H)$ <u>is bijective for all</u> $H \subset G$. <u>Suppose for</u> $C \in \pi_0(M^H)$ <u>and</u> $D \in \pi_0(N^H)$ <u>with</u> $f^H(C) \subset D$ <u>that</u> $(f^H|C)^* w_1(D)$ <u>is</u> $w_1(C)$. <u>Then the (non-equivariant) degree</u> $\deg(f^H|C : C \to D) \in \mathbb{Z}/\{\pm 1\}$ <u>is defined. Assume</u> $\deg(f^H|C) = \pm 1$ <u>and that</u> M <u>and</u> N <u>fulfill condition</u> (*).

<u>Then there is an</u> O(G)-<u>equivalence</u> $\omega: f^* tp_N \to tp_M$ <u>uniquely determined by the property that</u> $DEG(f, \omega) = 1$.

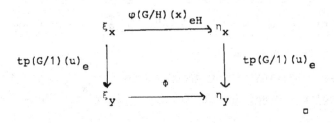

Proof. Denote by $M^H(x)$ the component of M^H containing x for $H \subset G$, $x \in M^H$. Let $\hat{M}^H(x)$ be $M^H(x)$ if $w_1(M^H(x))$ is zero and the orientation covering otherwise so that $\hat{M}^H(x)$ is an orientable connected manifold. Choose a lift $\hat{x} \in \hat{M}^H(x)$ of x and a lift $\hat{f}^H : \hat{M}^H(x) \to \hat{N}^H(fx)$ of $f|M^H(x) : M^H(x) \to N^H(fx)$. Write $\hat{y} = \hat{f}^H(\hat{x})$ and $y = f^H(x)$. Let $\psi(x,H) : (TN_y^c)^H \to (TM_x^c)^H$ be a (non-equivariant) map for $x \in M^H$, $H \subset G$ making the following diagram commutative where c denotes the collaps map and p the projection and n is $\dim M^H(x) = \dim N^H(fx)$.

$$\begin{array}{ccc}
H_n(\hat{M}^H(x)),\partial\hat{M}^H(x)) & \xrightarrow{\hat{f}^H_*} & H_n(\hat{N}^H(y)),\partial\hat{N}^H(y)) \\
c_* \downarrow \cong & & \cong \downarrow c_* \\
H_n(T\hat{M}^H(x)^c_{\hat{x}})_* & & H_n(T\hat{N}^H(y)^c_{\hat{y}})_* \\
(Tp^c_{\hat{x}}) \downarrow \cong & & \cong \downarrow (Tp^c_{\hat{y}}) \\
H_n((TM^c_x)^H) & \xleftarrow{\psi(x,H)_*} & H_n((TN^c_y)^H)
\end{array}$$

One easily checks that the homotopy class of $\psi(x,H)$ depends only on (x,H) but not on the choice of $\hat{x}$ and $\hat{f}^H$.

Lemma 7.5. *There is up to G_x-homotopy exactly one G_x-map $\Psi(x) : TN^c_{fx} \to TM^c_x$ for each x in M such that $\Psi(x)^H$ and $\psi(x,H)$ are non-equivariantly homotopic for all $H \subset G_x$. Each $\Psi(x)$ is a G_x-homotopy equivalence.*

Let m be the product $\pi|W_{G_x}H|$ running over $\{(H) \in \mathrm{Con}(G_x) \mid H \in \mathrm{ISO}(TN_y),\ W_{G_x}H \text{ finite}\}$. We get from [7] p. 173 + 174 the existence of G-maps

$\omega : TN_y \to TM_x$ and $\omega' : TM_x \to TN_y$ with $\deg((\omega' \circ \omega)^H) = 1 \bmod m$ for all $H \subset G$. Because of the equivariant Hopf theorem 3.5. and theorem 3.4. lemma 7.5. follows from:

7.6. The element $d = \{\deg(\omega^H \circ \psi(x,H)^{-1}) \mid (H) \in \mathrm{Con}(G_x)\} \in C(G_x)$ lies in the image of $\mathrm{DEG} : [TN_y, TN_y]^{G_x} \to C(G_x)$.

We firstly give the proof of 7.6. under the assumption that N^H is connected for all $H \subset G_x$. Consider $f : M \to N$ as a G_x-map so that x is a G_x-fixed point in M. As in the proof of theorem 7.3. ii) we get an $O(G_x)$-transformation $\varphi : f^* tp_N \to tp_M$ uniquely determined by the property that $\varphi(G_x/G_x)(x)$ is just ω. Note that for any $z \in M^H$, $H \subset G$ the case $w_1(M^H(z)) = 0$ and $w_1(N^H(z)) \neq 0$ never occurs because of $\deg(f^H | M^H(z)) = \pm 1$. Now one checks directly that $\mathrm{DEG}(f,\varphi) \in C(G_x)$ is just d. By theorem 5.1. we have $d \in \mathrm{image}(\mathrm{DEG} : [TN_y^c, TN_y^c]^{G_x} \to C(G_x))$.

In the general case one has the problem that ω does not determine an $O(G_x)$-transformation $\varphi : f^* tp_N \to tp_M$ if there is a non-connected N^H for some $H \subset G_x$. But we can restrict everything to the $O(G_x)$-subgroupoid $\Pi_x^{G_x} M$ of $\Pi^{G_x} M$ with $\Pi_x^{G_x} M(G_x/H) := \Pi(M^H(x))$ so that we consider only the component of M^H containing x. Then we get an $O(G_x)$-transformation $\varphi : f^* tp_N \mid \Pi_x^{G_x} M \to tp_M \mid \Pi_x^{G_x} M$ by ω as before. As in section two we can at least define $\mathrm{DEG}(f,\varphi)$ in $C(\Pi_x^{G_x} N) = C(G_x)$ and get $d = \mathrm{DEG}(f,\varphi)$. The same argument as in the proof of theorem 5.1. gives $d \in \mathrm{image}\ (\mathrm{DEG} : [TN_y^c, TN_y^c]^{G_x} \to C(G_x))$. This shows 7.6. and finishes the proof of Lemma 7.5.

Let $\varphi : f^* tp_N \to tp_M$ be defined by the property that $\varphi(G/H)(x)_{eH} : TN^c_{fx} \to TM^c_x$ is the restriction from G_x to H of $\psi(x)$. We leave it to the reader to check that φ is an well-defined O(G)-equivalence. By construction DEG(f,φ) = 1 holds. The uniqueness follows from theorem 4.1. This finishes the proof of theorem 7.4. □

We obtain as a corollary the homotopy invariance of the first equivariant Stiefel-Whitney class and the unstable version of the result in [13], corollary 2.4.

Corollary 7.7. Let $f : M \longrightarrow N$ be a G-map satisfying the assumption of Theorem 7.4. Then we have $f^* w_N = w_M$ and the spheres of the normal slices of M at x and N at fx are G_x - homotopy equivalent for all $x \in M$.

Proof We derive $f^* w_N = w_M$ and $TM^c_x \simeq_{G_x} TN_{fx}$ from theorem 7.4. Now apply theorem 3.5. □

If G_x is connected TM_x and TN_{fx} are even isomorphic as G_x-representations (see [22]). For finite G there are non-isomorphic G-representations V and W with $V^c \simeq_G W^c$ (see [6], p. 249).

Now we give a necessary condition for converting a G-map $f : M \to N$ between G-manifolds into a G-homotopy equivalence by surgery. Notice that this would imply the existence of a bordism appearing below. Theorem 7.8. motivates the approach to equivariant surgery given in [17].

Theorem 7.8. Let $(f,f,f_+) : (P,M,M_+) \to (Q,N,N_+)$ be a G-map between G-triads of G-manifolds such that f_+ is a G-homotopy equivalence. Assume that $M_+^H \to P^H$ is 1-connected for $H \subset G$.

Then we can find an O(G)-equivalence $\varphi : f^* tp_N \to tp_M$ with $DEG(f,\varphi) = 1$.

Proof. We get the existence of an O(G)-equivalence $\varphi_+ : f_+^* tp_{N_+} \to tp_{M_+}$ with $DEG(f_+,\varphi_+) = 1$ by corollary 7.7. Using the inward normal we get identifications $tp_P \mid M_+ = tp_{(TM_+ \oplus \mathbb{R})^c}$ and $tp_P \mid M = tp(TM \oplus \mathbb{R})^c$ and analogously for Q. For two G-representations V and N with $(V \oplus \mathbb{R})^c \simeq_G (W \oplus \mathbb{R})^c$ and $\dim V^G = \dim W^G \geq 1$ the suspension $[V^c,W^c]^G \to [V \oplus \mathbb{R}^c, W \oplus \mathbb{R}^c]^G$ is bijective (theorem 3.5.). Hence we obtain O(G)-equivalences $\Phi : F^* tp_Q \to tp_P$ and $\varphi : f^* tp_N \to tp_M$ such that $\Phi \mid M$ corresponds to φ and $\Phi \mid M_+$ to φ_+ under the identification above. Now apply the bordism invariance 2.3. □

References

[1] Atiyah, M. F. and Bott, R.: Lefschetz fixed point formula for elliptic complexes II. Applications. Ann. Math. 88, 451 - 491 (1968).

[2] Bredon, G. E.: Introduction to compact transformation groups, Academic Press, New York-London (1972).

[3] Bredon, G. E.: Fixed point sets of actions on Poincaré duality spaces. Topology 12 (1973), 159 - 175.

[4] Browder, W.: Pulling back fixed points, inv. math. 87, 331 - 342 (1987).

[5] tom Dieck, T.: The Burnside ring of a compact Lie group I. Math. Ann. 215, 235 - 250 (1975).

[6] tom Dieck, T.: Transformation groups and representation theory. Lect. notes in math. 766, Springer Verlag, Berlin-Heidelberg-New York (1979).

[7] tom Dieck, T.: Transformation groups, de Gruyter (1987).

[8] tom Dieck, T. and T. Petrie: Homotopy representations of finite groups, Publ. Math. IHES 56 (1982), 337 - 377.

[9] Dold, A.: Lectures on algebraic topology, Springer Verlag, Berlin-Heidelberg-New York (1972).

[10] Dovermann, K. H.: Addition of equivariant surgery obstructions, algebraic topology, Waterloo (1978), lecture notes in math. 741, Springer Verlag (1979), 244 - 271.

[11] Dovermann, K. H. and T. Petrie: G surgery II. Mem. of the AMS Vol. 37, no. 260 (1982).

[12] Ewing, J. and R. Stong: Group actions having one fixed point. Math. Zeitschrift 191, 159 - 164 (1986).

[13] Kawakubo, K.: Compact Lie group actions and fibre homotopy type. J. Math. Soc. Japan, Vol. 33, no. 2, 295 - 321 (1981).

[14] Laitinen, E.: Unstable homotopy theory of homotopy representations, in "transformation groups", Poznan, lect. notes in math., vol. 1217 (1986).

[15] Lück, W.: Seminarbericht "Transformationsgruppen und algebraische K-Theorie", Göttingen 1983.

[16] Lück, W.: Equivariant Eilenberg-MacLane spaces K($\mathfrak{g}$,μ,1) with possibly non-connected or empty fixed point sets, manuscr. math. 58, 67 - 75 (1987)

[17] Lück, W. and Madsen, I.: Equivariant L-theory, Aarhus preprint,(1988).

[18] Rubinsztein, R. L.: On the equivariant homotopy of spheres, preprint, Polish Academy of Science (1973).

[19] Switzer, R. M.: Algebraic topology - homology and homotopy, Springer Verlag, Berlin-Heidelberg-New York (1975).

[20] Thomason, R. W.: First quadrant spectral sequences in algebraic K-theory via homotopy colimit, Comm. in Algebra 10 (15), 1589 - 1668 (1982).

[21] Tornehave, J.: Equivariant maps of spheres with conjugate orthogonal actions, Can. Math. Soc. Conf. Proc., Vol. 2 part 2, 275-301, (1982).

[22] Traczyk, P.: On the G-homotopy equivalences of spheres of representations, Math. Zeitschrift 161, 257 - 261 (1978).

Wolfgang Lück
Mathematisches Institut
der Georg-August-Universität
Bunsenstraße 3 - 5

3400 Göttingen

Bundesrepublik Deutschland

SURGERY TRANSFER

by W. Lück and A. Ranicki

Introduction

Given a Hurewicz fibration $F \longrightarrow E \xrightarrow{p} B$ with fibre an n-dimensional geometric Poincaré complex F we construct algebraic transfer maps in the Wall surgery obstruction groups

$$p^! : L_m(\mathbb{Z}[\pi_1(B)]) \longrightarrow L_{m+n}(\mathbb{Z}[\pi_1(E)]) \quad (m \geqslant 0)$$

and prove that they agree with the geometrically defined transfer maps. In subsequent work we shall obtain specific computations of the composites $p^! p_!$, $p_! p^!$ with $p_! : L_m(\mathbb{Z}[\pi_1(E)]) \longrightarrow L_m(\mathbb{Z}[\pi_1(B)])$ the change of rings maps, and some vanishing results.

The construction of $p^!$ is most straightforward in the case when F is finite, with L_* the free L-groups L_*^h. In §9 we shall extend the definition of $p^!$ to finitely dominated F and the projective L-groups L_*^p, as well as to simple F and the simple L-groups L_*^s, and also to the intermediate cases.

There are two main sources of applications of the surgery transfer. The equivariant surgery obstruction groups of Browder and Quinn [1] were defined in terms of the geometric surgery transfer maps of the normal sphere bundles of the fixed point sets. An algebraic version will necessarily involve the algebraic surgery transfer maps. (In this connection see Lück and Madsen [8].) The recent work of Hambleton, Milgram, Taylor and Williams [3] on the evaluation of the surgery obstructions of normal maps of closed manifolds with finite fundamental group depends on the factorization of the assembly map by twisted product formulae which are closely related to the algebraic surgery transfer.

Our construction of the quadratic L-theory transfer maps is by a combination of the algebraic surgery theory of Ranicki [14],[19] and the method used by Lück [7] to define the algebraic K-theory transfer maps $p^!:K_m(\mathbb{Z}[\pi_1(B)])\longrightarrow K_m(\mathbb{Z}[\pi_1(E)])$ $(m=0,1)$ for a fibration with finitely dominated fibre F.

The algebraic surgery transfer maps $p^!$ for a fibration are a special case of transfer maps $(C,\alpha,U)^!:L_m(A)\longrightarrow L_{m+n}(B)$ $(m\geqslant 0)$ defined in abstract algebra. Here, A and B are rings with involution, C is an n-dimensional f.g. free B-module chain complex with a symmetric Poincaré duality chain equivalence $\alpha\simeq\alpha^*:C\longrightarrow C^{n-*}$, and $U:A\longrightarrow R=H_0(\mathrm{Hom}_B(C,C))^{op}$ is a morphism of rings with involution from A to the opposite of the ring of chain homotopy classes of B-module chain maps $f:C\longrightarrow C$, with the involution on R defined by $T(f)=\alpha^{-1}f^*\alpha$. An element of $L_{2i}(A)$ is represented by a nonsingular $(-)^i$-quadratic form $(M,\psi:M\longrightarrow M^*)$ on a f.g. free A-module $M=\underset{k}{\oplus}A$. We define

$(C,\alpha,U)^!(M,\psi)=(D,\theta)\in L_{n+2i}(B)$ to be the cobordism class of the $(n+2i)$-dimensional quadratic Poincaré complex (D,θ) given by

$$\theta_s=\begin{cases}U(\psi)(\underset{k}{\oplus}\alpha^{-1}) & \text{if } s=0\\ 0 & \text{if } s\neq 0\end{cases}:$$

$$D^{n+2i-r-s}=\underset{k}{\oplus}C^{n+i-r-s}\longrightarrow D_r=\underset{k}{\oplus}C_{r-i}\ .$$

There is a similar formula in the case $m=2i+1$, for which we refer to §4.

The algebraic transfer maps of fibration $F\longrightarrow E\xrightarrow{p}B$ with fibre an n-dimensional geometric Poincaré complex F are given by

$$p^!=(C(\tilde{F}),\alpha,U)^!\ :\ L_m(\mathbb{Z}[\pi_1(B)])\longrightarrow L_{m+n}(\mathbb{Z}[\pi_1(E)])$$

with $C(\tilde{F})$ the cellular $\mathbb{Z}[\pi_1(E)]$-module chain complex of the cover $\tilde{F}$ of F induced from the universal cover $\tilde{E}$ of E, $\alpha=([F]\cap-)^{-1}:C(\tilde{F})\longrightarrow C(\tilde{F})^{n-*}$ the Poincaré duality chain equivalence, and U determined by the fibre transport.

Here is the main idea in the identification of the algebraic and geometric surgery transfer. We know from the identification of the corresponding K-theory transfers in Lück [7] how to handle in algebra the lift of CW structures from the base to the total space of a fibration. We use the ultraquadratic L-theory of Ranicki [16,§7.8] both to encode the algebraic surgery data in the base spaces as CW structures, and to decode the algebraic surgery data from the lifted CW structures in the total spaces.

The paper was written during the second named author's visit in the academic year 1987/1988 to the Sonderforschungsbereich SFB170 in Göttingen, whose support is gratefully acknowledged.

The titles of the sections are:

Introduction
§1. The algebraic K-theory transfer
§2. Maps of L-groups
§3. The generalized Morita maps in L-theory
§4. The quadratic L-theory transfer
§5. The algebraic surgery transfer
§6. The geometric surgery transfer
§7. Ultraquadratic L-theory
§8. The connection
§9. Change of K-theory
Appendix 1. Fibred intersections
Appendix 2. A counterexample in symmetric L-theory
References

§1. The algebraic K-theory transfer

We recall from Lück [7] the construction of the algebraic K-theory transfer maps, and the connection with topology.

Given a ring R let R^{op} denote the opposite ring, with the same elements and additive structure but with the opposite multiplication.

Definition 1.1 A representation (A,U) of a ring R in an additive category $\mathbb{A}$ is an object A in $\mathbb{A}$ together with a morphism of rings $U:R \longrightarrow \mathrm{Hom}_{\mathbb{A}}(A,A)^{op}$.

□

Given an associative ring R with 1 let $\mathbb{B}(R)$ be the additive category of based f.g. free R-modules R^n $(n \geq 0)$. A morphism $f:R^n \longrightarrow R^m$ is an R-module morphism, corresponding to the $m \times n$ matrix $(a_{ij})_{1 \leq i \leq m, 1 \leq j \leq n}$ with entries $a_{ij} \in R$, such that

$$f = (a_{ij}) : R^n \longrightarrow R^m \; ; \; (x_j) \longrightarrow (\sum_{j=1}^{n} x_j a_{ij}) \; .$$

Example 1.2 The universal representation (R,U) of R in $\mathbb{B}(R)$ is defined by the ring isomorphism

$$U : R \longrightarrow \mathrm{Hom}_R(R,R)^{op} \; ; \; r \longrightarrow (\; s \longrightarrow sr \;) \; ,$$

which we shall use to identify $R = \mathrm{Hom}_R(R,R)^{op}$.

□

A functor of additive categories $F:\mathbb{A} \longrightarrow \mathbb{B}$ is required to preserve the additive structures.

Proposition 1.3 Given a ring R and an additive category $\mathbb{A}$ there is a natural one-one correspondence between functors $F:\mathbb{B}(R) \longrightarrow \mathbb{A}$ and representations (A,U) of R in $\mathbb{A}$.

Proof: Given a functor F define a representation (A,U) by

$$A = F(R) \ ,$$

$$U : R = \mathrm{Hom}_R(R,R)^{op} \longrightarrow \mathrm{Hom}_{\mathbb{A}}(A,A)^{op} \ ;$$

$$(\rho : R \longrightarrow R) \longrightarrow (F(\rho) : A \longrightarrow A) \ .$$

Conversely, given a representation (A,U) define a functor $F = -\otimes(A,U) : \mathbb{B}(R) \longrightarrow \mathbb{A}$ by

$$F(R^n) = A^n \ ,$$

$$F((a_{ij}) : R^n \longrightarrow R^m) = (U(a_{ij})) : A^n \longrightarrow A^m \ .$$

□

<u>Example</u> <u>1.4</u> A morphism of rings $f : R \longrightarrow S$ determines a representation (S,U) of R in $\mathbb{B}(S)$ with

$$U = f : R \longrightarrow \mathrm{Hom}_S(S,S)^{op} = S \ ,$$

such that $-\otimes(S,U) = f_! : \mathbb{B}(R) \longrightarrow \mathbb{B}(S)$ is the usual change of rings functor.

□

For any object A in an additive category $\mathbb{A}$ there is defined a representation $(A,1)$ of the ring $\mathrm{Hom}_{\mathbb{A}}(A,A)^{op}$ in $\mathbb{A}$. The corresponding functor is the full embedding

$$-\otimes(A,1) : \mathbb{B}(\mathrm{Hom}_{\mathbb{A}}(A,A)^{op}) \longrightarrow \mathbb{A} \ ;$$

$$\mathrm{Hom}_{\mathbb{A}}(A,A)^{op} \longrightarrow A \ .$$

The functor associated to a representation (A,U) of R in $\mathbb{A}$ is the composite

$$F = -\otimes(A,U) : \mathbb{B}(R) \xrightarrow{U_!} \mathbb{B}(\mathrm{Hom}_{\mathbb{A}}(A,A)^{op})$$

$$\xrightarrow{-\otimes(\mathbb{A},1)} \mathbb{A} .$$

Given chain complexes C,D in $\mathbb{A}$ let $\mathrm{Hom}_{\mathbb{A}}(C,D)$ be the abelian group chain complex defined by

$$d_{\mathrm{Hom}_{\mathbb{A}}(C,D)} : \mathrm{Hom}_{\mathbb{A}}(C,D)_r = \sum_{q-p=r} \mathrm{Hom}_{\mathbb{A}}(C_p,D_q)$$

$$\longrightarrow \mathrm{Hom}_{\mathbb{A}}(C,D)_{r-1} ; f \longrightarrow d_D f + (-)^q f d_C .$$

There is a natural one-one correspondence between chain maps $f:C\longrightarrow D$ and 0-cycles $f'\in\mathrm{Hom}_{\mathbb{A}}(C,D)_0$, with

$$f' = (-)^n f : C_n \longrightarrow D_n \quad (n\in\mathbb{Z}) .$$

Similarly for chain homotopies and 1-chains. Thus $H_0(\mathrm{Hom}_{\mathbb{A}}(C,D))$ is isomorphic to the additive group of chain homotopy classes of chain maps $C\longrightarrow D$.

A chain complex C is finite if $C_r=0$ for $r<0$ and there exists $n\geqslant 0$ such that $C_r=0$ for $r>n$.

<u>Definition 1.5</u> Given an additive category $\mathbb{A}$ let $\mathbb{D}(\mathbb{A})$ be the homotopy category of $\mathbb{A}$, the additive category of finite chain complexes in $\mathbb{A}$ and chain homotopy classes of chain maps with

$$\mathrm{Hom}_{\mathbb{D}(\mathbb{A})}(C,D) = H_0(\mathrm{Hom}_{\mathbb{A}}(C,D)) .$$

□

For a ring R we write $\mathbb{D}(\mathbb{B}(R))$ as $\mathbb{D}(R)$.

We refer to Ranicki [17],[18] for an account of the algebraic K-groups $K_m(\mathbb{A})$ $(m=0,1)$ of an additive category $\mathbb{A}$ with the split exact structure, and the application to chain complexes. In particular, the class of a finite chain complex C in $\mathbb{A}$ is defined by

$$[C] = \sum_{r=0}^{\infty} (-)^r [C_r] \in K_0(\mathbb{A}) ,$$

and the torsion of a self chain equivalence $f:C\longrightarrow C$ is defined by

$$\tau(f) = \tau(d+\Gamma : C(f)_{odd} \longrightarrow C(f)_{even}) \in K_1(\mathbb{A})$$

for any chain contraction $\Gamma : 0 \simeq 1 : C(f) \longrightarrow C(f)$ of the algebraic mapping cone $C(f)$.

<u>Definition 1.6</u> The <u>generalized Morita maps</u> $\mu : K_m(\mathbb{D}(\mathbb{A})) \longrightarrow K_m(\mathbb{A})$ $(m=0,1)$ are defined for any additive category $\mathbb{A}$ by:

for $m=0$ μ sends the class $[C] \in K_0(\mathbb{D}(\mathbb{A}))$ of an object C in $\mathbb{D}(\mathbb{A})$ to the class $[C] \in K_0(\mathbb{A})$,

for $m=1$ μ sends the torsion $\tau(f) \in K_1(\mathbb{D}(\mathbb{A}))$ of an automorphism $f : C \longrightarrow C$ in $\mathbb{D}(\mathbb{A})$ to the torsion $\tau(f) \in K_1(\mathbb{A})$ of any representative self chain equivalence.

□

A morphism in $\mathbb{D}(\mathbb{A})$ is a chain homotopy class and the definition of μ involves a choice of representative chain map. The generalized Morita maps μ are therefore not induced by a functor $\mathbb{D}(\mathbb{A}) \longrightarrow \mathbb{A}$.

<u>Example 1.7</u> (Lück [7]) A Hurewicz fibration $F \longrightarrow E \xrightarrow{p} B$ with the fibre F a CW complex determines a ring morphism

$$U : \mathbb{Z}[\pi_1(B)] \longrightarrow H_0(\mathrm{Hom}_{\mathbb{Z}[\pi_1(E)]}(C(\tilde{F}), C(\tilde{F})))^{op}$$

with $C(\tilde{F})$ the cellular based free $\mathbb{Z}[\pi_1(E)]$-module chain complex of the pullback $\tilde{F}$ to F of the universal cover $\tilde{E}$ of E, and U the chain homotopy action of $H_0(\Omega B) = \mathbb{Z}[\pi_1(B)]$ on $C(\tilde{F})$ determined by the homotopy action of the loop space ΩB on F. For finite F this defines a representation $(C(\tilde{F}), U)$ of $\mathbb{Z}[\pi_1(B)]$ in $\mathbb{D}(\mathbb{Z}[\pi_1(E)])$. For the identity map $p = 1 : E \longrightarrow B = E$ with $F = \{*\}$ this is the universal representation (R, U) of 1.2 for $R = \mathbb{Z}[\pi_1(B)] = \mathbb{Z}[\pi_1(E)]$.

□

The transfer map in the torsion groups associated to a representation (C,U) of a ring R in $\mathbb{D}(\mathbb{A})$ is the composite

$$(C,U)^! : K_1(R) = K_1(\mathbb{B}(R)) \xrightarrow{U_!} K_1(\mathbb{D}(\mathbb{A})) \xrightarrow{\mu} K_1(\mathbb{A})$$

of the map $U_!$ induced by the functor $(C,U)\otimes - : \mathbb{B}(R) \longrightarrow \mathbb{D}(\mathbb{A})$ and the generalized Morita map μ. The torsion $\tau(f) \in K_1(R)$ of an automorphism $f: R^k \longrightarrow R^k$ is sent by $(C,U)^!$ to the torsion $\tau(U(f)) \in K_1(\mathbb{A})$ of the self chain equivalence $U(f): \bigoplus_k C \longrightarrow \bigoplus_k C$.

The idempotent completion of an additive category $\mathbb{A}$ is the additive category $\hat{\mathbb{A}}$ with objects pairs

$$(A = \text{object of } \mathbb{A} \text{ , } p = p^2 : A \longrightarrow A)$$

and morphisms $f:(A,p) \longrightarrow (A',p')$ defined by morphisms $f: A \longrightarrow A'$ in $\mathbb{A}$ such that $p'fp = f : A \longrightarrow A'$. The evident functor $\mathbb{D}(\hat{\mathbb{A}}) \longrightarrow \hat{\mathbb{D}}(\mathbb{A})$ is an equivalence of additive categories, since every chain homotopy projection in $\hat{\mathbb{A}}$ splits (Lück and Ranicki [9]).

For any ring R the additive category $\mathbb{P}(R)$ of f.g. projective R-modules is equivalent to the idempotent completion $\hat{\mathbb{B}}(R)$ of the additive category $\mathbb{B}(R)$ of based f.g. free R-modules, with an equivalence

$$\hat{\mathbb{B}}(R) \longrightarrow \mathbb{P}(R) \text{ ; } (R^k, p) \longrightarrow \mathrm{im}(p) \text{ .}$$

For any representation (C,U) of a ring R in $\mathbb{D}(\hat{\mathbb{A}})$ the functor $(C,U)\otimes - : \mathbb{B}(R) \longrightarrow \mathbb{D}(\hat{\mathbb{A}})$ extends to a functor $\mathbb{P}(R) \longrightarrow \mathbb{D}(\hat{\mathbb{A}})$ (cf. Lemma 9.3), and so determines a transfer map in the class groups

$$(C,U)^! : K_0(R) = K_0(\mathbb{P}(R)) \xrightarrow{U_!} K_0(\mathbb{D}(\hat{\mathbb{A}}))$$

$$\xrightarrow{\mu} K_0(\hat{\mathbb{A}}) .$$

The class $[\mathrm{im}(p)] \in K_0(R)$ of a projection $p=p^2 : R^k \longrightarrow R^k$ is sent by $(C,U)^!$ to the projective class $[\underset{k}{\oplus} C, U(p)] \in K_0(\hat{\mathbb{A}})$ of the chain homotopy projection $U(p) \simeq U(p)^2 : \underset{k}{\oplus} C \longrightarrow \underset{k}{\oplus} C$.

<u>Example</u> <u>1.8</u> (Lück [7]) A representation (C,U) of a ring R in $\mathbb{D}(\mathbb{P}(S))$ induces algebraic K-theory transfer maps

$$(C,U)^! : K_m(R) = K_m(\mathbb{P}(R)) \longrightarrow K_m(S) = K_m(\mathbb{P}(S))$$

for $m=0,1$.

□

The algebraic K-theory transfer maps of a fibration $F \longrightarrow E \xrightarrow{p} B$ with finite (or finitely dominated) fibre F defined for $m=0,1$ by

$$p^! = (C(\tilde{F}),U)^! : K_m(\mathbf{Z}[\pi_1(B)]) \longrightarrow K_m(\mathbf{Z}[\pi_1(E)])$$

were shown in [7] to coincide with the geometric transfer maps using the following property of the functor

$$p^{\#} = -\otimes(C(\tilde{F}),U) : \mathbb{B}(\mathbf{Z}[\pi_1(B)]) \longrightarrow \mathbb{D}(\mathbf{Z}[\pi_1(E)]) .$$

<u>Proposition</u> <u>1.9</u> Let (X',X) be a relative CW pair such that X' is obtained from X by adjoining cells in dimensions $r,r+1$

$$X' = X \cup \bigvee_I e^r \cup \bigvee_J e^{r+1} .$$

Given a map $X' \longrightarrow B$ to a connected space B let $(\tilde{X}',\tilde{X})$ be the pullback to (X',X) of the universal cover $\tilde{B}$ of B, and let

$$d : C(\tilde{X}',\tilde{X})_{r+1} = \underset{J}{\oplus}\mathbf{Z}[\pi_1(B)] \longrightarrow C(\tilde{X}',\tilde{X})_r = \underset{I}{\oplus}\mathbf{Z}[\pi_1(B)]$$

be the differential in the cellular based free

$\mathbb{Z}[\pi_1(B)]$-module chain complex. Let $F \longrightarrow E \xrightarrow{p} B$ be a Hurewicz fibration such that the fibre F is a CW complex. Let $F \longrightarrow (Y',Y) \longrightarrow (X',X)$ be the fibration obtained from p by pullback along the map $X' \longrightarrow B$, with $(\tilde{Y}',\tilde{Y})$ the pullback to (Y',Y) of the universal cover $\tilde{E}$ of E. Then (Y',Y) is homotopy equivalent to a relative CW pair (also denoted by (Y',Y)) with cellular based free $\mathbb{Z}[\pi_1(E)]$-module chain complex

$$C(\tilde{Y}',\tilde{Y}) = S^r C(p^{\#}(d):\bigoplus_J C(\tilde{F}) \longrightarrow \bigoplus_I C(\tilde{F}))$$

the r-fold suspension of the algebraic mapping cone of a chain map in the chain homotopy class

$$p^{\#}(d) : p^{\#}(\bigoplus_J \mathbb{Z}[\pi_1(B)]) = \bigoplus_J C(\tilde{F})$$
$$\longrightarrow p^{\#}(\bigoplus_I \mathbb{Z}[\pi_1(B)]) = \bigoplus_I C(\tilde{F}) .$$

Proof: See Lück [7].

□

§2. Maps of L-groups

We refer to Ranicki [14],[19] for the definition of the quadratic L-groups $L_n(\mathbb{A})$ $(n\geq 0)$ of an additive category $\mathbb{A}$ with involution $*:\mathbb{A}\longrightarrow\mathbb{A}$, as the cobordism groups of n-dimensional quadratic Poincaré complexes $(C,\psi\in Q_n(C))$ in $\mathbb{A}$, and for the proof that these groups are 4-periodic, with $L_{2i}(\mathbb{A})$ (resp. $L_{2i+1}(\mathbb{A})$) the Witt group of nonsingular $(-)^i$-quadratic forms (resp. formations) in $\mathbb{A}$.

We now put an involution on the notions of §1.

Definition 2.1 An involution on an additive category $\mathbb{A}$ is a contravariant functor

$$* : \mathbb{A} \longrightarrow \mathbb{A} ; M \longrightarrow M^* ,$$

$$(f:M \longrightarrow N) \longrightarrow (f^*:N^* \longrightarrow M^*)$$

together with a natural equivalence

$$e : id_{\mathbb{A}} \longrightarrow ** : \mathbb{A} \longrightarrow \mathbb{A} ;$$

$$M \longrightarrow (e(M):M \longrightarrow M^{**})$$

such that

$$e(M^*) = (e(M)^{-1})^* : M^* \longrightarrow M^{***} .$$

□

We shall use the natural isomorphisms $e(M):M \longrightarrow M^{**}$ to identify $M^{**}=M$.

<u>Example 2.2</u> Given a ring R with involution

$$\bar{\ } : R \longrightarrow R ; r \longrightarrow \bar{r}$$

let the additive category

$$\mathbb{B}(R) = \{\text{based f.g. free } R\text{-modules}\}$$

have the duality involution

$$(R^n)^* = R^n \quad , \quad (a_{ij})^* = (\bar{a}_{ji}) \ ,$$

such that

$$L_n(\mathbb{B}(R)) = L_n(R) \quad (n \geqslant 0) .$$

By definition, a quadratic Poincaré complex over R is the same as a quadratic Poincaré complex in $\mathbb{B}(R)$.

□

<u>Notation 2.3</u> Let $\mathbb{A}$ be an additive category with involution.

i) A chain complex C in $\mathbb{A}$ is <u>n-dimensional</u> if $C_r=0$ for $r<0$ and $r>n$.

ii) The <u>n-dual</u> of an n-dimensional chain complex C is the n-dimensional chain complex C^{n-*} in $\mathbb{A}$ with

$$d_{C^{n-*}} = (-)^r (d_C)^* :$$

$$(C^{n-*})_r = C^{n-r} = (C_{n-r})^* \longrightarrow (C^{n-*})_{r-1} \ .$$

iii) For $n\geqslant 0$ let $\mathbb{D}_n(\mathbb{A})$ be the additive category of n-dimensional chain complexes in $\mathbb{A}$ and chain homotopy classes of chain maps, with the n-duality involution $T=n-*:\mathbb{D}_n(\mathbb{A})\longrightarrow\mathbb{D}_n(\mathbb{A});C\longrightarrow C^{n-*}$.

□

A functor of additive categories with involution $F:\mathbb{A}\longrightarrow\mathbb{B}$ is a functor of the underlying additive categories together with a natural equivalence $G:F*\longrightarrow *F:\mathbb{A}\longrightarrow\mathbb{B}$, such that for any object M in $\mathbb{A}$ there is defined a commutative diagram in $\mathbb{B}$

$$\begin{array}{ccc} F(M) & \xrightarrow{\ e_{\mathbb{B}}(F(M))\ } & F(M)^{**} \\ {\scriptstyle F(e_{\mathbb{A}}(M))}\Big\downarrow & & \Big\downarrow{\scriptstyle G(M)^*} \\ F(M^{**}) & \xrightarrow{\ G(M^*)\ } & F(M^*)^* \ . \end{array}$$

<u>Notation</u> <u>2.4</u> A functor $F:\mathbb{A}\longrightarrow\mathbb{B}$ of additive categories with involution induces morphisms of the quadratic L-groups which we write as

$$F_! \ : \ L_n(\mathbb{A}) \longrightarrow L_n(\mathbb{B}) \ ;$$

$$(C,\psi) \longrightarrow (F(C),F(\psi)) \quad (n\geqslant 0) \ .$$

□

<u>Example</u> <u>2.5</u> A morphism of rings with involution $f:R\longrightarrow S$

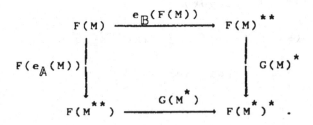

determines functors of additive categories with involution $f_!:\mathbb{B}(R)\longrightarrow\mathbb{B}(S)$ which induces change of rings morphisms in the quadratic L-groups $f_!:L_n(R)\longrightarrow L_n(S)$ $(n\geqslant 0)$.

□

<u>Definition</u> <u>2.6</u> Given a nonsingular symmetric form $(A,\alpha=\alpha^*:A\longrightarrow A^*)$ in an additive category with involution $\mathbb{A}$ let the ring $\mathrm{Hom}_{\mathbb{A}}(A,A)^{op}$ have the involution

$$\bar{\ } : \mathrm{Hom}_{\mathbb{A}}(A,A)^{op} \longrightarrow \mathrm{Hom}_{\mathbb{A}}(A,A)^{op} ;$$

$$(f:A\longrightarrow A) \longrightarrow (\alpha^{-1}f^*\alpha:A\longrightarrow A^*\longrightarrow A^*\longrightarrow A) .$$

□

By analogy with Definition 1.1:

<u>Definition</u> <u>2.7</u> A <u>symmetric</u> <u>representation</u> (A,α,U) of a ring with involution R in an additive category with involution $\mathbb{A}$ is a nonsingular symmetric form (A,α) in $\mathbb{A}$ together with a morphism of rings with involution $U:R\longrightarrow\mathrm{Hom}_{\mathbb{A}}(A,A)^{op}$.

□

In particular, (A,U) is a representation of R in the additive category $\mathbb{A}$ in the sense of 1.1.

By analogy with Example 1.2:

<u>Example</u> <u>2.8</u> The <u>universal</u> symmetric representation (R,α,U) of a ring with involution R in $\mathbb{B}(R)$ is defined by

$$\alpha : R \longrightarrow R^* ; r \longrightarrow (s \longrightarrow s\bar{r})$$

with U the isomorphism of rings with involution

$$U : R \longrightarrow \mathrm{Hom}_R(R,R)^{op} \; ; \; r \longrightarrow (\, s \longrightarrow sr \,) \; .$$

We shall use U as an identification of rings with involution $R = \mathrm{Hom}_R(R,R)^{op}$.

□

By analogy with Proposition 1.3:

Proposition 2.9 Given a ring with involution R and an additive category with involution $\mathbb{A}$ there is a natural one-one correspondence between functors of pairs of additive categories with involution $F : \mathbb{B}(R) \longrightarrow \mathbb{A}$ and symmetric representations (A, α, U) of R in $\mathbb{A}$.

Proof: Given a functor F define a symmetric representation (A, α, U) by

$$A = F(R) \; ,$$

$$\alpha = G(R) : F(R^*) = A \longrightarrow F(R)^* = A^* \; ,$$

$$U : R = \mathrm{Hom}_R(R,R)^{op} \longrightarrow \mathrm{Hom}_{\mathbb{A}}(A,A)^{op} \; ;$$

$$(\rho : R \longrightarrow R) \longrightarrow (F(\rho) : A \longrightarrow A) \; .$$

Conversely, given a symmetric representation (A, α, U) define a functor $F = - \otimes (A, \alpha, U) : \mathbb{B}(R) \longrightarrow \mathbb{A}$ by

$$F(R) = A \; ,$$

$$G(R) = \alpha : F(R^*) = A \longrightarrow F(R)^* = A^* \; ,$$

$$F((a_{ij}) : R^n \longrightarrow R^m) = (U(a_{ij})) : A^n \longrightarrow A^m \; .$$

□

By definition, a nonsingular symmetric form (C, α) in $\mathbb{D}_n(\mathbb{A})$ is an n-dimensional symmetric complex C in $\mathbb{A}$ together with a self dual chain homotopy class of chain equivalences $\alpha \simeq T\alpha : C \longrightarrow C^{n-*}$.

<u>Proposition</u> <u>2.10</u> A symmetric representation (C,α,U) of a ring with involution R in $\mathbb{D}_n(\mathbb{A})$ determines a functor $F=-\otimes(C,\alpha,U):\mathbb{B}(R)\longrightarrow\mathbb{D}_n(\mathbb{A})$ inducing morphisms in the quadratic L-groups

$$F_! = -\otimes(C,\alpha,U) : L_m(R) \longrightarrow L_m(\mathbb{D}_n(\mathbb{A})) \quad (m\geqslant 0) .$$

<u>Proof</u>: Immediate from 2.4 and 2.9.

□

Given a ring with involution S let $\mathbb{D}_n(S)=\mathbb{D}_n(\mathbb{B}(S))$, the additive category of n-dimensional chain complexes of based f.g. free S-modules and chain homotopy classes of chain maps with the n-duality involution $C\longrightarrow C^{n-*}$. A symmetric representation (C,α,U) of a ring with involution R in $\mathbb{D}_n(S)$ determines by 2.10 a functor $F=-\otimes(C,\alpha,U):\mathbb{B}(R)\longrightarrow\mathbb{D}_n(S)$ inducing morphisms in the quadratic L-groups

$$F_! = -\otimes(C,\alpha,U) : L_m(R) \longrightarrow L_m(\mathbb{D}_n(S)) \quad (m\geqslant 0) .$$

§3. <u>The generalized Morita maps in L-theory</u>

By analogy with the algebraic K-theory generalized Morita maps $\mu:K_m(\mathbb{D}(\mathbb{A}))\longrightarrow K_m(\mathbb{A})$ $(m=0,1)$ of §1 we define generalized Morita maps in the quadratic L-groups $\mu:L_m(\mathbb{D}_n(\mathbb{A}))\longrightarrow L_{m+n}(\mathbb{A})$ $(m,n\geqslant 0)$ by passing from nonsingular quadratic forms and formations in $\mathbb{D}_n(\mathbb{A})$ to quadratic Poincaré complexes in $\mathbb{A}$. The L-theory μ is the identity for $n=0$, since $\mathbb{D}_0(\mathbb{A})=\mathbb{A}$. For $n\geqslant 1$ the maps μ are not isomorphisms and are not induced by functors of additive categories with involution: a morphism in $\mathbb{D}_n(\mathbb{A})$ is a chain homotopy class and as in K-theory the definition of μ involves a choice of representative chain map.

<u>Proposition</u> <u>3.1</u> i) A nonsingular $(-)^i$-quadratic form in

$\mathbb{D}_n(\mathbb{A})$

$$(M,\theta\in \mathrm{coker}(1-(-)^iT:\mathrm{Hom}_{\mathbb{D}_n(\mathbb{A})}(M,M^*)\longrightarrow \mathrm{Hom}_{\mathbb{D}_n(\mathbb{A})}(M,M^*)))$$

is represented by an n-dimensional chain complex M in $\mathbb{A}$ together with a chain map $\theta:M\longrightarrow M^{n-*}$ such that $(1+(-)^iT)\theta=\theta+(-)^{n+i}\theta^*:M\longrightarrow M^{n-*}$ is a chain equivalence.

ii) The cobordism class $(C,\psi)\in L_{n+2i}(\mathbb{A})$ of the $(n+2i)$-dimensional quadratic Poincare complex in $\mathbb{A}$ (C,ψ) defined by $C=M^{n+i-*}$ and

$$\psi_s = \begin{cases} \theta & \text{if } s=0 \\ 0 & \text{if } s\geqslant 1 \end{cases} :$$

$$C^{n+2i-r-s} = M_{r-i-s} \longrightarrow C_r = M^{n+i-r}$$

depends only on the class

$$\theta \in \mathrm{coker}(1-(-)^iT:H_0(\mathrm{Hom}_{\mathbb{A}}(M,M^{n-*}))\longrightarrow H_0(\mathrm{Hom}_{\mathbb{A}}(M,M^{n-*})))$$

$$= \mathrm{coker}(1-(-)^iT:\mathrm{Hom}_{\mathbb{D}_n(\mathbb{A})}(M,M^{n-*})\longrightarrow \mathrm{Hom}_{\mathbb{D}_n(\mathbb{A})}(M,M^{n-*}))\ .$$

iii) Suppose given (C,ψ) as in ii), an n-dimensional chain complex L, a chain map $j:L\longrightarrow M$ and $\{\chi\in \mathrm{Hom}_A(L_r,L^{n+1-r})\mid r\geqslant 0\}$ defining a chain homotopy

$$\chi+(-)^{n+i+1}\chi^* \ :\ j^*\psi j \simeq 0 \ :\ L \longrightarrow L^{n-*}$$

such that the chain map $(j^*(1+T)\psi_0\ \ 0):C(j)\longrightarrow L^{n-*}$ is a chain equivalence, with $C(j)$ the algebraic mapping cone of j. Then $(C,\psi)=0\in L_{n+2i}(\mathbb{A})$.

<u>Proof</u>: i) Trivial.

ii) The isomorphism of abelian groups

$$Q_{(-)^i}(M) = \mathrm{coker}(1-(-)^iT:\mathrm{Hom}_{\mathbb{D}_n(\mathbb{A})}(M,M^*)\longrightarrow \mathrm{Hom}_{\mathbb{D}_n(\mathbb{A})}(M,M^*)) \longrightarrow Q_{n+2i}(C)\ ;$$

$$[\theta : M \longrightarrow M^{n-*}] \longrightarrow (\psi_s \in \mathrm{Hom}_{\mathbb{A}}(C^{n+2i-r-s}, C_r) \mid r,s \geqslant 0)$$

defined by

$$\psi_0 = \theta \ , \ \psi_s = 0 \text{ for } s \geqslant 1$$

sends the class of θ to the quadratic structure $\psi \in Q_{n+2i}(C)$.

iii) Define an $(n+2i+1)$-dimensional quadratic Poincaré pair in $\mathbb{A}$ $(f : C \longrightarrow D, (\delta\psi, \psi))$ by

$$f = j^* : C = M^{n+i-*} \longrightarrow D = L^{n+i-*} \ ,$$

$$\delta\psi_0 = \chi : D^{n+2i+1-r} = L_{r-i-1} \longrightarrow D_r = L^{n+i-r} \ ,$$

$$\delta\psi_s = 0 \text{ for } s \geqslant 1 \ .$$

□

We refer to §2 of Ranicki [19] for the definition of a nonsingular $(-)^i$-quadratic formation $(F,G) = (F, \begin{bmatrix}\gamma \\ \mu\end{bmatrix} G)$ in an additive category with involution $\mathbb{A}$, and for the result that $(F,G) = 0 \in L_{2i+1}(\mathbb{A})$ if and only if there there exist a $(-)^{i+1}$-quadratic form in $\mathbb{A}$ (H, ζ) and a morphism $j : F \longrightarrow H^*$ such that the morphism defined in $\mathbb{A}$ by

$$\begin{bmatrix} \mu^* & \gamma^* j^* \\ j & \zeta + (-)^{i+1} \zeta^* \end{bmatrix} : F \oplus H \longrightarrow G^* \oplus H^*$$

is an isomorphism.

<u>Proposition</u> <u>3.2</u> i) A nonsingular $(-)^i$-quadratic formation $(F, \begin{bmatrix}\gamma \\ \mu\end{bmatrix} G)$ in $\mathbb{D}_n(\mathbb{A})$ is represented by

n-dimensional chain complexes F, G in $\mathbb{A}$ together with chain maps $\gamma : G \longrightarrow F$, $\mu : G \longrightarrow F^{n-*}$, $\theta : G \longrightarrow G^{n-*}$ and a chain homotopy

$$\chi : \gamma^* \mu \simeq \theta + (-)^{n+i+1}\theta^* : G \longrightarrow G^{n-*}$$

such that the chain map

$$\begin{bmatrix} \chi + (-)^{n+i}\chi^* & \gamma^* \\ \gamma & 0 \end{bmatrix} : C(\mu^*)^{n+1-*} \longrightarrow C(\mu^* : F \longrightarrow G^{n-*})$$

is a chain equivalence.

ii) The cobordism class $(C,\psi) \in L_{n+2i+1}(\mathbb{A})$ of the $(n+2i+1)$-dimensional quadratic Poincaré complex $(C = S^i C(\mu^*), \psi)$ in $\mathbb{A}$ with

$$d_C = \begin{bmatrix} d_G^* & (-)^{r-1}T\mu \\ 0 & d_F \end{bmatrix} :$$

$$C_r = G^{n-r+i} \oplus F_{r-i-1} \longrightarrow C_{r-1} = G^{n-r-i+1} \oplus F_{r-i-2} ,$$

$$\psi_0 = \begin{bmatrix} (-)^{(n+1)(r-1)}\chi & 0 \\ (-)^{n(r-1)}\gamma & 0 \end{bmatrix} :$$

$$C^{n+2i+1-r} = G_{r-i-1} \oplus F^{n-r+i}$$

$$\longrightarrow C_r = G^{n-r+i} \oplus F_{r-i-1} ,$$

$$\psi_1 = \begin{bmatrix} (-)^{(n+1)r+1}\theta & 0 \\ 0 & 0 \end{bmatrix} :$$

$$C^{n+2i-r} = G_{r-i} \oplus F^{n-r+i-1}$$

$$\longrightarrow C_{r-i} = G^{n-r+i} \oplus F_{r-i-1} ,$$

$$\psi_s = 0 : C^{n+2i+1-r-s} \longrightarrow C_r \text{ for } s \geq 2 ,$$

depends only on the chain homotopy classes

$$\gamma \in H_0(\mathrm{Hom}_{\mathbb{A}}(G,F)) = \mathrm{Hom}_{\mathbb{D}_n(\mathbb{A})}(G,F) ,$$

$$\mu \in H_0(\mathrm{Hom}_{\mathbb{A}}(G,F^{n-*})) = \mathrm{Hom}_{\mathbb{D}_n(\mathbb{A})}(G,F^{n-*}) .$$

iii) Suppose given (C,ψ) as in i), an n-dimensional chain complex H in $\mathbb{A}$ and chain maps $j:F \longrightarrow H^{n-*}$, $\zeta:H \longrightarrow H^{n-*}$ such that the chain map

$$\begin{bmatrix} \mu^* & \gamma^* j^* \\ j & \zeta+(-)^{n+i+1}\zeta^* \end{bmatrix} : F \oplus H \longrightarrow G^{n-*} \oplus H^{n-*}$$

is a chain equivalence. Then $(C,\psi) = 0 \in L_{n+2i+1}(\mathbb{A})$.

<u>Proof</u>: i) The inclusion of the lagrangian $(G,0) \longrightarrow H_{(-)^i}(F)$ extends to an isomorphism of $(-)^i$-quadratic forms in $\mathbb{D}_n(\mathbb{A})$

$$\begin{bmatrix} \gamma & \tilde{\gamma} \\ \mu & \tilde{\mu} \end{bmatrix} : H_{(-)^i}(G) \longrightarrow H_{(-)^i}(F)$$

which is represented by a chain equivalence in $\mathbb{A}$

$$\begin{bmatrix} \gamma & \tilde{\gamma} \\ \mu & \tilde{\mu} \end{bmatrix} : G \oplus G^{n-*} \longrightarrow F \oplus F^{n-*}$$

with chain homotopy inverse

$$\begin{bmatrix} \gamma & \tilde{\gamma} \\ \mu & \tilde{\mu} \end{bmatrix}^{-1} = \begin{bmatrix} \tilde{\mu}^* & (-)^i \tilde{\gamma}^* \\ (-)^i \mu^* & \gamma^* \end{bmatrix} :$$

$$F \oplus F^{n-*} \longrightarrow G \oplus G^{n-*} .$$

For any representative chain map $\tilde{\mu}: G^{n-*} \longrightarrow F^{n-*}$ there exist a chain map $\nu: F \longrightarrow F^{n-*}$ and a chain homotopy

$$\eta : \mu\tilde{\mu}^{*} \simeq \nu + (-)^{n+i+1}\nu^{*} : F \longrightarrow F^{n-*} .$$

The chain maps in $\mathbb{A}$ defined by

$$f = \begin{bmatrix} \chi+(-)^{n+i}\chi^{*} & \gamma^{*} \\ \gamma & 0 \end{bmatrix} : C(\mu^{*})^{n+1-*} \longrightarrow C(\mu^{*})$$

$$g = \begin{bmatrix} 0 & \tilde{\mu}^{*} \\ \tilde{\mu} & \eta+(-)^{n+i}\eta^{*} \end{bmatrix} : C(\mu^{*}) \longrightarrow C(\mu^{*})^{n+1-*}$$

are such that there are defined chain homotopies

$$\begin{bmatrix} \epsilon & \tilde{\gamma}^{*} \\ 0 & \delta \end{bmatrix} : gf \simeq \begin{bmatrix} 1 & 0 \\ \beta & 1 \end{bmatrix} = \text{automorphism}$$

$$: C(\mu^{*})^{n+1-*} \longrightarrow C(\mu^{*})^{n+1-*} ,$$

$$\begin{bmatrix} \epsilon^{*} & 0 \\ \tilde{\gamma} & \delta^{*} \end{bmatrix} : fg \simeq \begin{bmatrix} 1 & \beta^{*} \\ 0 & 1 \end{bmatrix} = \text{automorphism}$$

$$: C(\mu^{*}) \longrightarrow C(\mu^{*})$$

with $\beta = \tilde{\mu}(\chi+(-)^{n+i}\chi^{*})+(\eta+(-)^{n+i}\eta^{*})\gamma$, and δ, ϵ chain homotopies

$$\delta : \tilde{\mu}\gamma^{*}+(-)^{i}\mu\tilde{\gamma}^{*} \simeq 1 : F^{n-*} \longrightarrow F^{n-*} ,$$

$$\epsilon : \tilde{\mu}^{*}\gamma+(-)^{i}\tilde{\gamma}^{*}\mu \simeq 1 : G \longrightarrow G .$$

Thus both fg and gf are chain equivalences, and f is a chain equivalence with chain homotopy inverse $(gf)^{-1}g \simeq g(fg)^{-1}$.

ii) With $\begin{bmatrix} \gamma & \tilde{\gamma} \\ \mu & \tilde{\mu} \end{bmatrix}$ as in i) there exist chain maps in $\mathbb{A}$

$$\tilde{\gamma} : G^{n-*} \longrightarrow F \ , \ \tilde{\mu} : G^{n-*} \longrightarrow F^{n-*} \ ,$$

$$\tilde{\theta} : G^{n-*} \longrightarrow G$$

and a chain homotopy

$$\tilde{\chi} : \tilde{\gamma}^*\tilde{\mu} \simeq \tilde{\theta} + (-)^{n+i+1}\tilde{\theta}^* : G^{n-*} \longrightarrow G$$

such that the chain map

$$\begin{bmatrix} \tilde{\chi}+(-)^{n+i}\tilde{\chi}^* & \tilde{\gamma}^* \\ \tilde{\gamma} & 0 \end{bmatrix} : C(\tilde{\mu}^*)^{n+1-*} \longrightarrow C(\tilde{\mu}^*)$$

is a chain equivalence in $\mathbb{A}$. Let $(\tilde{C},\tilde{\psi})$ be the $(n+2i+1)$-dimensional quadratic Poincaré complex derived from $(F,G^{n-*},\tilde{\gamma},\tilde{\mu},\tilde{\theta},\tilde{\chi})$ in the way (C,ψ) is derived from $(F,G,\gamma,\mu,\theta,\chi)$. Define an $(n+2i+2)$-dimensional quadratic Poincaré cobordism $((f\ \tilde{f}):C\oplus\tilde{C}\longrightarrow D,(\delta\psi,\psi\oplus-\tilde{\psi}))$ by

$$D = S^{i+1}F \ , \ \delta\psi = 0 \ ,$$

$$f = (0\ 1) : C_r = G^{n-r+i}\oplus F_{r-i-1} \longrightarrow D_r = F_{r-i-1}$$

$$\tilde{f} = (0\ 1) : \tilde{C}_r = G_{r-i}\oplus F_{r-i-1} \longrightarrow D_r = F_{r-i-1} \ .$$

Thus $(C,\psi)=(\tilde{C},\tilde{\psi})\in L_{n+2i+1}(\mathbb{A})$. Since $\tilde{\theta}$ and $\tilde{\chi}$ can be chosen independently of θ and χ it follows that the cobordism is independent of these choices also. Given $(F,G,\gamma,\mu,\theta,\chi)$ and chain equivalences $h:F\longrightarrow F'$, $k:G\longrightarrow G'$ it is possible to define $(F',G',\gamma',\mu',\theta',\chi')$ such that the corresponding quadratic Poincaré complex (C',ψ') is homotopy equivalent to (C,ψ), and so $(C',\psi')=(C,\psi)\in L_{n+2i+1}(\mathbb{A})$.

iii) Define an $(n+2i+2)$-dimensional quadratic Poincaré pair $(f:C\longrightarrow D,(\delta\psi,\psi))$ by

$D = H^{n+i+1-*}$,

$f = (0\ j)$:

$$C_r = G^{n-r+i} \oplus F_{r-i-1} \longrightarrow D_r = H^{n+i+1-r} ,$$

$\delta\psi_0 = \zeta$:

$$D^{n+2i+2-r} = H_{r-i-1} \longrightarrow D_r = H^{n+i+1-r} ,$$

$\delta\psi_s = 0$ for $s \geq 1$.

This is a quadratic Poincaré null-cobordism of (C,ψ), so that $(C,\psi) = 0 \in L_{n+2i+1}(\mathbb{A})$.

□

Definition 3.3 For any additive category with involution $\mathbb{A}$ define the generalized Morita maps

$$\mu : L_m(\mathbb{D}_n(\mathbb{A})) \longrightarrow L_{m+n}(\mathbb{A}) \quad (m,n \geq 0)$$

for $m=2i$ (resp. $2i+1$) by sending a nonsingular $(-)^i$-quadratic form (M,θ) (resp. formation (F,G)) in $\mathbb{D}_n(\mathbb{A})$ to the cobordism class of the $(m+n)$-dimensional quadratic Poincaré complex (C,ψ) in $\mathbb{A}$ defined in Proposition 3.1 ii) (resp. 3.2 ii)). The verification that the maps μ are well-defined is contained in Propositions 3.1 iii) (resp. 3.2 iii)).

□

For a ring with involution R apply 3.3 to $\mathbb{A} = \mathbb{B}(R)$ to obtain generalized Morita maps $\mu : L_m(\mathbb{D}_n(R)) \longrightarrow L_{m+n}(R)$ $(m,n \geq 0)$.

§4. The quadratic L-theory transfer

As before, let $\mathbb{A}$ be an additive category with

involution, and let $\mathbb{D}_n(\mathbb{A})$ be the chain homotopy category of n-dimensional chain complexes in $\mathbb{A}$ with the n-duality involution.

__Definition__ __4.1__ The __quadratic__ __L-theory__ __transfer__ __maps__ of a symmetric representation (C,α,U) of a ring with involution R in $\mathbb{D}_n(\mathbb{A})$

$$(C,\alpha,U)^! : L_m(R) \longrightarrow L_{m+n}(\mathbb{A}) \quad (m\geq 0)$$

are the composites

$$(C,\alpha,U)^! : L_m(R) = L_m(\mathbb{B}(R)) \xrightarrow{-\otimes(C,\alpha,U)} L_m(\mathbb{D}_n(\mathbb{A})) \xrightarrow{\mu} L_{m+n}(\mathbb{A})$$

of the maps $-\otimes(C,\alpha,U)$ of 2.10 and the generalized Morita maps μ of 3.3.

□

__Example__ __4.2__ Let $\mathbb{A}$ be the additive category $\mathbb{B}(S)$ of based f.g. free S-modules with the duality involution, for a ring with involution S. The transfer maps determined by an n-dimensional symmetric representation (C,α,U) of a ring with involution R in $\mathbb{D}_n(\mathbb{A})=\mathbb{D}_n(S)$ are morphisms of quadratic L-groups

$$(C,\alpha,U)^! : L_m(R) \longrightarrow L_{m+n}(\mathbb{A}) = L_{m+n}(S) \quad (m,n\geq 0) .$$

□

__Example__ __4.3__ Given a Hurewicz fibration $F \longrightarrow E \xrightarrow{p} B$ with the fibre F a finite n-dimensional geometric Poincaré complex we shall define in §5 below a symmetric representation $(C(\widetilde{F}),\alpha,U)$ of $\mathbf{Z}[\pi_1(B)]$ in $\mathbb{D}_n(\mathbf{Z}[\pi_1(E)])$, with $\widetilde{F}$ the pullback to F of the universal cover $\widetilde{E}$ of E and $\alpha=([F]\cap -)^{-1}:C(\widetilde{F})\longrightarrow C(\widetilde{F})^{n-*}$ the Poincaré duality chain equivalence. The algebraic surgery transfer maps will be defined in §5 to be

$$p^!_{alg} = (C(\tilde{F}),\alpha,U)^! :$$

$$L_m(\mathbb{Z}[\pi_1(B)]) \longrightarrow L_{m+n}(\mathbb{Z}[\pi_1(E)]) \quad (m\geq 0) .$$

In §6 we shall recall the definition via the lifting of normal maps of the geometric surgery transfer maps $p^!_{geo}$, which will be identified with $p^!_{alg}$ in §8.

□

Example 4.4 Given a morphism of rings with involution $f:R\longrightarrow S$ define a symmetric representation (C,α,U) of R in $\mathbb{D}_0(S)=\mathbb{B}(S)$ by

$$\alpha_0 : C_0 = S \longrightarrow C^0 = S^* ;$$

$$s \longrightarrow (t \longrightarrow t\bar{s}) ,$$

$$C_r = 0 \text{ for } r\neq 0 \quad ,$$

$$U = f : R \longrightarrow H_0(\mathrm{Hom}_S(C,C))^{op} = S .$$

In this case the transfer maps are just the change of rings morphisms $(C,\alpha,U)^! = f_! : L_m(R)\longrightarrow L_m(S)$. For $f=1:R\longrightarrow S=R$ (C,α,U) is the universal symmetric representation (2.8) of R in $\mathbb{B}(R)$.

□

Example 4.5 Given a ring with involution S and an integer $k\geq 1$ let $R=M_k(S)$ be the ring of $k\times k$ matrices $(s_{ij})_{1\leq i,j\leq k}$ with entries $s_{ij}\in S$, with the involution

$$^- : R \longrightarrow R ; (s_{ij}) \longrightarrow (\bar{s}_{ji}) .$$

Define a symmetric representation (C,α,U) of R in $\mathbb{D}_0(S)=\mathbb{B}(S)$ by

$$C_0 = \sum_1^k S \quad , \quad C_r = 0 \text{ for } r \neq 0 \ ,$$

$$\alpha : C_0 = \sum_1^k S \longrightarrow C^0 = (\sum_1^k S)^* \ ;$$

$$(s_1, s_2, \ldots, s_k) \longrightarrow$$

$$((t_1, t_2, \ldots, t_k) \longrightarrow t_1\bar{s}_1 + t_2\bar{s}_2 + \ldots + t_k\bar{s}_k) \ ,$$

$$U = 1 : R = M_k(S) \longrightarrow H_0(\mathrm{Hom}_S(C,C))^{op} = M_k(S) \ .$$

The generalized Morita maps $\mu : L_*(R) \longrightarrow L_*(S)$ in this case are just the usual Morita maps, which are isomorphisms for the projective and round L-groups. See Hambleton, Taylor and Williams [5] and Hambleton, Ranicki and Taylor [4] for Morita maps in quadratic L-theory.

□

<u>Example</u> <u>4.6</u> Let $F = \bigcup_k (*) \longrightarrow E \xrightarrow{p} B$ be a k-sheeted finite covering, so that $\pi_1(E)$ is a subgroup of $\pi_1(B)$ of index k. There are evident identifications of spaces

$$\tilde{F} = \pi_1(B) = \bigcup_k \pi_1(E) \subset \tilde{B} = \tilde{E} \ ,$$

and also of $\mathbb{Z}$-module chain complexes

$$C(\tilde{F}) = \mathbb{Z}[\pi_1(B)] = \bigoplus_k \mathbb{Z}[\pi_1(E)] \ .$$

The symmetric representation $(C(\tilde{F}), \alpha, U)$ of $\mathbb{Z}[\pi_1(B)]$ in $\mathbb{D}_0(\mathbb{Z}[\pi_1(E)]) = \mathbb{B}(\mathbb{Z}[\pi_1(E)])$ associated to $p: E \longrightarrow B$ (as in 4.3) is given by

$$U : \mathbb{Z}[\pi_1(B)] = H_0(\mathrm{Hom}_{\mathbb{Z}[\pi_1(B)]}(C(\tilde{F}), C(\tilde{F})))^{op}$$

$$\xrightarrow{\text{restriction}}$$

$$H_0(\mathrm{Hom}_{\mathbb{Z}[\pi_1(E)]}(C(\tilde{F}),C(\tilde{F})))^{op} = M_k(\mathbb{Z}[\pi_1(E)]) ,$$

$$\alpha = \underset{k}{\oplus}1 : C(\tilde{F}) = \underset{k}{\oplus}\mathbb{Z}[\pi_1(E)] \longrightarrow$$

$$\mathrm{Hom}_{\mathbb{Z}[\pi_1(E)]}(C(\tilde{F}),\mathbb{Z}[\pi_1(E)]) = \underset{k}{\oplus}\mathbb{Z}[\pi_1(E)]^* .$$

The algebraic transfer maps in this case are the composites

$$p^!_{alg} : L_m(\mathbb{Z}[\pi_1(B)]) \xrightarrow{U_!}$$

$$L_m(M_k(\mathbb{Z}[\pi_1(E)])) \xrightarrow{\mu} L_m(\mathbb{Z}[\pi_1(E)])$$

with $U_!$ induced by U as in 2.5 and μ the Morita maps of 4.5. In this case $p^!_{alg}$ can be described more directly by the restrictions of $\mathbb{Z}[\pi_1(B)]$-module actions to $\mathbb{Z}[\pi_1(E)]$-module actions, and it is clear that $p^!_{alg}=p^!_{geo}$.

□

<u>Example</u> <u>4.7</u> The algebraic S^1-bundle transfer maps of Munkholm and Pedersen [10] and Ranicki [16,§7.8] $p^!_{alg}:L_m(R)\longrightarrow L_{m+1}(S)$ are defined for any ring with involution S, with $R=S/(t-1)$ for a central element $t\in S$ such that $\bar{t}=t^{-1}$. (We are only dealing with the orientable case here). From our point of view these are the quadratic L-theory transfer maps $p^!_{alg}=(C,\alpha,U)^!$ of 4.1 with (C,α,U) the symmetric representation of R in $\mathbb{D}_1(S)$ given by

$$d = 1-t : C_1 = S \longrightarrow C_0 = S ,$$

$$\alpha = \begin{cases} -t : C_1 = S \longrightarrow C^0 = S \\ 1 : C_0 = S \longrightarrow C^1 = S . \end{cases}$$

For an S^1-bundle $S^1 \longrightarrow E \xrightarrow{p} B$ one takes $R=\mathbb{Z}[\pi_1(B)]$, $S=\mathbb{Z}[\pi_1(E)]$, $t=\text{fibre}\in\pi_1(E)$.

□

§5. The algebraic surgery transfer

A map $p:E\longrightarrow B$ of connected spaces with homotopy fibre of the homotopy type of a finite (or finitely dominated) CW complex F determines a representation of $\mathbb{Z}[\pi_1(B)]$ in $\mathbb{D}(\mathbb{Z}[\pi_1(E)])$

$$(C(\tilde{F}),U:\mathbb{Z}[\pi_1(B)]\longrightarrow H_0(\mathrm{Hom}_{\mathbb{Z}[\pi_1(E)]}(C(\tilde{F}),C(\tilde{F})))^{op})$$

as in 1.7. We shall now show that if F is a finite n-dimensional geometric Poincaré complex then for any choice of orientation map $w(B):\pi_1(B)\longrightarrow\mathbb{Z}_2$ in the base there is defined a symmetric representation $(C(\tilde{F}),\alpha,U)$ of $\mathbb{Z}[\pi_1(B)]$ in $\mathbb{D}_n(\mathbb{Z}[\pi_1(E)])$, and hence obtain from §4 quadratic L-theory transfer maps

$$p^!_{alg} = (C(\tilde{F}),\alpha,U)^! :$$

$$L_m(\mathbb{Z}[\pi_1(B)]) \longrightarrow L_{m+n}(\mathbb{Z}[\pi_1(E)]) \quad (m\geqslant 0) .$$

In §8 below we shall identify these algebraic surgery transfer maps with the geometric surgery transfer maps.

There is no loss of generality in assuming that $F\longrightarrow E\xrightarrow{p} B$ is a Hurewicz fibration with the fibre $F=p^{-1}(*)$ a finite CW complex F. If F is disconnected then $p:E\longrightarrow B$ is the composite of a Hurewicz fibration

$p':E \longrightarrow B'$ with connected fibre $p'^{-1}(*)$ and a finite covering $B' \longrightarrow B$. Since transfer theory is well-known for finite covers (cf. 4.6) there is no loss of generality in taking F to be connected. In fact, the algebraic transfer maps are defined in exactly the same way for disconnected F, and only the geometric treatment of the orientation maps has to be modified by using groupoids instead of groups.

Transport of the fibre along paths in the base space gives a map $\Omega B \longrightarrow F^F$ which on π_0 induces a group morphism $U:\pi_1(B) \longrightarrow [F,F]$ to the monoid of homotopy classes of self-maps of F (Whitehead [24,p.186]). Analogously, one has the pointed transport of the pointed fibre along paths in E, defining a morphism $U^+:\pi_1(E) \longrightarrow [F,F]^+$ to the monoid of pointed homotopy classes of pointed self-maps of F. Homotopy along a path defines a morphism $\pi_1(F) \longrightarrow [F,F]^+$ (Whitehead [24,p.98ff]).

<u>Proposition</u> <u>5.1</u> The transport maps define a morphism from an exact sequence of groups to an exact sequence of pointed sets

$$\begin{array}{ccccccc}
\pi_1(F) & \longrightarrow & \pi_1(E) & \xrightarrow{p_*} & \pi_1(B) & \longrightarrow & \{1\} \\
\| & & \downarrow U^+ & & \downarrow U & & \\
\pi_1(F) & \longrightarrow & [F,F]^+ & \longrightarrow & [F,F] & \longrightarrow & \{1\} .
\end{array}$$

□

We shall now use 5.1 in the case when F is a geometric Poincaré complex to lift an orientation map $w(B)$ for $\pi_1(B)$ to an orientation map $w(E)$ for $\pi_1(E)$.

__Definition 5.2__ An __orientation map__ for a group π is a morphism $w:\pi \longrightarrow \mathbb{Z}_2=\{\pm 1\}$. Let $\mathbb{Z}[\pi]^w$ denote the ring $\mathbb{Z}[\pi]$ with the __w-twisted involution__

$$\bar{\ } : \mathbb{Z}[\pi] \longrightarrow \mathbb{Z}[\pi] \ ; \ \sum_{g\in\pi} n_g g \longrightarrow \sum_{g\in\pi} n_g w(g) g^{-1} \ .$$

Given a chain complex C in $\mathbb{B}(\mathbb{Z}[\pi])$ let ${}^wC^{n-*}$ denote the n-dual chain complex C^{n-*} in $\mathbb{B}(\mathbb{Z}[\pi])$ defined using the w-twisted involution on $\mathbb{Z}[\pi]$. If w is trivial ${}^wC^{n-*}$ is written as C^{n-*}. Let $\mathbb{Z}^w$ denote the right $\mathbb{Z}[\pi]$-module with additive group $\mathbb{Z}$ and

$$\mathbb{Z}^w \times \mathbb{Z}[\pi] \longrightarrow \mathbb{Z}^w \ ; \ (m, \sum_{g\in\pi} n_g g) \longrightarrow m(\sum_{g\in\pi} w(g) n_g) \ .$$

Let ${}^w\mathbb{Z}$ denote the left $\mathbb{Z}[\pi]$-module defined in the same way.

□

When w is clear we abbreviate $\mathbb{Z}[\pi]^w$ to $\mathbb{Z}[\pi]$.

An n-dimensional geometric Poincaré complex X is a (connected) finite CW complex together with an orientation map $w(X):\pi_1(X) \longrightarrow \mathbb{Z}_2$ and a fundamental class

$$[X] \in H_n(X;\mathbb{Z}^{w(X)}) = H_n(\mathbb{Z}^w \otimes_{\mathbb{Z}[\pi_1(X)]} C(\tilde{X}))$$

such that the $\mathbb{Z}[\pi_1(X)]$-module chain map $[X]\cap - : {}^{w(X)}C(\tilde{X})^{n-*} \longrightarrow C(\tilde{X})$ is a chain equivalence, with $\tilde{X}$ the universal cover. See Wall [21] for the general theory.

The orientation map $w=w(X):\pi=\pi_1(X) \longrightarrow \mathbb{Z}_2$ of an n-dimensional geometric Poincaré complex X is determined by the topology of X, since the cap product with a fundamental class $[X]\in H_n(X;\mathbb{Z}^w)$ defines an isomorphism of $\mathbb{Z}[\pi]$-modules

$$[X]\cap - \; : \; H_0({}^wC(\tilde{X})^{n-*}) \longrightarrow H_0(\tilde{X}) = \mathbb{Z} \; .$$

If $H^n(\tilde{X})$ is defined to be $H_0(C(\tilde{X})^{n-*})$ using the untwisted involution $(\bar{g}=g^{-1})$ on $\mathbb{Z}[\pi]$ then we get $H^n(\tilde{X})\cong{}^w\mathbb{Z}$.

<u>Definition</u> <u>5.3</u> Let X be an n-dimensional geometric Poincaré complex.

i) The <u>degree</u> of a pointed self-map $f:X\longrightarrow X$ is the number $d(f)\in\mathbb{Z}$ such that

$$\tilde{f}^* \; : \; H^n(\tilde{X}) \longrightarrow H^n(\tilde{X}) \; ; \; 1 \longrightarrow d(f) \; ,$$

with $\tilde{f}:\tilde{X}\longrightarrow\tilde{X}$ a lift of f to a self map of the universal cover $\tilde{X}$.

ii) The <u>homotopy</u> <u>orientation</u> of X is the monoid morphism

$$\hat{w} = \hat{w}(X) \; : \; [X,X]^+ \longrightarrow \mathbb{Z}^\times \; ; \; f \longrightarrow d(f) \; ,$$

with $\mathbb{Z}^\times$ the monoid defined by $\mathbb{Z}$ and multiplication.

□

Let $f:X\longrightarrow X$ be a pointed self homotopy equivalence, inducing an automorphism $f_*:\pi\longrightarrow\pi$ of the fundamental group $\pi=\pi_1(X)$. A lift $\tilde{f}:\tilde{X}\longrightarrow\tilde{X}$ of f to the universal cover $\tilde{X}$ induces a $\mathbb{Z}$-module chain equivalence $\tilde{f}:C(\tilde{X})\longrightarrow C(\tilde{X})$ which is f_*-equivariant

$$\tilde{f}(gx) = f_*(g)(x) \in C(\tilde{X}) \quad (g\in\pi, x\in C(\tilde{X})) \ .$$

The induced isomorphism of additive groups $\tilde{f}^*: H^n(\tilde{X}) = {}^w\mathbb{Z} \longrightarrow H^n(\tilde{X}) = {}^w\mathbb{Z}$ is also f_*-equivariant. Hence we have

$$w = wf_* : \pi \xrightarrow{f_*} \pi \xrightarrow{w} \mathbb{Z}_2$$

and f_* defines an automorphism $f_*: \mathbb{Z}[\pi]^w \longrightarrow \mathbb{Z}[\pi]^w$ of the ring with involution $\mathbb{Z}[\pi]^w$. The $\mathbb{Z}$-module automorphism $f_*: H_n(X;\mathbb{Z}^w) = \mathbb{Z} \longrightarrow H_n(X;\mathbb{Z}^w) = \mathbb{Z}$ is such that $f_*([X]) = d(f)[X]$, with $d(f) = \hat{w}(f) \in \langle \pm 1\rangle = \mathbb{Z}_2 \subset \mathbb{Z}^\times$. In particular, it follows that the orientation map w and the homotopy orientation $\hat{w}$ are related by a commutative diagram of monoid morphisms

$$\begin{array}{ccc} \pi_1(X) & \longrightarrow & [X,X]^+ \\ {\scriptstyle w}\big\downarrow & & \big\downarrow{\scriptstyle \hat{w}} \\ \langle \pm 1\rangle & \longrightarrow & \mathbb{Z}^\times \end{array} \ .$$

<u>Proposition</u> <u>5.4</u> For any pointed self homotopy equivalence $f: X \longrightarrow X$ there is defined a chain homotopy commutative diagram of $\mathbb{Z}$-module chain complexes and chain equivalences

$$\begin{array}{ccc} {}^wC(\tilde{X})^{n-*} & \xrightarrow{d(f)(\tilde{f}^{-1})^*} & {}^wC(\tilde{X})^{n-*} \\ {\scriptstyle [X]\cap -}\big\downarrow & & \big\downarrow{\scriptstyle [X]\cap -} \\ C(\tilde{X}) & \xrightarrow{\tilde{f}} & C(\tilde{X}) \end{array}$$

with the horizontal chain maps f_*-equivariant, and the

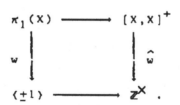

vertical chain maps $\pi_1(X)$-equivariant.

□

<u>Definition 5.5</u> An <u>n-dimensional Poincaré fibration</u> $F \longrightarrow E \xrightarrow{p} B$ is a Hurewicz fibration with the fibre F an n-dimensional geometric Poincaré complex, together with an orientation map $w(B):\pi_1(B) \longrightarrow \mathbb{Z}_2$. The <u>lift</u> of w(B) is the orientation map

$$p^! w(B) = w(E) \ : \ \pi_1(E) \longrightarrow \mathbb{Z}_2 \ ;$$

$$g \longrightarrow w(B)(p_*(g)).\hat{w}(F)(U^+(g))$$

with U^+ as in 5.1 and $\hat{w}$ as in 5.3.

□

<u>Proposition 5.6</u> An n-dimensional Poincaré fibration $F \longrightarrow E \xrightarrow{p} B$ determines a symmetric representation $(C(\tilde{F}),\alpha,U)$ of $\mathbb{Z}[\pi_1(B)]^{w(B)}$ in $\mathbb{D}_n(\mathbb{Z}[\pi_1(E)]^{w(E)})$ with $\alpha=([F]\cap-)^{-1}:C(\tilde{F})\longrightarrow C(\tilde{F})^{n-*}$ the Poincaré duality chain equivalence and $(C(\tilde{F}),U)$ the representation of $\mathbb{Z}[\pi_1(B)]$ in $\mathbb{D}_n(\mathbb{Z}[\pi_1(E)])$ associated to p.

<u>Proof</u>: We have to show that

$$U \ : \ \mathbb{Z}[\pi_1(B)]^{w(B)} \longrightarrow H_0(\mathrm{Hom}_{\mathbb{Z}[\pi_1(E)]}(C(\tilde{F}),C(\tilde{F})))^{op}$$

is a morphism of rings with involution, or equivalently that for every $g\in\pi_1(B)$ there is defined a chain homotopy commutative diagram of $\mathbb{Z}[\pi_1(E)]$-module chain complexes

$$\begin{array}{ccc} w(E)_{C(\tilde{F})^{n-*}} & \xrightarrow{U(g)^*} & w(E)_{C(\tilde{F})^{n-*}} \\ {\scriptstyle [F]\cap -}\downarrow & & \downarrow{\scriptstyle [F]\cap -} \\ C(\tilde{F}) & \xrightarrow{w(B)(g)U(g^{-1})} & C(\tilde{F}) \end{array} .$$

This follows from 5.4 and the $\pi_1(E)$-equivariant transport along the fibre $U:\pi_1(B) \longrightarrow [\tilde{F},\tilde{F}]_{\pi_1(E)}$ used to define the ring morphism U in Lück [7].

□

<u>Definition 5.7</u> The <u>algebraic</u> <u>surgery</u> <u>transfer</u> <u>maps</u> of an n-dimensional Poincaré fibration $F \longrightarrow E \xrightarrow{p} B$ are the quadratic L-theory transfer maps of 4.1 associated to the symmetric representation $(C(\tilde{F}),\alpha,U)$ of $\mathbb{Z}[\pi_1(B)]^{w(B)}$ in $\mathbb{D}_n(\mathbb{Z}[\pi_1(E)]^{w(E)})$ given by 5.6

$$p^!_{alg} = (C(\tilde{F}),\alpha,U)^! :$$

$$L_m(\mathbb{Z}[\pi_1(B)]) \longrightarrow L_{m+n}(\mathbb{Z}[\pi_1(E)]) \quad (m \geqslant 0) .$$

□

By definition, the algebraic surgery transfer maps are the composites

$$p^!_{alg} : L_m(\mathbb{Z}[\pi_1(B)]) \xrightarrow{(p^\#)_!} L_m(\mathbb{D}_n(\mathbb{Z}[\pi_1(E)]))$$

$$\xrightarrow{\mu} L_{m+n}(\mathbb{Z}[\pi_1(E)]) \quad (m \geqslant 0)$$

of the maps induced as in §2 by the functor of additive categories with involution

$$p^{\#} = -\otimes(C(\tilde{F}),\alpha,U) : \mathbb{B}(\mathbf{Z}[\pi_1(B)]) \longrightarrow \mathbb{D}_n(\mathbf{Z}[\pi_1(E)])$$

and the generalized Morita maps μ of §3.

§6. The geometric surgery transfer

Wall [22] defined the rel∂ surgery obstruction $\sigma_*(f,b)\in L_m(\mathbf{Z}[\pi_1(X)])$ for a normal map $(f,b):(M,\partial M)\longrightarrow(X,\partial X)$ from a compact m-manifold with boundary $(M,\partial M)$ to a finite m-dimensional geometric Poincaré pair $(X,\partial X)$ with $\partial f=f|:\partial M\longrightarrow\partial X$ a homotopy equivalence, and $b:\nu_M\longrightarrow\tilde{\nu}_X$ a map from the stable normal bundle of M to a topological reduction of the Spivak normal fibration ν_X of X, with the w(X)-twisted involution on $\mathbf{Z}[\pi_1(X)]$. The surgery obstruction has the property that $\sigma_*(f,b)=0$ if (and for $m\geqslant 5$ only if) (f,b) is normal bordant rel∂ to a homotopy equivalence of pairs. Given a connected space B with finitely presented $\pi_1(B)$, and given an orientation map $w(B):\pi_1(B)\longrightarrow\mathbf{Z}_2$, it is possible to realize every element $x\in L_m(\mathbf{Z}[\pi_1(B)])$ $(m\geqslant 5)$ as the surgery obstruction of an m-dimensional normal map $(f,b):(M,\partial M)\longrightarrow(X,\partial X)$ with a π_1-isomorphism reference map $X\longrightarrow B$ and orientation map $w(X):\pi_1(X)\longrightarrow\pi_1(B)\xrightarrow{w(B)}\mathbf{Z}_2$

$$x = \sigma_*(f,b) \in L_m(\mathbf{Z}[\pi_1(B)]) \ .$$

The total space E of an n-dimensional Poincaré fibration $F\longrightarrow E\xrightarrow{p}B$ over an m-dimensional geometric Poincaré complex B is homotopy equivalent to an (m+n)-dimensional geometric Poincaré complex, with the orientation map the lift $w(E)=p^!w(B):\pi_1(E)\longrightarrow\mathbf{Z}_2$ in the sense of 5.5 of the orientation map $w(B):\pi_1(B)\longrightarrow\mathbf{Z}_2$ (Quinn [12], Gottlieb [2]).

Quinn [11] used the realization theorem for

surgery obstructions to define geometric transfer maps in the quadratic L-groups for a fibre bundle (or even a block fibration) $F \longrightarrow E \xrightarrow{p} B$ with the fibre F a compact n-manifold

$$p^!_{geo} : L_m(\mathbb{Z}[\pi_1(B)]) \longrightarrow L_{m+n}(\mathbb{Z}[\pi_1(E)]) ;$$

$$\sigma_*((f,b):(M,\partial M)\longrightarrow(X,\partial X))$$

$$\longrightarrow \sigma_*((g,c):(N,\partial N)\longrightarrow(Y,\partial Y)) .$$

Here, $(g,c):(N,\partial N)\longrightarrow(Y,\partial Y)$ the $(m+n)$-dimensional normal map equipped with a reference map $Y\longrightarrow E$ obtained from the n-dimensional normal map $(f,b):M\longrightarrow X$ by the pullback of p along a reference map $X\longrightarrow B$.

The surgery obstruction of Wall [22] was defined using geometric intersection numbers on the homology remaining after surgery below the middle dimension. The theory of Ranicki [14],[15] associates an invariant in $L_m(\mathbb{Z}[\pi_1(X)])$ to a normal map $(f,b):(M,\partial M)\longrightarrow(X,\partial X)$ of m-dimensional geometric Poincaré pairs, with $b:\nu_M\longrightarrow\nu_X$ a map of the Spivak normal fibrations and $\partial f:\partial M\longrightarrow\partial X$ a homotopy equivalence. The quadratic kernel of (f,b) is an m-dimensional quadratic Poincaré complex $(C(f^!),\psi)$ over $\mathbb{Z}[\pi_1(X)]$. Here, $C(f^!)$ is the algebraic mapping cone of the Umkehr $\mathbb{Z}[\pi_1(X)]$-module chain map

$$f^! : C(\tilde{X},\partial\tilde{X}) \xrightarrow{([X]\cap-)^{-1}} C(\tilde{X})^{m-*} \xrightarrow{\tilde{f}^*} C(\tilde{M})^{m-*}$$

$$\xrightarrow{[M]\cap-} C(\tilde{M},\partial\tilde{M})$$

with $\tilde{X}$ the universal cover of X, $\tilde{f}:\tilde{M}\longrightarrow\tilde{X}$ a $\pi_1(X)$-equivariant lift of f to the pullback cover $\tilde{M}=f^*\tilde{X}$ of M. The Poincaré duality chain equivalence is given up to chain homotopy by the composite

$$(1+T)\psi_0 \ : \ C(f^!)^{m-*} \xrightarrow{e^*} C(\tilde{M},\partial\tilde{M})^{m-*}$$

$$\xrightarrow{[M]\cap -} C(\tilde{M}) \longrightarrow C(\tilde{M},\partial\tilde{M}) \xrightarrow{e} C(f^!)$$

with $e:C(\tilde{M},\partial\tilde{M}) \longrightarrow C(f^!)$ the inclusion. The quadratic signature of (f,b) is the cobordism class

$$\sigma_*(f,b) = (C(f^!),\psi) \in L_m(\mathbb{Z}[\pi_1(X)]) \ .$$

A normal map from a manifold to a geometric Poincaré complex determines a normal map of geometric Poincaré complexes with quadratic signature the surgery obstruction.

<u>Definition 6.1</u> The <u>geometric</u> <u>surgery</u> <u>transfer</u> <u>maps</u> of an n-dimensional Poincaré fibration $F \longrightarrow E \xrightarrow{p} B$ with finitely presented $\pi_1(B)$

$$p^!_{geo} \ : \ L_m(\mathbb{Z}[\pi_1(B)]) \longrightarrow L_{m+n}(\mathbb{Z}[\pi_1(E)]) \ ;$$

$$\sigma_*((f,b):M \longrightarrow X) \longrightarrow \sigma_*((g,c):N \longrightarrow Y) \quad (m \geqslant 5)$$

are defined using the quadratic signature of normal maps of geometric Poincaré complexes. Here, $(g,c):N \longrightarrow Y$ is the $(m+n)$-dimensional normal map obtained from an m-dimensional normal map $(f,b):M \longrightarrow X$ by the pullback of p along a reference map $X \longrightarrow B$.

□

<u>Theorem</u> <u>6.2</u> The geometric surgery transfer maps of an n-dimensional Poincaré fibration $F \longrightarrow E \xrightarrow{p} B$ coincide with the algebraic surgery transfer maps

$p^!_{geo} = p^!_{alg}$:

$$L_m(\mathbb{Z}[\pi_1(B)]) \longrightarrow L_{m+n}(\mathbb{Z}[\pi_1(E)]) \quad (m\geq 5) .$$

□

The proof of 6.2 is deferred to §8. The ideal proof would express the quadratic kernel of the pullback normal map of the total (m+n)-dimensional geometric Poincaré complexes $(g,c):N\longrightarrow Y$ as a twisted tensor product of the quadratic kernel of the normal map of the base m-dimensional geometric Poincaré complexes $(f,b):M\longrightarrow X$ and the symmetric Poincaré complex $(C(\tilde{F}),\phi)$. This would generalize the chain level proof of the surgery product formula in Ranicki [15] in the untwisted case p=projection:$E=B\times F\longrightarrow B$

$$\sigma_*((f,b)\times 1:M\times F\longrightarrow X\times F) = \sigma_*(f,b)\otimes\sigma^*(F)$$

$$\in L_{m+n}(\mathbb{Z}[\pi_1(B)\times\pi_1(F)])$$

which expressed the quadratic signature of a product $(f,b)\times 1$ as the tensor product of the quadratic signature of (f,b) and the symmetric signature $\sigma^*(F)=(C(\tilde{F}),\phi)\in L^n(\mathbb{Z}[\pi_1(F)])$. However, this would require the development of a fair amount of new technology, translating the homotopy action of ΩB on the geometric Poincaré complex F into a chain homotopy action of $C(\Omega B)$ on the symmetric Poincaré complex $(C(\tilde{F}),\phi)$ over $\mathbb{Z}[\pi_1(E)]$. For the purpose at hand we can assume by the realization theorem that the m-dimensional normal map $(f,b):M\longrightarrow X$ is $[(m-2)/2]$-connected. In the highly-connected case we can give a chain level geometric interpretation of both the element $U_!\sigma_*(f,b)\in L_m(\mathbb{D}_n(\mathbb{Z}[\pi_1(E)]))$ and its image under the generalized Morita map $\mu: L_m(\mathbb{D}_n(\mathbf{Z}[\pi_1(E)])) \longrightarrow L_{m+n}(\mathbf{Z}[\pi_1(E)])$. For a fibre bundle $F \longrightarrow E \xrightarrow{p} B$ it is possible to dispense with some of the algebra, using instead the fibred intersection theory of Hatcher and Quinn [6] as outlined in Appendix 1 below.

§7. Ultraquadratic L-theory

Ultraquadratic L-theory was developed in §7.8 of Ranicki [16] in connection with the algebraic theory of codimension 2 surgery. We use it here to recognize quadratic Poincaré complexes in the image of the generalized Morita maps $\mu: L_m(\mathbb{D}_n(\mathbb{A})) \longrightarrow L_{m+n}(\mathbb{A})$ of §3, providing a tool for the identification in §8 below of the algebraic and geometric surgery transfer maps.

Let $\mathbb{A}$ be an additive category with involution. As in Ranicki [15],[19] define for any finite chain complex C in $\mathbb{A}$ and $\epsilon = \pm 1$ the $\mathbf{Z}$-module chain complex

$$W_\% C = W \otimes_{\mathbf{Z}[\mathbf{Z}_2]} \mathrm{Hom}_{\mathbb{A}}(C^*, C) ,$$

with the generator $T \in \mathbf{Z}_2$ acting on $\mathrm{Hom}_{\mathbb{A}}(C^*, C)$ by the ϵ-transposition involution $T_\epsilon = \epsilon T$ and W the standard free $\mathbf{Z}[\mathbf{Z}_2]$-module resolution of $\mathbf{Z}$

$$W : \ldots \longrightarrow \mathbf{Z}[\mathbf{Z}_2] \xrightarrow{1-T} \mathbf{Z}[\mathbf{Z}_2] \xrightarrow{1+T} \mathbf{Z}[\mathbf{Z}_2] \xrightarrow{1-T} \mathbf{Z}[\mathbf{Z}_2] .$$

An m-chain $\psi \in (W_\% C)_m$ is a collection of morphisms

$$\psi = \{\psi_s \in \mathrm{Hom}_{\mathbb{A}}(C^*, C)_{m-s} \mid s \geqslant 0\}$$

such that for a cycle there is defined a chain map

$(1+T_\epsilon)\psi_0:C^{m-*}\longrightarrow C$. An m-dimensional ϵ-quadratic (Poincaré) complex (C,ψ) in $\mathbb{A}$ is an m-dimensional chain complex C in $\mathbb{A}$ together with an element $\psi\in Q_m(C,\epsilon)=H_m(W_\%C)$ (such that $(1+T_\epsilon)\psi_0:C^{m-*}\longrightarrow C$ is a chain equivalence). The skew-suspension isomorphisms

$$\bar{S}\ :\ Q_m(C,\epsilon)\longrightarrow Q_{m+2}(SC,-\epsilon)\ ;\ \psi\longrightarrow \bar{S}\psi$$

are defined by $(\bar{S}\psi)_s=\pm\psi_s$ $(s\geqslant 0)$, for any finite chain complex C in $\mathbb{A}$. The skew-suspension maps $\bar{S}:L_m(\mathbb{A},\epsilon)\longrightarrow L_{m+2}(\mathbb{A},-\epsilon)$ $(m\geqslant 0)$ in the $\pm\epsilon$-quadratic L-groups are also isomorphisms, so that

$$L_m(\mathbb{A},\epsilon)=L_{m+2}(\mathbb{A},-\epsilon)=L_{m+4}(\mathbb{A},\epsilon)\quad (m\geqslant 0)\ .$$

For $\epsilon=1$ we write $Q_m(C,1)=Q_m(C)$, $L_m(\mathbb{A},1)=L_m(\mathbb{A})$, and 1-quadratic = quadratic.

Ultraquadratic complexes are ϵ-quadratic complexes (C,ψ) with $\psi_s=0$ for $s\geqslant 1$.

For any finite chain complex C in $\mathbb{A}$ define the abelian group

$$\hat{Q}_m(C)=H_m(\mathrm{Hom}_{\mathbb{A}}(C^*,C))=H_0(\mathrm{Hom}_{\mathbb{A}}(C^{m-*},C))$$

of chain homotopy classes of chain maps $\hat{\psi}:C^{m-*}\longrightarrow C$.

<u>Definition</u> <u>7.1</u> An <u>m-dimensional</u> ϵ<u>-ultraquadratic</u> <u>(Poincaré)</u> <u>complex</u> <u>in</u> $\mathbb{A}$ $(C,\hat{\psi})$ is an m-dimensional chain complex C in $\mathbb{A}$ together with an element $\hat{\psi}\in\hat{Q}_m(C)$ (such that $(1+T_\epsilon)\hat{\psi}:C^{m-*}\longrightarrow C$ is a chain equivalence).

□

There is a corresponding notion of cobordism of

ϵ-ultraquadratic Poincaré complexes in $\mathbb{A}$, with the m-dimensional cobordism group denoted by $\hat{L}_m(\mathbb{A},\epsilon)$, and by $\hat{L}_m(\mathbb{A})$ for $\epsilon=+1$. The ϵ-ultraquadratic L-groups are 4-periodic, with

$$\hat{L}_m(\mathbb{A},\epsilon) = \hat{L}_{m+2}(\mathbb{A},-\epsilon) = \hat{L}_{m+4}(\mathbb{A},\epsilon) \quad (m\geq 0)$$

by skew-suspension isomorphisms, just like for the ϵ-quadratic L-groups L_*. If $\mathbb{A}=\mathbb{B}(R)$ for a ring with involution R we write $\hat{L}_m(\mathbb{A})$ as $\hat{L}_m(R)$.

Define a map $\hat{Q}_m(C)\longrightarrow Q_m(C,\epsilon);\hat{\psi}\longrightarrow\psi$ by $\psi_0=\hat{\psi}$, $\psi_s=0$ $(s\geq 1)$. An m-dimensional ϵ-ultraquadratic (Poincaré) complex $(C,\hat{\psi})$ determines an m-dimensional quadratic (Poincaré) complex (C,ψ). The forgetful maps in the cobordism groups

$$\hat{L}_m(\mathbb{A},\epsilon) \longrightarrow L_m(\mathbb{A},\epsilon) \;;\; (C,\hat{\psi}) \longrightarrow (C,\psi) \quad (m\geq 0)$$

are surjective for even m and injective for odd m.

The ultraquadratic L-group $\hat{L}_m(\mathbb{Z})$ was identified in §7.8 of [16] with the cobordism group C_{m-1} of knots $k:S^{m-1}\subset S^{m+1}$ $(m\geq 4)$. A Seifert surface for a knot $k:S^{m-1}\subset S^{m+1}$ is a codimension 1 framed submanifold $M^m\subset S^{m+1}$ with boundary $\partial M=k(S^{m-1})$. Inclusion defines an m-dimensional normal map $(f,b):(M,\partial M)\longrightarrow(D^{m+2},S^{m-1})$ with quadratic kernel $\sigma_*(f,b)=(C,\psi)$ such that $H_*(C)=H_{*+1}(D^{m+2},M)=\tilde{H}_*(M)$. The framing determines a map $M\longrightarrow S^{m+1}-M$ which induces a chain map $\hat{\psi}:C^{m-*}\longrightarrow C$, defining an m-dimensional ultraquadratic Poincaré complex $(C,\hat{\psi})$ over $\mathbb{Z}$. The knot complement $U=S^{m+1}-$(open nbhd. of $k(S^{m-1})$) has boundary $\partial U=S^{m-1}\times S^1$, and there is defined an $(m+1)$-dimensional normal map $(U,\partial U)\longrightarrow(D^{m+2},S^{m-1})\times S^1$ which is a $\mathbb{Z}$-homology equivalence. Let $(L^{m+1};M^m,zM^m)$ be the fundamental domain for the infinite cyclic cover $\bar{U}$ of U obtained by cutting U along M, and let

$$((e;f,zf),(a;b,zb)) :$$

$$(L^{m+1};M^m,zM^m) \longrightarrow D^{m+2}\times([0,1];\{0\},\{1\})$$

be the corresponding $(m+1)$-dimensional normal map of triads. The inclusions $j:M\longrightarrow L$, $k:zM\longrightarrow L$ induce $\mathbb{Z}$-module chain maps $j,k:C=C(f^!)\longrightarrow D=C(g^!)$ such that $j-k:C\longrightarrow D$ is a chain equivalence. The ultraquadratic structure $\hat{\psi}\in\hat{Q}_m(C)$ is determined by the symmetric structure $(1+T)\hat{\psi}:C^{m-*}\longrightarrow C$ and j,k, since up to chain homotopy

$$(j-k)^{-1}j = \hat{\psi}((1+T)\hat{\psi})^{-1} : C \longrightarrow C ,$$

$$(j-k)^{-1}k = -T\hat{\psi}((1+T)\hat{\psi})^{-1} : C \longrightarrow C .$$

More generally:

<u>Proposition</u> <u>7.2</u> Let (C,ψ) be an m-dimensional ϵ-quadratic Poincaré complex in $\mathbb{A}$. A cobordism $((j\ k):C\oplus C\longrightarrow D,(\delta\psi,\psi\oplus-\psi))$ with $j-k:C\longrightarrow D$ a chain equivalence determines an ϵ-ultraquadratic structure $\hat{\psi}\in\hat{Q}_m(C)$ with image $\psi\in Q_m(C,\epsilon)$, such that

$$(C,\psi) = \mu(C^{m-*},\hat{\psi}) \in \mathrm{im}(\mu:L_0(\mathbb{D}_m(\mathbb{A}),\epsilon)\longrightarrow L_m(\mathbb{A},\epsilon))$$

with $(C^{m-*},\hat{\psi})$ a nonsingular ϵ-quadratic form in $\mathbb{D}_m(\mathbb{A})$.

<u>Proof</u>: Define a morphism in $\mathbb{D}_m(\mathbb{A})$

$$h = (j-k)^{-1}j : C \xrightarrow{\ j\ } D \xrightarrow{\ (j-k)^{-1}\ } C .$$

By the chain homotopy invariance of the Q-groups we can replace $((j\ k),(\delta\psi,\psi\oplus-\psi))$ by a homotopy equivalent cobordism $((h\ h-1):C\oplus C \longrightarrow C, (\delta\psi, \psi\oplus -\psi) \in Q_{m+1}((h\ h-1),\epsilon))$. On the chain level

$$h_{\%}(\psi) - (h-1)_{\%}(\psi) = d(\delta\psi) \in (W_{\%}C)_m ,$$

so that there is defined a chain homotopy

$$(1+T_\epsilon)\delta\psi_0 : h(1+T_\epsilon)\psi_0 \simeq (1+T_\epsilon)\psi_0(1-h^*) :$$

$$C^{m-*} \longrightarrow C .$$

The m-dimensional ϵ-ultraquadratic Poincaré complex $(C,\hat{\psi})$ in $\mathbb{A}$ defined by the chain map

$$\hat{\psi} = h(1+T_\epsilon)\psi_0 : C^{m-*} \xrightarrow{(1+T_\epsilon)\psi_0} C \xrightarrow{h} C$$

is such that $\hat{\psi}+\epsilon\hat{\psi}^* \simeq (1+T_\epsilon)\psi_0 : C^{m-*} \longrightarrow C$. Define a chain $\chi \in (W_{\%}C)_{m+1}$ such that $\hat{\psi}-\psi = d(\chi+\delta\psi) \in (W_{\%}C)_m$ by

$$\chi_s = \begin{cases} 0 & \text{if } s=0 \\ hT_\epsilon\psi_{s-1} & \text{if } s\geq 1 \end{cases} : C^{m+1-r-s} \longrightarrow C_r .$$

Thus $\psi = \hat{\psi} \in Q_m(C,\epsilon)$ and

$$(C,\psi) = (C,\hat{\psi}) = \mu(C^{m-*},\hat{\psi}) \in L_m(\mathbb{A},\epsilon) .$$

□

<u>Corollary</u> <u>7.3</u> Let $(f,b):M \longrightarrow X$ be an $(i-1)$-connected normal map of $(n+2i)$-dimensional geometric Poincaré complexes, and let

$$((e;f,zf),(a;b,zb)) : (L;M,zM) \longrightarrow X\times([0,1];\{0\},\{1\})$$

be an $(i-1)$-connected normal bordism between (f,b) and a disjoint copy (zf,zb). If the $(i-1)$-connected normal map of $(n+2i+1)$-dimensional geometric Poincaré complexes

$$(e/(f=zf),a/(b=zb)) :$$

$$L/(M=zM) \longrightarrow X\times([0,1]/0=1) = X\times S^1$$

is a $\mathbf{Z}[\pi_1(X)]$-homology equivalence then the bordism determines an $(-)^i$-ultraquadratic structure $\hat{\psi}\in\hat{Q}_n(S^{-i}C(f^!))$ with image the quadratic kernel structure $\psi\in Q_n(S^{-i}C(f^!),(-)^i)=Q_{n+2i}(C(f^!))$. The nonsingular $(-)^i$-quadratic form $(S^{-i}C(f^!)^{n+2i-*},\hat{\psi})$ in $\mathbb{D}_n(\mathbf{Z}[\pi_1(X)])$ is such that

$$\sigma_*(f,b) = (C(f^!),\psi) = \mu(S^{-i}C(f^!)^{n+2i-*},\hat{\psi})$$

$$\in \operatorname{im}(\mu:L_0(\mathbb{D}_n(\mathbf{Z}[\pi_1(X)]),(-)^i)\longrightarrow L_n(\mathbf{Z}[\pi_1(X)],(-)^i))$$

$$= \operatorname{im}(\mu:L_{2i}(\mathbb{D}_n(\mathbf{Z}[\pi_1(X)]))\longrightarrow L_{n+2i}(\mathbf{Z}[\pi_1(X)])) .$$

<u>Proof</u>: The kernel $\mathbf{Z}[\pi_1(X)]$-module chain complexes

$$C = C(f^!:C(\tilde{X})\longrightarrow C(\tilde{M})) , D = C(g^!:C(\tilde{X}\times[0,1])\longrightarrow C(\tilde{L}))$$

are i-fold suspensions of n-dimensional chain complexes (up to chain equivalence). The inclusions $M\longrightarrow L$, $zM\longrightarrow L$ induce $\mathbf{Z}[\pi_1(X)]$-module chain maps $j:C\longrightarrow D$, $k:C\longrightarrow D$ such that $j-k:C\longrightarrow D$ is a chain equivalence. Let $h=(j-k)^{-1}j:C\longrightarrow C$ for any chain homotopy inverse $(j-k)^{-1}:D\longrightarrow C$. The quadratic kernel

$$\sigma_*((e;f,zf),(a;b,zb)) = ((j\ \ k):C\oplus C\longrightarrow D,(\delta\psi,\psi\oplus-\psi))$$

is the i-fold skew-suspension of a cobordism of n-dimensional $(-)^i$-quadratic Poincaré complexes over

$\mathbb{Z}[\pi_1(X)]$ satisfying the hypothesis of 7.2. It follows that $\psi \in Q_{n+2i}(C)$ is the image of the element $\hat{\psi} \in H_0(\mathrm{Hom}_{\mathbb{A}}(C^{n+2i-*},C))$ defined by the composite chain map

$$\hat{\psi} : C^{n+2i-*} \xrightarrow{\phi_0} C \xrightarrow{j} D \xrightarrow{(j-k)^{-1}} C$$

with $\phi_0 = [M]\cap - : C^{n+2i-*} \longrightarrow C$ the Poincaré duality chain equivalence. The nonsingular $(-)^i$-quadratic form $(S^{-i}C^{n+2i-*},\hat{\psi})$ in $\mathbb{D}_n(\mathbb{Z}[\pi_1(X)])$ is such that

$$\sigma_*(f,b) = (C,\psi) = \mu(S^{-i}C^{n+2i-*},\hat{\psi})$$

$$\in \mathrm{im}(\mu : L_0(\mathbb{D}_n(\mathbb{Z}[\pi_1(X)]),(-)^i) \longrightarrow L_n(\mathbb{Z}[\pi_1(X)],(-)^i))$$

$$= \mathrm{im}(\mu : L_{2i}(\mathbb{D}_n(\mathbb{Z}[\pi_1(X)])) \longrightarrow L_{n+2i}(\mathbb{Z}[\pi_1(X)])) .$$

□

<u>Proposition</u> <u>7.4</u> Let $((j\ \ j') : C\oplus C' \longrightarrow D, (\delta\psi, \psi\oplus -\psi'))$ be a cobordism of m-dimensional ϵ-quadratic Poincaré complexes in $\mathbb{A}$, such that D, $C(j)$ and $C(j')$ are the suspensions of $(m-1)$-dimensional chain complexes (up to chain equivalence), with $m \geq 1$. The chain homotopy classes of the chain maps

$$\gamma = \text{inclusion} : G = S^{-1}D \longrightarrow S^{-1}C(j') = F$$

$$\mu = \text{inclusion} :$$

$$G = S^{-1}D \longrightarrow S^{-1}C(j) \simeq C(j')^{m-*} = F^{m-1-*}$$

are the components of a morphism of ϵ-symmetric forms in $\mathbb{D}_{m-1}(\mathbb{A})$

$$\begin{bmatrix}\gamma\\ \mu\end{bmatrix} : (G,0) \longrightarrow H^{\epsilon}(F) = (F\oplus F^{m-1-*}, \begin{bmatrix}0 & 1\\ \epsilon & 0\end{bmatrix})$$

such that $\gamma^*\mu=(1+T_{-\epsilon})\theta_0:G\longrightarrow G^{m-1-*}$ for a certain element $\theta\in Q_{m-1}(G^{m-1-*},-\epsilon)$ determined by $(\delta\psi,\psi\oplus-\psi)$.

If the morphism $\begin{bmatrix}\gamma\\ \mu\end{bmatrix}:G\longrightarrow F\oplus F^{m-1-*}$ is a split injection in $\mathbb{D}_{m-1}(\mathbb{A})$ and if $\theta\in \mathrm{im}(\hat{Q}_{m-1}(G^{m-1-*})\longrightarrow Q_{m-1}(G^{m-1-*},-\epsilon))$ then G is a lagrangian of the hyperbolic ϵ-quadratic form

$$H_{\epsilon}(F)=(F\oplus F^{m-1-*},\begin{bmatrix}0 & 1\\ 0 & 0\end{bmatrix})$$

and (F,G) is a nonsingular ϵ-quadratic formation in $\mathbb{D}_{m-1}(\mathbb{A})$ such that

$$(C,\psi) = \mu(F,G) \in \mathrm{im}(\mu:L_1(\mathbb{D}_{m-1}(\mathbb{A}),\epsilon)\longrightarrow L_m(\mathbb{A},\epsilon)) .$$

<u>Proof</u>: Let (D^{m+1-*},θ) be the $(m+1)$-dimensional ϵ-quadratic complex in $\mathbb{A}$ (not in general Poincaré) defined by the algebraic Thom construction, the image of $(\delta\psi/\psi\oplus-\psi')\in Q_{m+1}(C(j\ j'),\epsilon)$ under the isomorphism

$$((1+T_{\epsilon})(\delta\psi_0,\psi_0\oplus-\psi_0')_{\%})^{-1} :$$

$$Q_{m+1}(C(j\ j'),\epsilon) \longrightarrow$$

$$Q_{m+1}(D^{m+1-*},\epsilon) = Q_{m-1}(G^{m-1-*},-\epsilon) .$$

Up to chain homotopy

$$\gamma^*\mu : G = S^{-1}D \xrightarrow{\text{inclusion}}$$

$$S^{-1}C(j\ j') \simeq D^{m-*} = G^{m-1-*} ,$$

so that there exists a chain homotopy

$$\gamma^*\mu \simeq (1+T_{-\epsilon})\theta_0 \; : \; G \longrightarrow G^{m-1-*} ,$$

and

$$(\gamma^* \; \mu^*)\begin{pmatrix} 0 & 1 \\ \epsilon & 0 \end{pmatrix}\begin{pmatrix} \gamma \\ \mu \end{pmatrix} = \gamma^*\mu + \epsilon\mu^*\gamma \simeq 0 \; : \; G \longrightarrow G^{m-1-*}$$

as required for G to be a lagrangian in $H^{\epsilon}(F)$. If $\theta \in Q_{m-1}(G^{m-1-*},-\epsilon)$ is the image of $\hat{\theta} \in \hat{Q}_{m-1}(G^{m-1-*})$ then $(G,\hat{\theta})$ is the hessian $(-\epsilon)$-quadratic form in $\mathbb{D}_{m-1}(\mathbb{A})$ required for G to be a lagrangian in $H_{\epsilon}(F)$. The algebraic Thom construction defines a one-one correspondence between the homotopy equivalence classes of $(m+1)$-dimensional ϵ-quadratic Poincaré pairs in $\mathbb{A}$ and $(m+1)$-dimensional ϵ-quadratic complexes in $\mathbb{A}$ (Proposition 3.4 of Ranicki [14]). Thus $((j \; j') : C \oplus C' \longrightarrow D, (\delta\psi, \psi \oplus \psi'))$ is homotopy equivalent to the $(m+1)$-dimensional ϵ-quadratic Poincaré pair $((0 \; \pm 1) : \partial D \longrightarrow D, (0, \partial\hat{\theta}))$ defined by

$$d_{\partial D} = \begin{bmatrix} d_D^* & (-)^r(1+T_{-\epsilon})\hat{\theta} \\ 0 & (-)^r d_D \end{bmatrix} :$$

$$\partial D_r = D^{m-r} \oplus D_r \longrightarrow \partial D_{r-1} = D^{m-r+1} \oplus D_{r-1} ,$$

$$\partial\hat{\theta}_0 = \begin{bmatrix} 0 & 0 \\ 1 & 0 \end{bmatrix} :$$

$$\partial D^{m-r} = D_r \oplus D^{m-r} \longrightarrow \partial D_r = D^{m-r} \oplus D_r ,$$

$$\partial\hat{\theta}_1 = \begin{pmatrix} (-)^{m-r+s}\hat{\theta} & 0 \\ 0 & 0 \end{pmatrix} :$$

$$\partial D^{m-r-1} = D_{r+1} \oplus D^{m-r-1} \longrightarrow \partial D_r = D^{m-r} \oplus D_r ,$$

$$\partial\hat{\theta}_s = 0 : \partial D^{m-r-s} \longrightarrow \partial D_r \quad (s \geqslant 2) .$$

Up to chain homotopy

$$\mu^* : F = S^{-1}C(j') \xrightarrow{\text{inclusion}} S^{-1}C(j\ j') \simeq D^{m-*} = G^{m-1-*} ,$$

so that there is defined a chain equivalence $f : C \longrightarrow C(\mu^*)$. Choosing a representative chain map $\hat{\theta} : D \longrightarrow D^{m+1-*}$ and a chain homotopy $\chi : \gamma^* \mu \simeq (1+T_\epsilon)\hat{\theta} : D \longrightarrow D^{m+1-*}$ define a chain map $g : \partial D \longrightarrow C(\mu^*)$ by

$$g = \begin{pmatrix} 1 & \chi \\ 0 & \gamma \end{pmatrix} :$$

$$\partial D_r = D^{m-r} \oplus D_r \longrightarrow C(\mu^*)_r = G^{m-r-1} \oplus F_{r-1}$$

such that

$$f_\%(\psi) = g_\%(\partial\hat{\theta}) \in Q_m(C(\mu^*), \epsilon) .$$

Now $(C(\mu^*), g_\%(\partial\hat{\theta}))$ is the m-dimensional ϵ-quadratic Poincaré complex in $\mathbb{A}$ constructed in 3.2 from the nonsingular ϵ-quadratic formation (F,G) in $\mathbb{D}_{m-1}(\mathbb{A})$, so that

$$(C,\psi) = (C(\mu^*), f_\%(\psi)) = (C(\mu^*), g_\%(\partial\hat{\theta}))$$

$$= \mu(F,G) \in \mathrm{im}(\mu : L_1(\mathbb{D}_{m-1}(\mathbb{A}),\epsilon) \longrightarrow L_m(\mathbb{A},\epsilon)) .$$

□

§8. The connection

We now connect the algebra and the geometry, verifying the claim of Theorem 6.2 that the geometric surgery transfer maps for an n-dimensional Poincaré fibration $F \longrightarrow E \xrightarrow{p} B$ coincide with the algebraic surgery transfer maps

$$p^!_{geo} = p^!_{alg} :$$

$$L_m(\mathbf{Z}[\pi_1(B)]) \longrightarrow L_{m+n}(\mathbf{Z}[\pi_1(E)]) \quad (m \geqslant 0) .$$

We know from 1.9 how a CW complex structure behaves under transfer on the cellular chain level. The strategy is to encode the L-theory data in CW complex structures, and to decode the lifted L-theory data from the CW lifts using the ultraquadratic L-theory of §7.

We consider first the case m=2i. By Chapter 5 of Wall [22] every element $x \in L_{2i}(\mathbf{Z}[\pi_1(B)])$ $(i \geqslant 3)$ is the Witt class of the nonsingular $(-)^i$-quadratic form in $\mathbb{B}(\mathbf{Z}[\pi_1(B)])$

$$(K_i(M) , \lambda : K_i(M) \times K_i(M) \longrightarrow \mathbf{Z}[\pi_1(B)] ,$$

$$\mu : K_i(M) \longrightarrow \mathbf{Z}[\pi_1(B)]/\langle a-(-)^i\bar{a} \mid a \in \mathbf{Z}[\pi_1(B)]\rangle)$$

on the kernel $\mathbf{Z}[\pi_1(B)]$-module

$$K_i(M) = \pi_{i+1}(f) = H_i(f^!) = \ker(\tilde{f}_* : H_i(\tilde{M}) \longrightarrow H_i(\tilde{X}))$$

of an (i-1)-connected normal map $(f,b) : (M,\partial M) \longrightarrow (X,\partial X)$

from a 2i-dimensional manifold with boundary $(M,\partial M)$ to a 2i-dimensional geometric Poincaré pair $(X,\partial X)$, with $\partial f:\partial M \longrightarrow \partial X$ a homotopy equivalence, and with a π_1-isomorphism reference map $X \longrightarrow B$ such that $w(X):\pi_1(X) \longrightarrow \pi_1(B) \xrightarrow{w(B)} \mathbb{Z}_2$. The adjoint of λ defines an isomorphism in $\mathbb{B}(\mathbb{Z}[\pi_1(B)])$

$$\lambda : K_i(M) \longrightarrow K_i(M)^* ;$$

$$u \longrightarrow (v \longrightarrow \lambda(u,v)) .$$

$(K_i(M),\lambda,\mu)$ can be viewed as a nonsingular $(-)^i$-quadratic form $(K_i(M)^*,\psi)$ over $\mathbb{Z}[\pi_1(B)]$, with ψ an equivalence class of $\mathbb{Z}[\pi_1(B)]$-module morphisms $\hat{\psi}:K_i(M)^* \longrightarrow K_i(M)$ such that

$$\hat{\psi}+(-)^i\hat{\psi}^* = \lambda^{-1} : K_i(M)^* \longrightarrow K_i(M) ,$$

$$\hat{\psi}(\lambda(v))(\lambda(v)) = \mu(v) \quad (v\in K_i(M)) ,$$

with $\hat{\psi}$ equivalent to $\hat{\psi}+\chi+(-)^{i+1}\chi^*$ for any $\mathbb{Z}[\pi_1(B)]$-module morphism $\chi:K_i(M)^* \longrightarrow K_i(M)$. The surgery obstruction is thus given by

$$x = \sigma_*(f,b) = (K_i(M),\lambda,\mu) = (K_i(M)^*,\hat{\psi})$$

$$\in L_{2i}(\mathbb{Z}[\pi_1(B)]) .$$

We shall be regarding modules as 0-dimensional chain complexes, and for any $q\in\mathbb{Z}$ we write S^qC for the q-fold suspension of a chain complex C, with

$$d_{S^qC} = d_C : (S^qC)_r = C_{r-q} \longrightarrow (S^qC)_{r-1} = C_{r-q-1} .$$

The quadratic kernel $\sigma_*(f,b)=(C(f^!),\psi)$ of the

$(i-1)$-connected $2i$-dimensional normal map $(f,b):M\longrightarrow X$ is an $(i-1)$-connected $2i$-dimensional quadratic Poincaré complex over $\mathbf{Z}[\pi_1(B)]$ which is homotopy equivalent to $(S^iK_i(M),\hat{\psi})$. Thus we can identify $\psi_0=\hat{\psi}$, and up to chain homotopy

$$(1+T)\psi_0 = \hat{\psi}+(-)^i\hat{\psi}^* = \lambda^{-1} :$$

$$C(f^!)^{2i-*} = S^iK_i(M)^* \longrightarrow C(f^!) = S^iK_i(M) .$$

The quadratic structure $\psi\in Q_{2i}(C(f^!))$ is the equivalence class of $\mathbf{Z}[\pi_1(B)]$-module morphisms $\hat{\psi}:K_i(M)^*\longrightarrow K_i(M)$ described above. A choice of representative $\hat{\psi}$ is a choice of ultraquadratic structure $\hat{\psi}\in\hat{Q}_{2i}(C(f^!))$ for the quadratic structure $\psi\in Q_{2i}(C(f^!))$. We now fix a choice of $\hat{\psi}$.

Let $\langle v_1,v_2,\ldots,v_k\rangle$ be a base for the f.g. free $\mathbf{Z}[\pi_1(B)]$-module $K_i(M)$, and use the dual to define a base for $K^i(M)=K_i(M)^*$. The functor of additive categories with involution

$$p^{\#} = -\otimes(C(\widetilde{F}),\alpha,U) :$$

$$\mathbb{B}(\mathbf{Z}[\pi_1(B)]) \longrightarrow \mathbb{D}_n(\mathbf{Z}[\pi_1(E)])$$

sends the morphisms in $\mathbb{B}(\mathbf{Z}[\pi_1(B)])$

$$\hat{\psi}\ ,\ \hat{\psi}^* :$$

$$K_i(M)^* = \bigoplus_k \mathbf{Z}[\pi_1(B)] \longrightarrow K_i(M) = \bigoplus_k \mathbf{Z}[\pi_1(B)]$$

to chain homotopy classes of $\mathbf{Z}[\pi_1(E)]$-module chain maps

$p^{\#}(\hat{\psi})\ ,\ p^{\#}(\hat{\psi}^*)\ :$

$$p^{\#}(K_i(M)^*) = \underset{k}{\oplus}C(\tilde{F}) \longrightarrow p^{\#}(K_i(M)) = \underset{k}{\oplus}C(\tilde{F})$$

such that there is defined a chain homotopy commutative diagram

$$\begin{array}{ccc} \underset{k}{\oplus}C(\tilde{F})^{n-*} & \xrightarrow{\ p^{\#}(\hat{\psi})^*\ } & \underset{k}{\oplus}C(\tilde{F})^{n-*} \\ {\scriptstyle \oplus[F]\cap-}\downarrow & & \downarrow{\scriptstyle \oplus[F]\cap-} \\ \underset{k}{\oplus}C(\tilde{F}) & \xrightarrow{\ p^{\#}(\hat{\psi}^*)\ } & \underset{k}{\oplus}C(\tilde{F}) \end{array}$$

and such that

$$p^{\#}(\hat{\psi}) + (-)^i p^{\#}(\hat{\psi}^*) = p^{\#}(\lambda^{-1})\ :\ \underset{k}{\oplus}C(\tilde{F}) \longrightarrow \underset{k}{\oplus}C(\tilde{F})$$

is a chain equivalence.

In Lemmas 8.1,8.3 below we shall show that the quadratic kernel $\sigma_*(g,c)=(C(g^!),\eta)$ of the pullback $(n+i-1)$-connected $(n+2i)$-dimensional normal map $(g,c):(N,\partial N)\longrightarrow(Y,\partial Y)$ is homotopy equivalent to the $(n+2i)$-dimensional quadratic Poincaré complex (D,η) defined by the $\mathbf{Z}[\pi_1(E)]$-module chain complex

$$D = S^i p^{\#}(K_i(M)) = \underset{k}{\oplus}S^i C(\tilde{F})$$

with the (ultra)quadratic structure

$$\eta_0\ :\ D^{n+2i-*} = \underset{k}{\oplus}S^i C(\tilde{F})^{n-*} \xrightarrow{\ \oplus[F]\cap-\ } \underset{k}{\oplus}S^i C(\tilde{F})$$

$$\xrightarrow{\ p^{\#}(\hat{\psi})\ } D = \underset{k}{\oplus}S^i C(\tilde{F})\ ,$$

$$\eta_s = 0 : D^{n+2i-r-s} \longrightarrow D_r \quad (s \geq 1) .$$

It will follow that the nonsingular $(-)^i$-quadratic form $(p^{\#}(K_i(M)^*),(\oplus[F]\cap-)^{-1}p^{\#}(\hat{\psi}))$ in $\mathbb{D}_n(\mathbf{Z}[\pi_1(E)])$ defined by

$$(\oplus[F]\cap-)^{-1}p^{\#}(\hat{\psi}) :$$

$$p^{\#}(K_i(M)^*) = \underset{k}{\oplus}C(\tilde{F}) \xrightarrow{p^{\#}(\hat{\psi})} p^{\#}(K_i(M)) = \underset{k}{\oplus}C(\tilde{F})$$

$$\xrightarrow{\oplus([F]\cap-)^{-1}} \underset{k}{\oplus}C(\tilde{F})^{n-*} = (p^{\#}(K_i(M)^*))^{n-*}$$

is such that

$$p^!_{geo}\sigma_*(f,b) = \sigma_*(g,c) =$$

$$\mu(p^{\#}(K_i(M)^*),(\oplus[F]\cap-)^{-1}p^{\#}(\hat{\psi}))$$

$$= p^!_{alg}(K_i(M)^*,\hat{\psi}) = p^!_{alg}\sigma_*(f,b)$$

$$\in \mathrm{im}(\mu:L_0(\mathbb{D}_n(\mathbf{Z}[\pi_1(E)]),(-)^i)\longrightarrow L_n(\mathbf{Z}[\pi_1(E)],(-)^i))$$

$$= \mathrm{im}(\mu:L_{2i}(\mathbb{D}_n(\mathbf{Z}[\pi_1(E)]))\longrightarrow L_{n+2i}(\mathbf{Z}[\pi_1(E)])) ,$$

verifying that $p^!_{geo}=p^!_{alg}$ in the case $m=2i$.

For the symmetric structure of $\sigma_*(g,c)=(C(g^!),\eta)$ we have:

<u>Lemma 8.1</u> The symmetric kernel $\sigma^*(g,c)=(C(g^!),(1+T)\eta)$ is such that up to chain homotopy

$$(1+T)\eta_0 :$$

$$C(g^!)^{n+2i-*} = S^i p^{\#}(K_i(M)^*)^{n-*} = \bigoplus_k S^i C(\tilde{F})^{n-*}$$

$$\xrightarrow{\oplus[F]\cap -} \bigoplus_k S^i C(\tilde{F}) = S^i p^{\#}(K_i(M)^*)$$

$$\xrightarrow{p^{\#}((1+T)\psi_0)} C(g^!) = S^i p^{\#}(K_i(M)) = \bigoplus_k S^i C(\tilde{F}) \ .$$

<u>Proof</u>: Represent the base elements $v_j \in K_i(M)$ $(1 \leqslant j \leqslant k)$ by framed immersions $v_j : S^i \longrightarrow \mathrm{int}(M^{2i})$ with nullhomotopies in X, and with $\pi_1(B)$-equivariant lifts $\tilde{v}_j : \tilde{S}^i = \pi_1(B) \times S^i \longrightarrow \tilde{M}$. Replace $f : M \longrightarrow X$ by the inclusion of M in the CW complex $M \cup \bigcup_k e^{i+1}$ homotopy equivalent to X (which is also denoted by X), so that

$$C(f^!) \simeq S^{-1} C(\tilde{X}, \tilde{M}) = S^i K_i(M) = \bigoplus_k S^i \mathbf{Z}[\pi_1(B)] \ .$$

In the total spaces of the pullbacks $g : N \longrightarrow Y$ is replaced by the inclusion of N in the CW complex $N \cup \bigcup_k F \times e^{i+1}$, so that

$$C(g^!) \simeq S^{-1} C(\tilde{Y}, \tilde{N}) = S^i p^{\#}(K_i(M)) = \bigoplus_k S^i C(\tilde{F}) \ .$$

The Poincaré duality chain equivalence is given up to chain homotopy by the composite

$$(1+T)\eta_0 : C(g^!)^{n+2i-*} \xrightarrow{e^*} C(\tilde{N}, \partial\tilde{N})^{n+2i-*}$$

$$\xrightarrow{[N]\cap -} C(\tilde{N}) \longrightarrow C(\tilde{N}, \partial\tilde{N}) \xrightarrow{e} C(g^!)$$

with e the inclusion.

For a sufficiently large number $q \geqslant 0$ the framed immersions can be approximated by framed embeddings

$v_j : S^i \longrightarrow \mathrm{int}(M^{2i} \times D^q)$ with nullhomotopies in X. Let V_j be a regular neighbourhood of $v_j(S^i)$ in $M \times D^q$, and let $P_j = \mathrm{closure}(M \times D^q - V_j)$, so that

$$M \times D^q = V_j \cup_{\partial V_j} P_j \quad , \quad \partial P_j = \partial V_j \cup \partial(M \times D^q) \ ,$$

$$(V_j, \partial V_j) = v_j(S^i) \times (D^{i+q}, S^{i+q-1}) \ .$$

The intersection number

$$\lambda_{j,j'} = \lambda(\tilde{v}_j, \tilde{v}_{j'}) \in \mathbf{Z}[\pi_1(B)] \quad (1 \leqslant j, j' \leqslant k)$$

is the image of $1 \in \mathbf{Z}[\pi_1(B)]$ under the composite $\mathbf{Z}[\pi_1(B)]$-module morphism

$$\begin{aligned} H_i(\tilde{S}^i) &= \mathbf{Z}[\pi_1(B)] \\ &\xrightarrow{\tilde{v}_{j'}} H_i(\tilde{M}) \cong H_{i+q}(\tilde{M} \times D^q, \tilde{M} \times S^{q-1}) \\ &\longrightarrow H_{i+q}(\tilde{M} \times D^q, \tilde{P}_j) \cong H_{i+q}(\tilde{V}_j, \partial \tilde{V}_j) \\ &= H_0(\tilde{S}^i) = \mathbf{Z}[\pi_1(B)] \ , \end{aligned}$$

which can also be expressed as

$$H_i(\tilde{S}^i) = \mathbf{Z}[\pi_1(B)] \xrightarrow{\tilde{v}_{j'}} H_i(\tilde{M}, \partial \tilde{M})$$

$$\xrightarrow{([M] \cap -)^{-1}} H^i(\tilde{M}) \xrightarrow{\tilde{v}_j^*} H^i(\tilde{S}^i) = \mathbf{Z}[\pi_1(B)] \ .$$

The pullbacks from the n-dimensional Poincaré fibration $F \longrightarrow E \xrightarrow{p} B$ define framed Poincaré immersions $w_j : F^n \times S^i \longrightarrow N^{n+2i}$ with nullhomotopies in Y, and with $\pi_1(E)$-equivariant lifts $\tilde{w}_j : \tilde{F} \times S^i \longrightarrow \tilde{N}$ $(1 \leqslant j \leqslant k)$. Let

$W_j, Q_j \subset N\times D^q$ be the total spaces of the fibrations over $V_j, P_j \subset M\times D^q$, so that

$$N\times D^q = W_j \cup_{\partial W_j} Q_j \quad , \quad \partial Q_j = \partial W_j \cup \partial(N\times D^q) \ ,$$

$$(W_j, \partial W_j) = w_j(F\times S^i)\times(D^{i+q}, S^{i+q-1}) \ .$$

For any embedding $D^{2i+q} \subset \mathrm{int}(V_j) \subset M^{2i}\times D^q$ the pair

$$(\ (M\times D^q - \mathrm{int}(D^{2i+q})) \cup_{v_{j'}\times 1} D^{i+1}\times D^q \ , \ P_j \cup D^{i+1}\times S^{q-1}\ )$$

has a relative CW structure with one (i+q)-cell and one (i+q+1)-cell, such that the cellular chain complex in $\mathbb{B}(\mathbf{Z}[\pi_1(B)])$ is $\lambda_{j,j'}:\mathbf{Z}[\pi_1(B)] \longrightarrow \mathbf{Z}[\pi_1(B)]$. By 1.9 the chain homotopy class of the $\mathbf{Z}[\pi_1(E)]$-module chain map $p^{\#}(\lambda_{j,j'}):C(\tilde{F}) \longrightarrow C(\tilde{F})$ coincides with the composite

$$C(\tilde{F}) \longrightarrow S^{-i}C(\tilde{F}\times S^i) \xrightarrow{\tilde{w}_{j'}}$$

$$S^{-i}C(\tilde{N}, \partial\tilde{N}) \simeq S^{-i-q}C(\tilde{N}\times D^q, \partial(\tilde{N}\times D^q))$$

$$\longrightarrow S^{-i-q}C(\tilde{N}\times D^q, \tilde{Q}_j) \simeq S^{-i-q}C(\tilde{W}_j, \partial\tilde{W}_j) \simeq C(\tilde{F}\times S^i)$$

$$\longrightarrow C(\tilde{F}) \ ,$$

and hence also with the composite

$$C(\tilde{F}) \longrightarrow S^{-i}C(\tilde{F}\times S^i) \xrightarrow{\tilde{w}_{j'}} S^{-i}C(\tilde{N}, \partial\tilde{N})$$

$$\xrightarrow{([N]\cap -)^{-1}} S^{i}C(\tilde{N})^{n-*} \xrightarrow{\tilde{w}_j^*} S^{i}C(\tilde{F}\times S^i)^{n-*}$$

$$\longrightarrow C(\tilde{F})^{n-*} \xrightarrow{[F]\cap -} C(\tilde{F}) \ .$$

The (j,j')-component of the $\mathbf{Z}[\pi_1(E)]$-module chain equivalence

$$((1+T)\eta_0)^{-1} :$$

$$C(g^!) \simeq \bigoplus_k S^i C(\tilde{F}) \longrightarrow C(g^!)^{n+2i-*} \simeq \bigoplus_k S^i C(\tilde{F})^{n-*}$$

is thus the composite

$$S^i C(\tilde{F}) \xrightarrow{p^{\#}(\lambda_{j,j'})} S^i C(\tilde{F})$$

$$\xrightarrow{([F]\cap-)^{-1}} S^i C(\tilde{F})^{n-*}$$

and up to chain homotopy

$$(1+T)\eta_0 : C(g^!)^{n+2i-*} = \bigoplus_k S^i C(\tilde{F})^{n-*} \xrightarrow{\oplus[F]\cap-}$$

$$\bigoplus_k S^i C(\tilde{F}) \xrightarrow{p^{\#}(\lambda^{-1})} C(g^!) = \bigoplus_k S^i C(\tilde{F}) .$$

□

We extend the description of the symmetric structure of $\sigma_*(g,c)$ given by 8.1 to the quadratic structure, using the ultraquadratic L-theory of §7. A choice of ultraquadratic structure $\hat{\psi}:K_i(M)^* \longrightarrow K_i(M)$ for $\sigma_*(f,b)$ is used to construct a normal bordism between $(f,b):M \longrightarrow X$ and a copy $(zf,zb):zM \longrightarrow zX$ which encodes the quadratic self-intersection form μ in the CW structure. The quadratic structure of $\sigma_*(g,c)$ is then decoded from the CW structure of the pullback normal bordism between $(g,c):N \longrightarrow Y$ and a copy $(zg,zc):zN \longrightarrow zY$, using 1.9 and 7.3. The construction of the bordism is motivated by the way in which the infinite cyclic cover of a knot complement can be obtained by cutting along a Seifert surface.

__Lemma__ __8.2__ A choice of ultraquadratic structure $\hat{\psi}$ for $(K_i(M),\lambda,\mu)$ can be realized by an (i-1)-connected (2i+1)-dimensional normal bordism

$$((e;f,zf),(a;b,zb)) : (L;M,zM) \longrightarrow X\times([0,1];\langle 0\rangle,\langle 1\rangle)$$

between $(f,b):M\longrightarrow X$ and a disjoint copy $(zf,zb):zM\longrightarrow zX$, such that the difference of the $\mathbb{Z}[\pi_1(B)]$-module morphisms $j,k:K_i(M)\longrightarrow K_i(L)$ induced by the inclusions $j:M\longrightarrow L$, $k:zM\longrightarrow L$ is an isomorphism $j-k:K_i(M)\longrightarrow K_i(L)$ with

$$(j-k)^{-1}j = \hat{\psi}\lambda : K_i(M) \longrightarrow K_i(M) ,$$

$$(j-k)^{-1}k = (-)^{i+1}\hat{\psi}^*\lambda : K_i(M) \longrightarrow K_i(M) .$$

The (i-1)-connected (2i+1)-dimensional normal map

$$(e/(f=zf),a/(b=zb)) : (L/(M=zM),\partial M\times S^1)$$

$$\longrightarrow (X,\partial X)\times([0,1]/0=1) = (X,\partial X)\times S^1$$

is a $\mathbb{Z}[\pi_1(B)]$-homology equivalence, with the homotopy equivalence $\partial f\times 1:\partial M\times S^1\longrightarrow\partial X\times S^1$ on the boundary.

__Proof__: Every based f.g. free lagrangian of the $(-)^i$-quadratic form $(K_i(M),\lambda,\mu)\oplus(K_i(M),-\lambda,-\mu)$ can be realized by disjoint framed embeddings of S^i in $M\cup_{\partial M\times[0,1]}zM$ with nullhomotopies in X, such that the trace of the surgeries on these framed embedded i-spheres defines a normal bordism between (f,b) and (zf,zb). The realization of the lagrangian

$$\mathrm{im}\left(\begin{bmatrix} \hat{\psi}\lambda \\ (-)^{i+1}\hat{\psi}^*\lambda \end{bmatrix} : K_i(M) \longrightarrow K_i(M)\oplus K_i(M)\right)$$

has the required properties. (This lagrangian is a direct complement of the diagonal lagrangian $\mathrm{im}(\begin{bmatrix} 1 \\ 1 \end{bmatrix}: K_i(M) \longrightarrow K_i(M)\oplus K_i(M))$. The realization of the diagonal lagrangian is the product $(2i+1)$-dimensional normal map

$$(f,b)\times 1 \ : \ M\times([0,1];\{0\},\{1\}) \longrightarrow X\times([0,1];\{0\},\{1\}) \ .$$

The required normal map (e,a) can also be obtained from $(f,b)\times 1$ by surgeries on i-spheres in the interior of $M\times[0,1]$ representing a base of $K_i(M\times[0,1])=K_i(M)$.)

□

We can now extend Lemma 8.1 to the quadratic structure:

<u>Lemma</u> <u>8.3</u> The quadratic kernel $\sigma_*(g,c)=(C(g^!),\eta)$ is such that up to chain homotopy

$$\eta_0 \ : \ C(g^!)^{n+2i-*} = S^i p^{\#}(K_i(M)^*)^{n-*} = \bigoplus_k S^i C(\tilde{F})^{n-*}$$

$$\xrightarrow{\oplus[F]\cap -} \bigoplus_k S^i C(\tilde{F}) = S^i p^{\#}(K_i(M)^*)$$

$$\xrightarrow{p^{\#}(\hat{\psi})} C(g^!) = S^i p^{\#}(K_i(M)) = \bigoplus_k S^i C(\tilde{F}) \ ,$$

$$\eta_s = 0 \ : \ C(g^!)^{n+2i-r-s} \longrightarrow C(g^!)_r \quad (s\geqslant 1) \ .$$

<u>Proof</u>: Let $\hat{\psi},(e,a),(f,b),j,k$ be as in 8.2, and let

$$((h;g,zg),(d;c,zc)) :$$

$$(P;N,zN) \longrightarrow Y\times([0,1];\{0\},\{1\})$$

be the $(n+i-1)$-connected $(n+2i+1)$-dimensional normal bordism between $(g,c):N\longrightarrow Y$ and a disjoint copy $(zg,zc):zN\longrightarrow zY$ obtained from $((e;f,zf),(a;b,zb))$ by pullback from $F\longrightarrow E\xrightarrow{p}B$ along the reference map $X\longrightarrow B$. The $(n+i-1)$-connected $(n+2i+1)$-dimensional normal map

$$(h/(g=zg),d/(c=zc)) :$$

$$P/(N=zN) \longrightarrow Y\times([0,1]/0=1) = Y\times S^1$$

is a $\mathbb{Z}[\pi_1(E)]$-homology equivalence. By 7.3 the quadratic kernel $\sigma_*(g,c)$ is determined by the chain homotopy classes of the $\mathbb{Z}[\pi_1(E)]$-module chain maps $C(g^!)\longrightarrow C(h^!)$, $C(zg^!)\longrightarrow C(h^!)$ and the Poincaré duality chain equivalence $C(g^!)^{n+2i-*}\longrightarrow C(g^!)$. We shall now arrange CW structures for (e,a) in such a way that only cells in dimensions $i,i+1$ occur in the relevant pairs, and 1.9 applies to obtain the $\mathbb{Z}[\pi_1(E)]$-module chain homotopy data in the total spaces of the pullbacks from $F\longrightarrow E\xrightarrow{p}B$ as the algebraic transfers of $\mathbb{Z}[\pi_1(B)]$-module data.

L is the trace of surgeries on $(i-1)$- and i-spheres in M, so that (L,M) has a relative CW structure with i- and $(i+1)$-cells, with the cellular chain complex in $\mathbb{B}(\mathbb{Z}[\pi_1(B)])$ given by

$$d = j : C(\tilde{L},\tilde{M})_{i+1} = K_i(M) \longrightarrow C(\tilde{L},\tilde{M})_i = K_i(L) .$$

Replacing $e:L\longrightarrow X\times[0,1]$ by the inclusion of L in the mapping cylinder it may be assumed that L is a subcomplex of X, such that (X,L) and (X,M) have cellular chain complexes in $\mathbb{B}(\mathbb{Z}[\pi_1(B)])$

$$C(\tilde{X},\tilde{L}) = S^{i+1}K_i(L) ,$$

$$d = (j\ \ 1) : C(\tilde{X},\tilde{M})_{i+1} = K_i(M)\oplus K_i(L)$$

$$\longrightarrow C(\tilde{X},\tilde{M})_i = K_i(L) \ .$$

The kernel chain complexes $C(f^!)$, $C(e^!)$ are chain equivalent to $S^{-1}C(\tilde{X},\tilde{M})$, $S^{-1}C(\tilde{X},\tilde{L})$ respectively. Replacing the inclusion $C(f^!)\longrightarrow C(e^!)$ by the inclusion $S^{-1}C(\tilde{X},\tilde{M})\longrightarrow S^{-1}C(\tilde{X},\tilde{L})$ of the chain equivalent complexes corresponds in the pullbacks to replacing $C(g^!)\longrightarrow C(h^!)$ by $S^{-1}C(\tilde{Y},\tilde{N})\longrightarrow S^{-1}C(\tilde{Y},\tilde{P})$, and by 1.9

$$C(\tilde{Y},\tilde{N}) = p^{\#}C(\tilde{X},\tilde{M}) \quad , \quad C(\tilde{Y},\tilde{P}) = p^{\#}C(\tilde{X},\tilde{L}) \ .$$

Thus up to chain homotopy the inclusion $C(g^!)\longrightarrow C(h^!)$ may be identified with the $\mathbf{Z}[\pi_1(E)]$-module chain map

$$p^{\#}(j) \ : \ C(g^!) = S^i p^{\#}(K_i(M)) \longrightarrow$$
$$C(h^!) = S^i p^{\#}(K_i(L)) \ .$$

Similarly, up to chain homotopy $C(zg^!)\longrightarrow C(h^!)$ may be identified with

$$p^{\#}(k) \ : \ C(zg^!) = S^i p^{\#}(K_i(M)) \longrightarrow$$
$$C(h^!) = S^i p^{\#}(K_i(L)) \ .$$

The $\mathbf{Z}[\pi_1(B)]$-module isomorphism $j-k:K_i(M)\longrightarrow K_i(L)$ such that $(j-k)^{-1}j=\hat{\psi}\lambda:K_i(M)\longrightarrow K_i(M)$ lifts to (the chain homotopy class of) a $\mathbf{Z}[\pi_1(E)]$-module chain equivalence

$$p^{\#}(j-k) = p^{\#}(j) - p^{\#}(k) \ : \ C(g^!) \longrightarrow C(h^!)$$

such that up to chain homotopy

$$p^{\#}(j-k)^{-1}p^{\#}(j) = p^{\#}(\hat{\psi}\lambda) \ :$$
$$C(g^!) = \bigoplus_k S^i C(\tilde{F}) \longrightarrow \bigoplus_k S^i C(\tilde{F}) \ .$$

Applying 8.1 and 7.3 we have that the quadratic kernel $\sigma_*(g,c)$ is homotopy equivalent to the (n+2i)-dimensional quadratic Poincaré complex $(\bigoplus_k S^i C(\tilde{F}),\eta)$ over $\mathbb{Z}[\pi_1(E)]$ with

$$(1+T)\eta_0 \ : \ C(g^!)^{n+2i-*} = \bigoplus_k S^i C(\tilde{F})^{n-*} \xrightarrow{\oplus[F]\cap -}$$

$$\bigoplus_k S^i C(\tilde{F}) \xrightarrow{p^{\#}(\lambda^{-1})} C(g^!) = \bigoplus_k S^i C(\tilde{F}) \ ,$$

$$\eta_0 = p^{\#}(\hat{\psi}\lambda)(1+T)\eta_0 \ :$$

$$C(g^!)^{n+2i-*} = \bigoplus_k S^i C(\tilde{F})^{n-*} \xrightarrow{\oplus[F]\cap -} \bigoplus_k S^i C(\tilde{F})$$

$$\xrightarrow{p^{\#}(\hat{\psi})} C(g^!) = \bigoplus_k S^i C(\tilde{F}) \ ,$$

$$\eta_s = 0 \ : \ C(g^!)^{n+2i-r-s} \longrightarrow C(g^!)_r \quad (s\geq 1) \ .$$

□

This completes the proof of Theorem 6.2 in the case m=2i, and we proceed to the case m=2i+1.

By Chapter 6 of Wall [22] every element $x\in L_{2i+1}(\mathbb{Z}[\pi_1(B)])$ $(i\geq 2)$ is the Witt class of the kernel nonsingular $(-)^i$-quadratic formation over $\mathbb{Z}[\pi_1(B)]$

$$(F,G) = (K_{i+1}(U,\partial U),K_{i+1}(M_0,\partial U))$$

of an (i-1)-connected (2i+1)-dimensional normal map $(f,b):(M,\partial M)\longrightarrow(X,\partial X)$ with $\partial f:\partial M\longrightarrow\partial X$ a homotopy equivalence, and with a π_1-isomorphism reference map $X\longrightarrow B$ such that $w(X):\pi_1(X)\longrightarrow\pi_1(B)\xrightarrow{w(B)}\mathbb{Z}_2$. Here, U is the connected sum of a sufficiently large number $k\geq 0$ of framed embeddings $S^i\subset \mathrm{int}(M)$ with nullhomotopies in X to generate the f.g. $\mathbb{Z}[\pi_1(B)]$-module $K_i(M)$, and M_0=closure(M-U). Thus $F=K_{i+1}(U,\partial U)$ is a based f.g. free $\mathbb{Z}[\pi_1(B)]$-module, and $G=K_{i+1}(M_0,\partial U)$ is a based f.g. free lagrangian of the hyperbolic $(-)^i$-quadratic form $H_{(-)^i}(F)=(F\oplus F^*,\begin{pmatrix}0&1\\0&0\end{pmatrix})$, with

$$F = G = \bigoplus_k \mathbf{Z}[\pi_1(B)] .$$

The inclusion $\begin{pmatrix}\gamma\\ \mu\end{pmatrix}:G\longrightarrow F\oplus F^*$ extends to an isomorphism of hyperbolic $(-)^i$-quadratic forms

$$\begin{pmatrix}\gamma & \tilde{\gamma}\\ \mu & \tilde{\mu}\end{pmatrix} : H_{(-)^i}(G) \longrightarrow H_{(-)^i}(F) .$$

Surgery on the framed embedded i-spheres in U defines an (i-1)-connected (2i+2)-dimensional normal map of triads

$$((e;f,f'),(a;b,b')) : (L^{2i+2};M^{2i+1},M'^{2i+1})$$

$$\longrightarrow X\times([0,1];\{0\},\{1\})$$

with (F^*,G) the kernel nonsingular $(-)^i$-quadratic formation of (f',b'), and

$$G = K_{i+1}(L) \ , \ F = K_{i+1}(L,M') \ , \ F^* = K_{i+1}(L,M) \ ,$$

$$C(e^!) = S^{i+1}G \ , \ C(e^!,f'^!) = S^{i+1}F \ ,$$

$$C(e^!,f^!) = S^{i+1}F^* .$$

The quadratic kernel $\sigma_*((e;f,f'),(a;b,b'))$ is an (i-1)-connected (2i+2)-dimensional quadratic Poincaré cobordism over $\mathbf{Z}[\pi_1(B)]$ with algebraic Thom complex homotopy equivalent to the i-connected (2i+2)-dimensional quadratic complex $(S^{i+1}G^*,\theta)$ corresponding to the $(-)^{i+1}$-quadratic hessian form (G,θ) such that $\gamma^*\mu=\theta+(-)^{i+1}\theta^*:G\longrightarrow G^*$. The base elements of the f.g. free $\mathbf{Z}[\pi_1(B)]$-module $G=\pi_{i+1}(e)$ can be represented by immersed (i+1)-spheres in $\mathrm{int}(L^{2i+2})$ with nullhomotopies in X, so that the form (G,θ) can be expressed in terms of geometric intersection and self-intersection numbers exactly as in Chapter 5 of Wall [22].

The pullback of $((e;f,f'),(a;b,b'))$ from $F \longrightarrow E \xrightarrow{p} B$ along the reference map $X \longrightarrow B$ is an $(n+i-1)$-connected normal map of $(n+2i+2)$-dimensional geometric Poincaré triads

$$((h;g,g'),(d;c,c')) : (P^{n+2i+2};N^{n+2i+1},N'^{n+2i+1})$$

$$\longrightarrow X\times([0,1];\{0\},\{1\}) \ .$$

The $\mathbb{Z}[\pi_1(E)]$-module chain maps $C(h^!) \longrightarrow C(h^!,g'^!)$, $C(h^!) \longrightarrow C(h^!,g^!)$ defined by projections are given up to chain homotopy by

$$p^{\#}(\gamma) : C(h^!) = S^{i+1}p^{\#}G = \bigoplus_k S^{i+1}C(\tilde{F})$$

$$\longrightarrow C(h^!,g'^!) = S^{i+1}p^{\#}F = \bigoplus_k S^{i+1}C(\tilde{F}) \ ,$$

$$(\oplus[F]\cap -)^{-1}p^{\#}(\mu) : C(h^!) = S^{i+1}p^{\#}G = \bigoplus_k S^{i+1}C(\tilde{F})$$

$$\xrightarrow{p^{\#}(\mu)} S^{i+1}p^{\#}(F^*) = \bigoplus_k S^{i+1}C(\tilde{F})$$

$$\xrightarrow{(\oplus[F]\cap -)^{-1}}$$

$$\bigoplus_k S^{i+1}C(\tilde{F})^{n-*} = C(h^!,g'^!)^{n+2i+2-*} \simeq C(h^!,g^!) \ .$$

The quadratic kernel $\sigma_*((h;g,g'),(d;c,c'))$ is the i-fold skew-suspension of an $(n+1)$-dimensional $(-)^i$-quadratic Poincaré cobordism over $\mathbb{Z}[\pi_1(E)]$ satisfying the hypotheses of Proposition 7.4, with

$$\begin{bmatrix} p^{\#}(\gamma) \\ (\oplus[F]\cap -)^{-1}p^{\#}(\mu) \end{bmatrix} : (p^{\#}G,0) \longrightarrow H_{(-)^i}(p^{\#}F)$$

the inclusion of a lagrangian of the $(-)^i$-quadratic hyperbolic form $H_{(-)^i}(p^{\#}F)$ in $\mathbb{D}_n(\mathbf{Z}[\pi_1(E)])$ because it extends to an isomorphism of $(-)^i$-quadratic forms

$$\begin{bmatrix} p^{\#}(\gamma) & p^{\#}(\tilde{\gamma})(\oplus[F]\cap -) \\ (\oplus[F]\cap -)^{-1}p^{\#}(\mu) & (\oplus[F]\cap -)^{-1}p^{\#}(\tilde{\mu})(\oplus[F]\cap -) \end{bmatrix}$$

$$: H_{(-)^i}(p^{\#}G) \longrightarrow H_{(-)^i}(p^{\#}F) \ .$$

Working as in the proof of Lemma 8.1 the hessian $(-)^{i+1}$-quadratic form in $\mathbb{D}_n(\mathbf{Z}[\pi_1(E)])$ may be expressed as $(p^{\#}G,(\oplus[F]\cap -)^{-1}p^{\#}(\hat{\theta}))$, with $\hat{\theta}:G\longrightarrow G^*$ an $(-)^{i+1}$-ultraquadratic structure for (G,θ). By Proposition 7.4 the nonsingular $(-)^i$-quadratic formation $(p^{\#}F,p^{\#}G)$ in $\mathbb{D}_n(\mathbf{Z}[\pi_1(E)])$ is such that

$$p^!_{geo}\sigma_*(f,b) = \sigma_*(g,c) = \mu(p^{\#}F,p^{\#}G)$$

$$= p^!_{alg}(F,G) = p^!_{alg}\sigma_*(f,b)$$

$$\in \mathrm{im}(\mu:L_1(\mathbb{D}_n(\mathbf{Z}[\pi_1(E)]),(-)^i) \longrightarrow L_{n+1}(\mathbf{Z}[\pi_1(E)],(-)^i))$$

$$= \mathrm{im}(\mu:L_{2i+1}(\mathbb{D}_n(\mathbf{Z}[\pi_1(E)])) \longrightarrow L_{n+2i+1}(\mathbf{Z}[\pi_1(E)])) \ .$$

This verifies $p^!_{geo}=p^!_{alg}$ also in the case $m=2i+1$, completing the proof of Theorem 6.2.

We can now write the surgery transfer maps unambiguously as

$$p^! : L_m(\mathbf{Z}[\pi_1(B)]) \longrightarrow L_{m+n}(\mathbf{Z}[\pi_1(E)]) \quad (m\geqslant 0) \ .$$

§9. Change of K-theory

We now extend the definition of the algebraic surgery transfer maps $(C,\alpha,U)^!:L_m(R)\longrightarrow L_{m+n}(S)$ to the intermediate L-groups, and show that they are compatible with the Rothenberg exact sequences.

An involution $R\longrightarrow R;r\longrightarrow\bar{r}$ on a ring R determines a duality involution $*:\mathbb{P}(R)\longrightarrow\mathbb{P}(R);P\longrightarrow P^*=\mathrm{Hom}_R(P,R)$ on the additive category $\mathbb{P}(R)$ of f.g. projective R-modules by

$$R\times P^*\longrightarrow P^*\ ;\ (r,f)\longrightarrow(x\longrightarrow f(x).\bar{r})\ ,$$

$$e(P)\ :\ P\xrightarrow{\ \simeq\ }P^{**}\ ;\ x\longrightarrow(f\longrightarrow\overline{f(x)})\ .$$

The duality involution on $\mathbb{P}(R)$ determines involutions on the algebraic K-groups

$$*\ :\ K_0(R)\longrightarrow K_0(R)\ ;\ [P]\longrightarrow[P^*]\ ,$$

$$*\ :\ K_1(R)\longrightarrow K_1(R)\ ;$$

$$\tau(f:P\longrightarrow Q)\longrightarrow\tau(f^*:Q^*\longrightarrow P^*)$$

and also on the reduced K-groups

$$\tilde{K}_i(R)=\mathrm{coker}(K_i(\mathbb{Z})\longrightarrow K_i(R))\quad(i=0,1)\ .$$

The intermediate quadratic L-groups $L^X_*(R)$ of a ring with involution R are defined for *-invariant subgroups $X\subseteq\tilde{K}_i(R)$ $(i=0,1)$, such that $x^*\in X$ for all $x\in X$. The intermediate L-groups for $X=\{0\},\tilde{K}_i(R)$ are written as

$$L_*^{\tilde{K}_0(R)}(R)=L^p_*(R)\quad,\quad L_*^{\{0\}\subseteq\tilde{K}_1(R)}(R)=L^s_*(R)\quad,$$

$$L_*^{\{0\}\subseteq\tilde{K}_0(R)}(R)=L_*^{\tilde{K}_1(R)}(R)=L^h_*(R)=L_*(R)\ .$$

For $*$-invariant subgroups $X \subseteq X' \subseteq \tilde{K}_i(R)$ there is defined a Rothenberg exact sequence

$$\ldots \longrightarrow L_n^X(R) \longrightarrow L_n^{X'}(R) \longrightarrow \hat{H}^n(\mathbb{Z}_2; X'/X) \longrightarrow L_{n-1}^X(R) \longrightarrow \ldots$$

with

$$\hat{H}^n(\mathbb{Z}_2; X'/X) = \{a \in X'/X \mid a^* = (-)^n a\} / \{b + (-)^n b^* \mid b \in X'/X\} .$$

See Ranicki [13],[14] for further details.

We consider first the torsion case $X \subseteq \tilde{K}_1(R)$.

A representation (C,U) of R in $\mathbb{D}(S)$ determines a transfer map in the absolute torsion groups $(C,U)^! : K_1(R) \longrightarrow K_1(S)$ (Example 1.8), and also in the reduced torsion groups $(C,U)^! : \tilde{K}_1(R) \longrightarrow \tilde{K}_1(S)$. By definition, $\mathbb{D}(S)$ is the homotopy category of finite chain complexes of based f.g. free S-modules. We shall now make use of the bases.

<u>Proposition</u> <u>9.1</u> Let (C,α,U) be a symmetric representation of R in $\mathbb{D}_n(S)$, for some rings with involution R,S.
i) For any $*$-invariant subgroups $X \subseteq \tilde{K}_1(R)$, $Y \subseteq \tilde{K}_1(S)$ such that $(C,U)^!(X) \subseteq Y$ and $\tau(\alpha : C \longrightarrow C^{n-*}) \in Y$ there are defined transfer maps in the intermediate torsion L-groups

$$(C,\alpha,U)^! \ : \ L_m^X(R) \longrightarrow L_{m+n}^Y(S) \quad (n \geqslant 0) .$$

ii) For any $*$-invariant subgroups $X \subseteq X' \subseteq \tilde{K}_1(R)$, $Y \subseteq Y' \subseteq \tilde{K}_1(S)$ such that $(C,U)^!(X) \subseteq Y$, $(C,U)^!(X') \subseteq Y'$, $\tau(\alpha) \in Y$ there is defined a morphism of Rothenberg exact sequences

$$\begin{array}{ccccccc} \cdots \longrightarrow & L^X_m(R) & \longrightarrow & L^{X'}_m(R) & \longrightarrow & \hat{H}^m(\mathbb{Z}_2;X'/X) & \longrightarrow \cdots \\ & \Big\downarrow (C,\alpha,U)^! & & \Big\downarrow (C,\alpha,U)^! & & \Big\downarrow (C,U)^! & \\ \cdots \longrightarrow & L^Y_{m+n}(S) & \longrightarrow & L^{Y'}_{m+n}(S) & \longrightarrow & \hat{H}^{m+n}(\mathbb{Z}_2;Y'/Y) & \longrightarrow \cdots . \end{array}$$

<u>Proof</u>: The transfer map in the reduced torsion groups $(C,U)^! : \tilde{K}_1(R) \longrightarrow \tilde{K}_1(S)$ is such that

$$*(C,U)^! = (-)^n (C,U)^! * \ : \ \tilde{K}_1(R) \longrightarrow \tilde{K}_1(S) \ .$$

Let $m=2i$. For any nonsingular $(-)^i$-quadratic form (M,ψ) on a based f.g. free R-module $M=R^k$ the n-dimensional $(-)^i$-quadratic Poincaré complex $(\underset{k}{\oplus}C,\theta)$ representing $(C,U)^!(M,\psi)$ has reduced torsion

$$\begin{aligned} \tau((1+T)\theta_0 : \underset{k}{\oplus}C^{n-*} \longrightarrow \underset{k}{\oplus}C) \\ = (C,U)^! \tau(\psi+(-)^i\psi^* : M \longrightarrow M^*) \in \tilde{K}_1(S) \end{aligned}$$

the image of $\tau(\psi+(-)^i\psi^*) \in \tilde{K}_1(R)$. Similarly for $m=2i+1$ and formations.

□

Next, we consider the projective case $X \subseteq \tilde{K}_0(R)$. It is more convenient to work with the preimage of X in $K_0(R)$, so we regard X as a $*$-invariant subgroup of $K_0(R)$ such that $[R] \in X$.

Given a ring S let $\mathbb{E}(S) = \mathbb{D}(\mathbb{P}(S))$, the homotopy category of finite-dimensional f.g. projective S-module chain complexes. A representation (C,U) of a ring R in $\mathbb{E}(S)$ determines transfer maps in the algebraic K-groups

$$(C,U)^! \ : \ K_i(R) = K_i(\mathbb{P}(R)) \longrightarrow$$

$$K_i(S) = K_i(\mathbb{P}(S)) \quad (i=0,1)$$

(Example 1.8). For $n \geq 0$ let $\mathbb{E}_n(S) = \mathbb{D}_n(\mathbb{P}(S))$, the full subcategory of $\mathbb{E}(S)$ with objects n-dimensional f.g. projective S-module chain complexes. An involution on S determines the n-duality involution $C \longrightarrow C^{n-*}$ on $\mathbb{E}_n(S)$.

Proposition 9.2 Let (C,α,U) be a symmetric representation of R in $\mathbb{E}_n(S)$, for some rings with involution R,S.

i) For any $*$-invariant subgroups $X \subseteq K_0(R)$, $Y \subseteq K_0(S)$ such that $[R] \in X$, $[S] \in Y$, $(C,U)^!(X) \subseteq Y \subseteq K_0(S)$ there are defined transfer maps in the intermediate class L-groups

$$(C,\alpha,U)^! : L^X_m(R) \longrightarrow L^Y_{m+n}(S) \quad (n \geq 0) .$$

ii) For any $*$-invariant subgroups $X \subseteq X' \subseteq K_0(R)$, $Y \subseteq Y' \subseteq K_0(S)$ such that $[R] \in X$, $[S] \in Y$, $(C,U)^!(X) \subseteq Y$, $(C,U)^!(X') \subseteq Y'$ there is defined a morphism of Rothenberg exact sequences

$$\begin{array}{ccccccccc}
\dots & \longrightarrow & L^X_m(R) & \longrightarrow & L^{X'}_m(R) & \longrightarrow & \hat{H}^m(\mathbb{Z}_2;X'/X) & \longrightarrow & \dots \\
 & & \Big\downarrow (C,\alpha,U)^! & & \Big\downarrow (C,\alpha,U)^! & & \Big\downarrow (C,U)^! & & \\
\dots & \longrightarrow & L^Y_{m+n}(S) & \longrightarrow & L^{Y'}_{m+n}(S) & \longrightarrow & \hat{H}^{m+n}(\mathbb{Z}_2;Y'/Y) & \longrightarrow & \dots .
\end{array}$$

□

The proof of 9.2 is somewhat more involved than that of 9.1.

A splitting (B,r,i) in $\mathbb{A}$ of an object (A,p) in the idempotent completion $\hat{\mathbb{A}}$ is an object B in $\mathbb{A}$ together with morphisms $r:A \longrightarrow B$, $i:B \longrightarrow A$ in $\mathbb{A}$ such that

$$ri = 1 : B \longrightarrow B \ , \ ir = p : A \longrightarrow A \ .$$

Lemma 9.3 A functor of additive categories $F:\mathbb{A} \longrightarrow \mathbb{B}$

extends to a functor $\hat{F}:\hat{\mathbb{A}}\longrightarrow\mathbb{B}$ if and only if for each object (A,p) in $\hat{\mathbb{A}}$ the object $(F(A),F(p))$ in $\hat{\mathbb{B}}$ has a splitting in $\mathbb{B}$. Any two such extensions of F are naturally equivalent.

Proof: It is clear that the splitting condition is necessary for F to extend to $\hat{F}$, so we need only prove that it is sufficient. For each object (A,p) in $\hat{\mathbb{A}}$ choose a splitting (B,r,i) of the object $(F(A),F(p))$ in $\hat{\mathbb{B}}$, and set $\hat{F}(A,p)=B$, with $(B,r,i)=(F(A),1,1)$ for $p=1:A\longrightarrow A$. For a morphism $f:(A,p)\longrightarrow(A',p')$ let

$$\hat{F}(f) : \hat{F}(A) = B \xrightarrow{\ i\ } A \xrightarrow{\ f\ } A' \xrightarrow{\ r'\ } \hat{F}(A') = B' .$$

□

An additive category $\mathbb{A}$ is idempotent complete if the functor $\mathbb{A}\longrightarrow\hat{\mathbb{A}};A\longrightarrow(A,1)$ is an equivalence of categories. Applying 9.3 to $1:\mathbb{A}\longrightarrow\mathbb{A}$ we have that $\mathbb{A}$ is idempotent complete if and only if every object (A,p) in $\hat{\mathbb{A}}$ splits in $\mathbb{A}$. If $\mathbb{B}$ is idempotent complete every functor $F:\mathbb{A}\longrightarrow\mathbb{B}$ extends to a functor $\hat{F}:\hat{\mathbb{A}}\longrightarrow\mathbb{B}$, namely the composite of $\hat{F}:\hat{\mathbb{A}}\longrightarrow\hat{\mathbb{B}}$ and an equivalence $\hat{\mathbb{B}}\longrightarrow\mathbb{B}$.

For any ring S the additive category $\mathbb{P}(S)$ of f.g. projective S-modules is idempotent complete, with every object (A,p) in $\hat{\mathbb{P}}(S)$ split by the triple (B,r,i) defined by

$$r : A \longrightarrow B = \mathrm{im}(p) \ ;\ x \longrightarrow p(x) ,$$

$$i = \text{inclusion} : B \longrightarrow A .$$

This is the special case $n=0$ of:

Lemma 9.4 For any ring S and any $n\geqslant 0$ the homotopy category $\mathbb{E}_n(S)$ of n-dimensional f.g. projective S-module chain complexes is idempotent complete.

Proof: For every chain homotopy projection $p\simeq p^2:D\longrightarrow D$ of an object D in $\mathbb{E}_n(S)$ there exists by Lemma 3.4 of Lück [7] an $(n+1)$-dimensional infinitely generated projective S-module chain complex C with chain maps $r:D\longrightarrow C$, $i:C\longrightarrow D$ and chain homotopies

$$ri \simeq 1 : C \longrightarrow C \quad , \quad ir \simeq p : D \longrightarrow D \ .$$

Since C is dominated by an object in $\mathbb{E}_n(S)$ (namely D) it is chain equivalent to an object in $\mathbb{E}_n(S)$, by Proposition 3.1 of Ranicki [17].

□

The idempotent completion of an additive category $\mathbb{A}$ with an involution $*:\mathbb{A}\longrightarrow\mathbb{A}$ is an additive category $\hat{\mathbb{A}}$ with the involution

$$* : \hat{\mathbb{A}} \longrightarrow \hat{\mathbb{A}} \ ; \ (A,p) \longrightarrow (A^*,p^*) \ .$$

For a ring with involution R the functor $\hat{\mathbb{B}}(R)\longrightarrow\mathbb{P}(R);(R^k,p)\longrightarrow im(p)$ is an equivalence of additive categories with involution. Both 9.3 and 9.4 have evident versions for additive categories with involution.

<u>Definition</u> <u>9.5</u> Let R,S be rings with involution. The <u>projective</u> <u>surgery</u> <u>transfer</u> <u>maps</u> of a symmetric representation (C,α,U) of R in $\mathbb{E}_n(S)$

$$(C,\alpha,U)^! \ : \ L^p_m(R) \longrightarrow L^p_{m+n}(S) \quad (m\geqslant 0)$$

are the composites

$$(C,\alpha,U)^! \ : \ L^p_m(R) = L_m(\mathbb{P}(R)) \xrightarrow{\hat{F}} L_m(\mathbb{E}_n(S)) \xrightarrow{\mu} L^p_{m+n}(S)$$

with μ the generalized Morita maps of 3.3 for $\mathbb{A}=\mathbb{P}(S)$ and $\hat{F}$ induced by the functor of additive categories with involution $\hat{F}:\hat{\mathbb{B}}(R)\simeq\mathbb{P}(R)\longrightarrow\mathbb{E}_n(S)$ associated by 9.4 to the functor

$$F = -\otimes(C,\alpha,U) : \mathbb{B}(R) \longrightarrow \mathbb{E}_n(S) \; ; \; R \longrightarrow C \; .$$

□

The proof of 9.2 is now completed by observing that the transfer map in the projective class groups $(C,U)^!:K_0(R)\longrightarrow K_0(S)$ is such that

$$*(C,U)^! = (-)^n(C,U)^!* : K_0(R) \longrightarrow K_0(S) \; .$$

<u>Remark 9.6</u> Our methods also apply to construct algebraic surgery transfer maps in the round L-groups $L_*^{rX}(R)$ of Hambleton, Ranicki and Taylor [4], which are defined for $*$-invariant subgroups $X\subseteq K_1(R)$. For any symmetric representation (C,α,U) of R in $\mathbb{E}_n(S)$ and any $*$-invariant subgroup $X\subseteq K_1(R)$, $Y\subseteq K_1(S)$ such that $(C,U)^!(X)\subseteq Y$ there are defined round L-theory transfer maps

$$(C,\alpha,U)^! : L_m^{rX}(R) \longrightarrow L_{m+n}^{rY}(S) \quad (m\geq 0)$$

which are compatible with the round L-theory Rothenberg exact sequences.

□

<u>Remark 9.7</u> The connection established in §8 between the algebraic and geometric surgery transfer maps extends to the intermediate cases, and also to round L-theory.

□

<u>Remark 9.8</u> Our algebraic constructions apply also to the ϵ-quadratic L-groups $L_*(R,\epsilon)$, which are defined for a ring with involution R and a central unit $\epsilon\in R$ such that $\bar{\epsilon}\epsilon=1$. $L_{2i}(R,\epsilon)$ (resp. $L_{2i+1}(R,\epsilon)$) is the Witt group of nonsingular $(-)^i\epsilon$-quadratic forms (resp. formations) over R. A symmetric representation (C,α,U) of R in $\mathbb{D}_n(S)$ such that $U(\epsilon)=\eta:C\longrightarrow C$ for a central unit

$\eta \in S$ with $\bar{\eta}\eta = 1$ induces transfer maps

$$(C,\alpha,U)^{!} \; : \; L_m(R,\epsilon) \longrightarrow L_{m+n}(S,\eta) \quad (m \geqslant 0) \; .$$

Hitherto we considered the case $\epsilon = 1 \in R$ for which $L_*(R,1) = L_*(R)$, with $\eta = 1 \in S$.

□

Appendix 1. Fibred intersections

The proof of $p^{!}_{geo} = p^{!}_{alg}$ in §8 makes heavy use of the algebraic properties of the L-groups. For a fibre bundle $F \longrightarrow E \xrightarrow{p} B$ with the fibre F a compact n-dimensional manifold it is possible to verify that the algebraic and geometric surgery transfer maps coincide more directly, using the bordism intersection theory of Hatcher and Quinn [6] to obtain fibred versions of the geometric intersection forms (resp. formations) used by Wall [22] to define the surgery obstruction of a highly-connected even (resp. odd-) dimensional normal map. The quadratic kernel of the pullback normal map is the fibred intersection form (resp. formation) both algebraically and geometrically. We now sketch the argument for the intersection pairing λ in the even-dimensional case, leaving the self-intersection function μ and the odd-dimensional case to the interested reader.

Given two maps $v_i : Q_i \longrightarrow M$ $(i=1,2)$ let $E(v_1,v_2)$ be the pointed space of triples (x_1,x_2,ω) defined by points $x_i \in Q_i$ and a path $\omega : [0,1] \longrightarrow M$ from $\omega(0) = v_1(x_1)$ to $\omega(1) = v_2(x_2)$, so that there is defined a homotopy fibre square

$$\begin{array}{ccc} E(v_1,v_2) & \longrightarrow & Q_1 \\ \Big\downarrow & & \Big\downarrow{\scriptstyle v_1} \\ Q_2 & \xrightarrow{\;v_2\;} & M \; . \end{array}$$

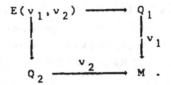

Given a stable vector bundle η over a space M let $\Omega_n^{fr}(M,\eta)$ be the bordism group of n-manifolds N equipped with a map $N \longrightarrow M$ and a compatible stable bundle map $\nu_N \longrightarrow \eta$. For trivial η this is the usual framed cobordism group $\Omega_n^{fr}(M)=\pi_n^S(M \cup \{*\})$. For $v_1=v_2:Q_1=Q_2=\{*\} \longrightarrow M$ the homotopy pullback is the loop space, $E(*,*)=\Omega M$.

Now suppose that M is an m-manifold, and that $v_i:Q_i \longrightarrow M$ is an immersion of a q_i-manifold Q_i $(i=1,2)$ such that $v_1(Q_1)$ intersects $v_2(Q_2)$ in general position. Let $Q_1 \cap Q_2$ denote the corresponding (q_1+q_2-m)-dimensional submanifold of M. The bordism invariant of the intersection ([6,2.1]) is the bordism class

$$\lambda(v_1,v_2) = [Q_1 \cap Q_2]$$

$$\in \Omega^{fr}_{q_1+q_2-m}(E(v_1,v_2),\nu_{Q_1}\oplus\nu_{Q_2}\oplus\tau_M) \ .$$

If Q_1 and Q_2 are (q_1+q_2-m+1)-connected the map $E(*,*)=\Omega M \longrightarrow E(v_1,v_2)$ induces an isomorphism ([6,3.1])

$$\Omega^{fr}_{q_1+q_2-m}(E(*,*)) = \Omega^{fr}_{q_1+q_2-m}(\Omega M) \longrightarrow$$

$$\Omega^{fr}_{q_1+q_2-m}(E(v_1,v_2),\nu_{Q_1}\oplus\nu_{Q_2}\oplus\tau_M)$$

which is used as an identification.

Let $(f,b):M \longrightarrow X$ be an $(i-1)$-connected $2i$-dimensional normal map with a π_1-isomorphism reference map $X \longrightarrow B$, with the surgery obstruction $\sigma_*(f,b)=(K_i(M),\lambda,\mu)\in L_{2i}(\mathbf{Z}[\pi_1(B)])$ defined as in Chapter 5 of Wall [22]. Let $v_1,v_2,\ldots,v_k$ be a base of the kernel f.g. free $\mathbf{Z}[\pi_1(B)]$-module $K_i(M)=\pi_{i+1}(f)$. Represent each $v_j\in K_i(M)$ by a pointed framed immersion $v_j:S^i \longrightarrow M$ with a nullhomotopy in X. The values taken by the $(-)^i$-symmetric form $(K_i(M),\lambda)$ on the base elements are just the bordism intersections

$$\lambda(v_j,v_{j'}) \in \Omega_0^{fr}(E(v_j,v_{j'}),\nu_{S^i}\oplus\nu_{S^i}\oplus\tau_M)$$
$$= \Omega_0^{fr}(\Omega M) = H_0(\Omega M) = \mathbb{Z}[\pi_1(B)]$$
$$(1\leqslant j,j'\leqslant k) \ .$$

Now let $(g,c):N\longrightarrow Y$ be the $(i-1)$-connected $(n+2i)$-dimensional normal map with a π_1-isomorphism reference map $Y\longrightarrow E$ obtained from $(f,b):M\longrightarrow X$ by the pullback of the fibre bundle $F\longrightarrow E\xrightarrow{p} B$ along $X\longrightarrow B$. The pointed framed immersions $v_j:S^i\longrightarrow M$ $(1\leqslant j\leqslant k)$ with nullhomotopies in X lift to pointed framed immersions $w_j:S^i\times F\longrightarrow N$ with nullhomotopies in Y. On the chain level this corresponds to lifting the kernel $\mathbb{Z}[\pi_1(B)]$-module chain complex $C(f^!)=S^iK_i(M)=\bigoplus_k S^i\mathbb{Z}[\pi_1(B)]$ to the kernel $\mathbb{Z}[\pi_1(E)]$-module chain complex $C(g^!)=\bigoplus_k S^iC(\tilde{F})$. The bordism intersections

$$\lambda(w_j,w_{j'}) \in \Omega_n^{fr}(E(w_j,w_{j'}),\nu_{S^i\times F}\oplus\nu_{S^i\times F}\oplus\tau_N)$$
$$= \Omega_n^{fr}(\Omega M\times F,\nu_F) \quad (1\leqslant j,j'\leqslant k)$$

are the images of the bordism intersections $\lambda(v_j,v_{j'})$ under the geometric bordism transfer map

$$p^! = -\times F \ : \ \Omega_0^{fr}(\Omega M) \longrightarrow \Omega_n^{fr}(\Omega M\times F,\nu_F) \ ;$$
$$\lambda \longrightarrow \lambda\times F \ .$$

The Poincaré duality isomorphism of based f.g. free $\mathbb{Z}[\pi_1(B)]$-modules

$$(\lambda(v_j,v_{j'})) \ :$$
$$C(f^!)^{2i-*} = S^iK_i(M)^* \longrightarrow C(f^!) = S^iK_i(M)$$

is lifted to the Poincaré duality chain equivalence of

chain complexes of based f.g. free $\mathbb{Z}[\pi_1(E)]$-modules

$$(\lambda(w_j,w_{j'})) \; : \; C(g^!)^{n+2i-*} = \bigoplus_k S^i C(\tilde{F})^{n-*}$$
$$\longrightarrow C(g^!) = \bigoplus_k S^i C(\tilde{F}) \; .$$

Using the Poincaré duality $\mathbb{Z}[\pi_1(E)]$-module chain equivalence $[F]\cap - : C(\tilde{F})^{n-*} \longrightarrow C(\tilde{F})$, the action of ΩM on the $\pi_1(E)$-equivariant homotopy type of $\tilde{F}$ and Hurewicz maps there is defined a commutative diagram

$$\begin{array}{ccccc}
\Omega_0^{fr}(\Omega M) & \xrightarrow{\ \simeq\ } & H_0(\Omega M) & \xrightarrow{\ \simeq\ } & \mathbb{Z}[\pi_1(B)] \\
 & & & & \big\downarrow U \\
p^! \big\downarrow & & -\otimes[F] \big\downarrow & & H_0(\mathrm{Hom}_{\mathbb{Z}[\pi_1(E)]}(C(\tilde{F}),C(\tilde{F}))^{op} \\
 & & & & \big\downarrow [F]\cap - \\
\Omega_n^{fr}(\Omega M\times F,\nu_F) & \longrightarrow & H_n(\Omega M\times F) & \longrightarrow & H_n(C(\tilde{F})\otimes_{\mathbb{Z}[\pi_1(E)]}C(\tilde{F}))
\end{array} \; .$$

The anticlockwise composition gives the geometric surgery transfer $p^!_{geo}$ on the level of intersections, while the clockwise composition gives the algebraic surgery transfer $p^!_{alg}$.

Appendix 2. A counterexample in symmetric L-theory

An n-dimensional Poincaré fibration $F \longrightarrow E \xrightarrow{p} B$ does not in general induce transfer maps in the symmetric L-groups $p^! : L^m(\mathbb{Z}[\pi_1(B)]) \longrightarrow L^{m+n}(\mathbb{Z}[\pi_1(E)])$, either algebraically or geometrically. It is not possible to define $p^!$ geometrically since the symmetric L-groups are not geometrically realizable (Ranicki [16,7.6.8]). There are two obstructions to an algebraic definition of $p^!$, which requires the lifting of an m-dimensional symmetric Poincaré complex (C,ϕ) over $\mathbb{Z}[\pi_1(B)]$ representing an element $(C,\phi) \in L^m(\mathbb{Z}[\pi_1(B)])$ to an $(m+n)$-dimensional symmetric Poincaré complex $(C^!,\phi^!)$ over $\mathbb{Z}[\pi_1(E)]$ representing the putative transfer $p^!(C,\phi) = (C^!,\phi^!) \in L^{m+n}(\mathbb{Z}[\pi_1(E)])$. The symmetric L-groups are not 4-periodic, so it cannot be assumed that (C,ϕ)

is highly-connected as in the quadratic case. In the following discussion we assume that the fibre F is finite, and that the chain complex C consists of based f.g. free $\mathbb{Z}[\pi_1(B)]$-modules. The two obstructions to lifting (C,ϕ) to $(C^!,\phi^!)$ are given by:

i) it may not be possible to lift C to a based f.g. free $\mathbb{Z}[\pi_1(E)]$-module chain complex $C^!$ with a filtration $F_0C^! \subseteq F_1C^! \subseteq \ldots \subseteq F_mC^! = C^!$ such that the connecting chain maps between successive filtration quotients are given up to chain homotopy by

$$\partial = p^{\#}(d_C) : F_rC^!/F_{r-1}C^! = S^r p^{\#}(C_r) \longrightarrow$$

$$S(F_{r-1}C^!/F_{r-2}C^!) = S^r p^{\#}(C_{r-1}) \quad (1 \leq r \leq m)$$

where S^r denotes the r-fold dimension shift and $p^{\#}$ is the functor of §1

$$p^{\#} = -\otimes(C(\tilde{F}),U) : \mathbb{B}(\mathbb{Z}[\pi_1(B)]) \longrightarrow \mathbb{D}_n(\mathbb{Z}[\pi_1(E)]) ,$$

ii) even if $C^!$ exists, it may not be possible to lift the m-dimensional symmetric Poincaré structure ϕ on C to an (m+n)-dimensional symmetric Poincaré structure $\phi^!$ on $C^!$.

If C can be assembled over B in the sense of Ranicki and Weiss [20] then it can be lifted to $C^!$, but in general it is not possible to assemble $\mathbb{Z}[\pi_1(B)]$-module chain complexes, so already i) presents a non-trivial obstruction to the existence of transfer in symmetric L-theory. Even if the obstruction of i) vanishes (e.g. if B is an Eilenberg-MacLane space $K(\pi_1(B),1)$) then ii) may present a non-trivial obstruction. This is illustrated by the following example, which exhibits the failure of a projection of rings with involution $p:S \longrightarrow R = S/(1-t)$ (t = central unit $\in$ S, $\bar{t} = t^{-1} \in S$) to induce an S^1-bundle symmetric L-theory transfer map $p^! : L^0(R) \longrightarrow L^1(S)$ analogous to the S^1-bundle quadratic L-theory transfer map $p^! : L_0(R) \longrightarrow L_1(S)$ (cf. 4.7). The transfer $p^!(C,\phi)=(C^!,\phi^!)$ of a 0-dimensional symmetric Poincaré complex (= nonsingular symmetric form) (C,ϕ) over R with $C_0=R^k$ is defined if the symmetric $k\times k$ matrix

$$\phi_0 = (\phi_0)^* \in M_k(R)$$

can be lifted to a $k\times k$ matrix $\phi_0^! \in M_k(S)$ such that $p(\phi_0^!)=\phi_0 \in M_k(R)$ and

$$t\phi_0^! - (\phi_0^!)^* = (1-t)\phi_1^! \in M_k(S)$$

for some symmetric $k\times k$ matrix $\phi_1^!=(\phi_1^!)^* \in M_k(S)$, so that $(C^!,\phi^!)$ is a 1-dimensional symmetric Poincaré complex over S with $C^!=C(1-t:S^k\longrightarrow S^k)$. In particular, for

$$S = \mathbb{Z}_2[\mathbb{Z}_2\times\mathbb{Z}_2] = \mathbb{Z}_2[t,u]/(t^2-1,u^2-1) ,$$

$$\bar{t} = t , \bar{u} = t+u+1 ,$$

$$p : S \longrightarrow R = \mathbb{Z}_2[\mathbb{Z}_2] = \mathbb{Z}_2[u]/(u^2-1) ;$$

$$t \longrightarrow 1 , u \longrightarrow u$$

the transfer is not defined for the 0-dimensional symmetric Poincaré complex $(C,\phi)=(R,u)$ over R, for although C can be lifted to $C^!$ and ϕ_0 can be lifted to $\phi_0^!$ there does not exist a symmetric $\phi_1^!$. Both the obstructions to i) and ii) vanish for the visible symmetric L-groups $VL^*(\mathbb{Z}[\pi])$ of Weiss [23] provided that B is an Eilenberg-MacLane space $K(\pi_1(B),1)$, in which case there are defined transfer maps $p^!:VL^m(\mathbb{Z}[\pi_1(B)])\longrightarrow VL^{m+n}(\mathbb{Z}[\pi_1(E)])$.

REFERENCES

[1] W.Browder and F.Quinn
A surgery theory for G-manifolds and stratified sets
Proceedings 1973 Tokyo Conference on

Manifolds, Tokyo Univ. Press, 27-36 (1974)

[2] D.Gottlieb
Poincaré duality and fibrations
Proc. A.M.S. 76, 148-150 (1979)

[3] I.Hambleton, J.Milgram, L.Taylor and B.Williams
Surgery with finite fundamental group
Proc. Lond. Math. Soc. (3) 56, 349-379 (1988)

[4] I.Hambleton, A.Ranicki and L.Taylor
Round L-theory
J. Pure and Appl. Alg. 47, 131-154 (1987)

[5] I.Hambleton, L.Taylor and B.Williams
Maps between surgery obstruction groups
Proc. 1982 Arhus Topology Conf.,
Springer Lecture Notes 1051, 149-227 (1984)

[6] A.Hatcher and F.Quinn
Bordism invariants of intersections of submanifolds
Trans. A.M.S. 200, 326-344 (1974)

[7] W.Lück
The transfer maps induced in the algebraic K_0- and K_1-groups by a fibration I.
Math. Scand. 59, 93-121 (1986)

[8] W.Lück and I.Madsen
Equivariant L-theory II.
to appear

[9] W.Lück and A.Ranicki
Chain homotopy projections
to appear in J. of Algebra

[10] H.Munkholm and E.Pedersen
The S^1-transfer in surgery theory
Trans. A.M.S. 280, 277-302 (1983)

[11] F.Quinn
A geometric formulation of surgery
Princeton Ph.D.thesis (1969)

[12] Surgery on Poincaré and normal spaces
Bull. A.M.S. 78, 262-267 (1972)

[13] A.Ranicki
Algebraic L-theory I. Foundations
Proc. Lond. Math. Soc. (3) 27, 101-125 (1973)

[14] The algebraic theory of surgery I. Foundations
Proc. Lond. Math. Soc. (3) 40, 87-192 (1980)

[15] The algebraic theory of surgery II. Applications to topology
Proc. Lond. Math. Soc. (3) 40, 193-283 (1980)

[16] Exact sequences in the algebraic theory of surgery
Mathematical Notes 26, Princeton (1981)

[17] The algebraic theory of finiteness obstruction
Math. Scand. 57, 105-126 (1985)

[18] The algebraic theory of torsion I. Foundations
Algebraic and Geometric Topology,
Springer Lecture Notes 1126, 199-237 (1985)

[19] Additive L-theory
Mathematica Gottingensis 12 (1988)

[20] A.Ranicki and M.Weiss
Chain complexes and assembly
Mathematica Gottingensis 28 (1987)

[21] C.T.C.Wall
Poincaré complexes
Ann. of Maths. 86, 213-245 (1970)

[22] Surgery on compact manifolds
Academic Press (1970)

[23] M.Weiss
On the definition of the symmetric L-groups
preprint

[24] G.W.Whitehead
Elements of homotopy theory
Springer (1978)

W.Lück: Mathematisches Institut,
Georg-August Universität,
Bunsenstr. 3-5,
34 Göttingen,
Bundesrepublik Deutschland.

A.Ranicki: Mathematics Department,
Edinburgh University,
Edinburgh EH9 3JZ,
Scotland, UK.

SOME REMARKS ON THE KIRBY-SIEBENMANN CLASS

R. J. Milgram

In this note we study the relations that hold between the Kirby-Siebenmann class $\{KS\} \in H^4(B_{STOP};\ \mathbf{Z}/2)$ and the first Pontrajagin class.

The first result is that that the natural map $p_0 :\ B_{STOP} \longrightarrow B_{SG}$ does not detect $\{KS\}$ no matter what coefficients might be used. However, the homology dual of $\{KS\}$ is in the image of the Hurewicz map

$$\pi_4(B_{STOP}) \longrightarrow H_4(B_{STOP};\ \mathbf{Z}/2).$$

In fact there is a unique non-zero element $[KS] \in \pi_4(B_{STOP})$ of order 2, and $p_0([KS]) \neq 0 \in \pi_4(B_{SG})$. In particular this implies that $w_4 + \{KS\}$ is a mod(24) fiber-homotopy invariant of SPIN-TOP bundles. However, it is interesting to ask what happens when w_2 is non-zero. To understand this we introduce an intermediate classifying space, B_{TSG} for which we have a factorization

$$p_0 = p \cdot f, \quad B_{STOP} \xrightarrow{f} B_{TSG} \xrightarrow{p} B_{SG}.$$

B_{TSG} is univeral for the vanishing of transversality obstructions through dimension 5. Additionally, B_{TSG} is built out of finite groups ($\mathbf{Z}/2$-extensions of the symmetric groups S_n) in the same way that B_{SG} is constructed from the S_n. As a result, explicit construction of homotopy classes of maps into B_{TSG} is often possible.

We show that $H_4(B_{TSG};\ \mathbf{Z}/2) \ = \ \mathbf{Z}/2 \oplus \mathbf{Z}/48$ and that the homology dual of the Kirby-Siebenmann class maps to 24 times the second generator. Thus, this transversality theory does detect $\{KS\}$. But note also the $\mathbf{Z}/48$. Our main question is the extent to which it gives rise to a fiber homotopy invariant of topological $\mathbf{R}^n$-bundles. The general result is

Theorem I: *Let ξ, ψ be two stable $\mathbf{R}^n$-bundles over X, and suppose they are fiber homotopy equivalent. Then there is $\alpha \in H^2(X;\ \mathbf{Z}/2)$ and*

$$24\alpha^2 + P_1(\xi) + 24\{KS(\xi)\} \ = \ P_1(\psi) + 24\{KS(\psi)\}$$

in $H^4(X;\ \mathbf{Z}/48)$ where $P_1(\xi)$ is the $\mathbf{Z}/48$ reduction of the first Pontrajagin class.

In other words, there is an element $A \in H^4(B_{TSG};\ \mathbf{Z}/48)$ with $f^*(A) \ = \ P_1 + 24\{KS\}$, and (I) gives the effect of different liftings of a map $p_0 \cdot g :\ X \longrightarrow B_{STOP} \longrightarrow B_{SG}$ on A.

$H^2(B_{STOP};\ \mathbf{Z}/2) \ = \mathbf{Z}/2$ with generator w_2, so the possible factorizations of p_0 through B_{TSG} differ in their effect on A only by $24w_2^2$. In particular this gives

Corollary: *If M^4 is a compact closed topological manifold with even index, and ν is its stable normal bundle, then $w_2^2 = 0 \in H^2(M;\ \mathbf{Z}/2)$ and*

$$\nu^* f^*(A) = P_1(\nu) + 24\{KS(\nu)\}$$

is independent of the choice of f factoring p_0.

This note came about in answer to a question of Frank Quinn. He pointed out that in [M-M] the exact structure of B_{STOP}, and the various surgery maps in dimension 4 were never worked out. But currently it appears very useful to understand them. Of course, we do not attempt to work out explicit geometric methods for evaluating the new invariants. But knowing what they are and how they fit together should make that fairly direct.

The homotopy types of B_{SO}, B_{SG} in dimension ≤ 7

A Postnikov system for B_{SO} through dimension 7 is given by

(1) $$B_{SO} \longrightarrow K(\mathbf{Z}/2,\ 2) \longrightarrow K(\mathbf{Z},\ 5)$$

with K-invariant $2\{Sq^2Sq^1(\iota_2) + \iota_2 \cdot Sq^1(\iota_2)\}$. (Note that $H^5(K(\mathbf{Z}/2,\ 2);\ \mathbf{Z}) = \mathbf{Z}/4$ with generator having mod(1) reduction γ and

(2) $$\gamma = Sq^2Sq^1(\iota_2) + \iota_2 \cdot Sq^1(\iota_2).$$

Moreover, $\beta_4(\iota_2^2) = \gamma$.)

The stable homotopy of spheres is given in the first 6 dimensions by

(3) $$\pi_i^s(S^0) = \begin{cases} \mathbf{Z} & i = 0 \\ \mathbf{Z}/2 & i = 1,\ \text{generator } \eta \\ \mathbf{Z}/2 & i = 2,\ \text{generator } \kappa_1 \\ \mathbf{Z}/24 & i = 3,\ \text{generator } \nu \\ 0 & i = 4,\ 5 \\ \mathbf{Z}/2 & i = 6,\ \text{generator } \kappa_2 = \nu^2 \end{cases}$$

and we will use the same names for the corresponding elements in $\pi_{i+1}(B_{SG}) \cong \pi_i^s(S^0)$. One relation that should be kept in mind is $\eta\kappa_1 = 12\nu$, since it also holds in $\pi_*(B_{SG})$, though the relation $\eta^2 = \kappa_1$ which holds stably does not hold in $\pi_*(B_{SG})$.

Lemma (4): *A Postnikov system for B_{SG} through 7 is given by*

$$K(\mathbf{Z}/2,\ 2) \times K(\mathbf{Z}/2,\ 3) \times K(\mathbf{Z}/2,\ 7) \longrightarrow K(\mathbf{Z}/24,\ 5)$$

where the K-invariant is $2\{Sq^2Sq^1(\iota_2) + \iota_2 \cdot Sq^1(\iota_2)\} + 4\{Sq^2(\iota_3)\}$.

Proof: With $\mathbf{Z}/24$-coefficients the K-invariant for B_{SG} maps back to the image of the corrisponding K-invariant for B_{SO}. Hence, the class in (2) must appear in the K-invariant. Also, the kernel of the map $H^5(K(\mathbf{Z}/2,\ 2,3);\ \mathbf{Z}/24) \longrightarrow H^5(K(\mathbf{Z}/2,\ 3);\ \mathbf{Z}/24)$ is generated by $4Sq^2(\iota_3)$. It follows that $4Sq^2(\iota_3)$ is the only term which can be added to the K-invariant. But, in fact, this term must be involved in the K-invariant because there is the homotopy relation which we have already noted $\eta\kappa_1 = 12\nu$, since η is detected by Sq^2.

In order to understand the integral homology of B_{SG}, B_{STOP}, and the intermediate space B_{TSG} which we will introduce shortly, we need a method for obtaining Bochstein information from K-invariants. The following result will suffice.

Lemma (5): *Let $K(\mathbf{Z}/2^i,\ j) \times K(\mathbf{Z}/2,\ j+1) \xrightarrow{\kappa} K(\mathbf{Z}/2^s,\ j+1)$ be given with*

$$\kappa = 2^w\beta(\iota_j) + 2^{s-1}(\iota_{j+1}),$$

then the fiber E of the map κ is $K(\mathbf{Z}/2^{i+s-w-1} \times \mathbf{Z}/2^w)$.

Proof: The homotopy exact sequence of the fibration in dimensions j, $j+1$ is

(6) $$0 \longrightarrow \pi_{j+1}(E) \longrightarrow \mathbf{Z}/2 \xrightarrow{\kappa_*} \mathbf{Z}/2^s \xrightarrow{\partial} \pi_j(E) \longrightarrow \mathbf{Z}/2^j \longrightarrow 0$$

But the term $2^{s-1}\iota_{j+1}$ in $\kappa^*(\iota_{j+1})$ implies that κ_* is injective in (6). Thus E is a $K(\pi,\ j)$ and π is given as an extension in the sequence

$$0 \longrightarrow \mathbf{Z}/2^{s-1} \longrightarrow \pi_j(E) \longrightarrow \mathbf{Z}/2^i \longrightarrow 0.$$

The type of this extension is determined by the term $2^w(\beta(\iota_j))$ in $\kappa^*(\iota_{j+1})$. From this (5) follows.

(4) and (5) imply that there is a mod(8) Bochstein

$$\beta_8(\iota_2^2) = \{Sq^2(\iota_3)\} \text{ in } H^*(B_{SG};\ \mathbf{Z}/2).$$

Additionally, the Hurewicz image of ν is $\{w_4^*\} + 2\{\iota_2^{2,*}\}$ since this is already true in B_{SO}, where it is well known. As a consequence $H_4(B_{SG};\ \mathbf{Z}) = \mathbf{Z}/2 \oplus \mathbf{Z}/24$ with generators $\{w_4^*\}$, $\{w_2^{2,*}\}$ respectively, and 12ν is in the kernel of the Hurewicz map.

The structure of B_{STOP} through dimension 7

From the fiberings

$$(7) \qquad \begin{array}{ccccc} G/O & \longrightarrow & B_{SO} & \longrightarrow & B_{SG} \\ \downarrow & & \downarrow & & \downarrow \\ G/TOP & \longrightarrow & B_{STOP} & \longrightarrow & B_{SG} \end{array}$$

and the well known result of Kirby-Siebenmann that $\pi_4(G/TOP) = \pi_4(G/O) = \mathbf{Z}$, but that the map between them is multiplication by 2, we get the diagram of extensions in π_4,

$$(8) \qquad \begin{array}{ccccccccc} 0 & \longrightarrow & \mathbf{Z} & \xrightarrow{\cdot 24} & \mathbf{Z} & \longrightarrow & \mathbf{Z}/24 & \longrightarrow & 0 \\ & & \downarrow{\scriptstyle \cdot 2} & & \downarrow & & \downarrow{\scriptstyle =} & & \\ 0 & \longrightarrow & \mathbf{Z} & \longrightarrow & \pi_4(B_{STOP}) & \longrightarrow & \mathbf{Z}/24 & \longrightarrow & 0 \end{array}$$

The only way this diagram can commute is if $\pi_4(B_{STOP}) = \mathbf{Z}/2 \oplus \mathbf{Z}$ with the element of order 2 mapping to $12 \cdot \nu$, and the generator of the $\mathbf{Z}$-summand mapping to ν.

Lemma (9): $\pi_i(B_{STOP}) = \begin{cases} \mathbf{Z}/2 & i = 2 \\ \mathbf{Z} \oplus \mathbf{Z}/2 & i = 4 \\ 0 & 4 < i < 8. \end{cases}$ *Moreover, a Postnikov system for B_{STOP} through this range is given by*

$$(10) \qquad K(\mathbf{Z}/2,\ 2) \times K(\mathbf{Z}/2,\ 4) \longrightarrow K(\mathbf{Z},\ 5)$$

with K-invariant $2\{Sq^2Sq^1(\iota_2) + \iota_2 \cdot Sq^1(\iota_2)\}$.

(This is clear.)

In particular, the class $\{KS^*\} \in H_4(B_{STOP};\ \mathbf{Z})$ which is in the Hurewicz image of the element of order 2, must go to zero in $H_4(B_{SG};\ \mathbf{Z})$, since, in homotopy, it goes to 12ν. This shows that $\{KS^*\}$ has no homology (or cohomology) relations implied by the map into B_{SG}. However, in homotopy, the fact that it maps to 12ν should have some consequeences.

The space B_{TSG}

The failure to detect the Kirby-Siebenmann class in $H_*(B_{SG};\ \mathbf{Z})$ is the influence of the first exotic class ι_3. In fact, the term $4Sq^2(\iota_3)$ in the 5-dimensional K-invariant (4) is exactly the difficulty. (For example, if we kill w_2 but leave ι_3 in $H^*(B_{SG};\ \mathbf{Z}/2)$ the resulting space has only $\mathbf{Z}/4$-torsion in $H_4(\ \ ;\ \mathbf{Z})$.) Hence it is natural to consider the classifying space B_{TSG} obtained from B_{SG} by killing the exotic class ι_3. For definiteness, recall that ι_3 is detected with 0-indeterminacy in the Thom-complex MSG by applying the twisted secondary operation associated to the relation $(w_2 + Sq^2)(w_2 + Sq^2)$ to the Thom class, and using the Thom isomorphism to bring the class back to B_{SG}. For details see [R].

We have the fibration sequence

$$K(\mathbf{Z}/2,\ 2) \longrightarrow B_{TSG} \xrightarrow{p} B_{SG} \xrightarrow{\iota_3} K(\mathbf{Z}/2,\ 3) \tag{11.}$$

with K-invariant ι_3. This is the universal space for fiber homotopy transversality to hold in the Thom space, at least through dimension 5 (Compare [B-M]). Indeed, a fiber homotopy sphere bundle $\xi \longrightarrow X$ and reduction to B_{TSG} is equivalent to the condition $\iota_3(\xi) = 0 \in H^3(X;\ \mathbf{Z}/2)$, together with a specific choice of 2-dimensional cochain c so

$$\delta c \ = \ f^{\#}(\iota_3)$$

where $f\ :\ X \longrightarrow B_{SG}$ classifies ξ. This situation is very close, but certainly not the same as the situation studied in [F-K]. Also, there is a factorization of the canonical map $B_{STOP} \longrightarrow B_{SG}$ as

$$B_{STOP} \longrightarrow B_{TSG} \longrightarrow B_{SG}.$$

Precisely, there are exactly two such factorizations differing by a map

$$B_{STOP} \longrightarrow K(\mathbf{Z}/2,\ 2).$$

Now, we look at the 6-skeleton of B_{TSG}. This is the 6-skeleton of the 2-stage Postnikov system

$$K(\mathbf{Z}/2,\ 2) \times K(\mathbf{Z}/3,\ 4) \longrightarrow K(\mathbf{Z}/8,\ 5)$$

with K-invariant $2\{Sq^2Sq^1(\iota_2) + \iota_2 \cdot Sq^1(\iota_2)\}$. From (5) the resulting space has 4th integral homology group given as

$$H_4(B_{TSG};\ \mathbf{Z}) \ = \ \mathbf{Z}/2 \oplus \mathbf{Z}/48$$

with generators $(w_4)^*$, $(w_2^2)^*$ respectively. Here, w_2 can be identified with ι_2. Note that this implies that the Kirby-Siebenmann class maps *non-trivially* to $24((w_2^2)^*)$.

The proof of theorem (I)

Lemma (12): *Let $X \xrightarrow{f} B_{TSG}$ be given and suppose f' is the composite*

$$X \xrightarrow{(\alpha, f)} K(\mathbf{Z}/2,\ 2) \times B_{TSG} \xrightarrow{\mu} B_{TSG}$$

where μ is the principal bundle map $K(\mathbf{Z}/2,\ 2) \times B_{TSG} \longrightarrow B_{TSG}$, then

$$f'^{,*}\{w_2^2\} = f^*\{w_2^2\} + 24\alpha^2 \in H^4(X;\ \mathbf{Z}/48).$$

Proof: $H^4(K(\mathbf{Z},\ 2) \times B_{TSG};\ \mathbf{Z}/16) = (\mathbf{Z}/2)^2 \oplus \mathbf{Z}/4 \oplus \mathbf{Z}/16$ with generators

$$8(\iota_2 \otimes w_2),\ 8(1 \otimes w_4) \text{ of order } 2,\ (4\iota_2^2 \otimes 1) \text{ of order } 4, \text{ and } (1 \otimes w_2^2) \text{ of order } 16.$$

We will show that $\mu^*(w_2^2) = 8(\iota_2^2 \otimes 1) + 1 \otimes w_2^2$. We first note, by naturality and the primitivity of w_2^2 in $H^4(B_{SO};\ \mathbf{Z})$ that $8(\iota_2 \otimes w_2)$ is not in this image. Next, we look at the cohomology Serre spectral sequence of the fibering

$$K(\mathbf{Z}/2,\ 2) \longrightarrow B_{TSG} \longrightarrow B_{SG}$$

with $\mathbf{Z}/16$-coefficients. $E_2^{0\ 4} = H^4(K(\mathbf{Z}/2,_2;\ \mathbf{Z}/16) = \mathbf{Z}/4$, with generator $4\iota_2^2$. Also, $E_2^{4\ 0} = H^4(B_{SG};\ \mathbf{Z}/16) = \mathbf{Z}/2 \oplus \mathbf{Z}/8$ with generators $8w_4$, $2(w_2^2)$, and

$$E_2^{5\ 0} = H^5(B_{SG};\ \mathbf{Z}/16) = (\mathbf{Z}/2)^3 + \mathbf{Z}/8.$$

Here, only the $\mathbf{Z}/8$ is of interest. It has generator $Sq^2(\iota_3)$, so $d_5(4\iota_2^2) = 4Sq^2(\iota_3)$, and at $E_\infty^{i,j}$, $i + j = 4$, only $E^{0\ 4} = \mathbf{Z}/2$, $E^{4\ 0} = \mathbf{Z}/8 \oplus \mathbf{Z}/2$ are non-zero. Thus there is a non-trivial extension for $H^4(B_{TSG};\ \mathbf{Z}/16)$

$$0 \longrightarrow \mathbf{Z}/8 \text{ (generator } 2w_2^2) \longrightarrow \mathbf{Z}/16 \longrightarrow \mathbf{Z}/2 \text{ (generator } 8\iota_2^2) \longrightarrow 0.$$

But this forces the result.

Theorem (I) is direct from (12). The corollary follows, also, since the assumption of even index implies that $w_2(M^4)^2 = 0 \pmod 2$. Hence, either lifting gives the same map in cohomology with $\mathbf{Z}/48$-coefficients.

Concluding remarks

From Quillen's work we know that $B_{SG} \otimes \hat{\mathbf{Z}}_2$ can be identified with $B(B^+(SO(\mathbf{F}_3)))$ in dimensions ≤ 6, and as $B(B^+(S_\infty))$ in all dimensions. Here, S_∞ is the infinite symmetric group. Similarly we can describe B_{TSG} as $B(B^+(\tilde{S}O(\mathbf{F}_3)))$ in this same range. Moreover, B_{TSG} can be given as $B(B^+(\tilde{S}_\infty))$ in all dimensions. Here, these new groups are described by central extensions

$$\mathbf{Z}/2 \longrightarrow \tilde{S}O(\mathbf{F}_3) \longrightarrow SO(\mathbf{F}_3) \longrightarrow 0$$

$$\mathbf{Z}/2 \longrightarrow \tilde{S}_\infty \longrightarrow S_\infty \longrightarrow 0$$

where, for S_∞ the extension is the (unique) non-trivial one for which the transposition (1, 2) continues to have order 2. This might be very useful in understanding Casson's recent results on the Rochlin invariant.

It seems direct to use the description above of B_{TSG} by finite models to calculate the order of the classes which carry the remaining Pontrajagin classes. I hope to return to this later.

Also, there is a second factorizing space for the map $B_{STOP} \longrightarrow B_{SG}$, namely the space where we kill *all* the exotic classes $\sigma(e_{2^i-1,2^i-1})$. The precise structure of these classes is not entirely known, but there is considerable information in [R]. So it should be possible to understand the higher torsion in the cohomology and homology of this intermediate classifying space. Moreover, it is likely that it is the universal space for the vanishing of transversality obstructions.

Bibliography

[B-M] G. Brumfiel-J.Morgan, *Homotopy theoretic consequences of N. Levitts obstruction theory to transversality for spherical fibrations*, Pac. J. Math (1976) 1-100

[F-K] M. Freedman-R. Kirby, *A geometric proof of Rochlin's theorem*, Algebraic and Geometric Topology, A.M.S. Proceedings of Symposia in Pure Mathematics, Vol. XXXII(1) (1978) 85-98

[M-M] Ib Madsen-R.J.Milgram, Classifying Spaces for Surgery and Cobordism of Manifolds, Ann. of Math Studies #92, Princeton U. Press (1979)

[R] Doug Ravenal, Thesis, Brandeis University (1970)

November, 1987
Sonderforschungsbereich 170
Göttingen Universität

The Fixed-Point Conjecture for p-Toral Groups

by

Dietrich Notbohm

1. Introduction

Suppose that X is a space with an action of the topological group G. Let X^G and X^{hG} denote the fixed-point set respectively the homotopy fixed-point set of this action.
We define

$$X^{hG} := map_G(EG,X)$$

as the space of G-maps in the category *Top* of topological spaces and maps. As model for EG any acyclic G-complex is possible (Here complex always means CW-complex.) X^{hG} is then unique up to homotopy.

The definition is not given in the category *S* of semisimplicial sets, as it happens in [DZ] and [M] for finite groups. For topological groups the space EG, constructed as nerve over a category, is not a simplicial set, but a semisimplicial object over the category *Top* . Therefore the same is true for the space

$$map_G(EG,X) \quad ,$$

where X is interpreted as the singular chain complex of the topological space X. For finite groups both definitions agree up to weak homotopy [BK; chapter VIII].

There are two other interpretations of the homotopy fixed-point set. The first one is as section space

$$\Gamma(EG\times_G G\to BG)$$

of the fibration

$$EG\times_G X \longrightarrow BG \quad ,$$

the second one is as fixed-point set

$$map(EG,X)^G \quad ,$$

where G operates canonically on map(EG,X) .
Let p be a prime, for all time fixed. $X^\wedge_p$ denotes the $\mathbb{Z}/p$-completion in the sense of Bousfield and Kan [BK].
X is called $\mathbb{Z}/p$-good, if $X^\wedge_p$ is p-complete [BK; I,5].
Especially nilpotent and other "nice" spaces are $\mathbb{Z}/p$-good.
Look at [BK;VII].

The unique G-map $EG \to *$, where $*$ is the one point set with a trivial G-action, induces a map

$$X^G = map_G(*,X) \longrightarrow map_G(EG,X) \quad .$$

Functoriality of the composition gives a composite map

$$X^{G\wedge}_{p} \longrightarrow X^{\wedge G}_{p} \longrightarrow X^{\wedge hG}_{p}$$

which fits into a commutative diagram

$$\begin{array}{ccc} X^G & \longrightarrow & X^{hG} \\ \downarrow & & \downarrow \\ X^{G\wedge}_{p} & \longrightarrow & X^{\wedge G}_{p} \end{array}$$

Definition: A topological group N is called a p-toral group, iff there exists an exact sequence

$$1 \longrightarrow T \longrightarrow N \longrightarrow P \longrightarrow 1 \ ,$$

where T is a Torus and P a finite p-group.

Theorem: If N is a p-toral group and X a $\mathbb{Z}/p$-good connected finite N-complex, then the map

$$X^{N\wedge}_{p} \longrightarrow X^{\wedge hN}_{p}$$

is a weak homotopy equivalence.

Remark: The analogue theorem for finite p-groups, but without the technical condition $\mathbb{Z}/p$-good, is proved by H. Miller in [M]. It is the foundation of the rest of the paper. For this result J. Lannes found another proof.

It is a pleasure to thank J. McClure for valuable discussions about the book of Bousfield and Kan.

2. Proof of the Theorem

We need some remarks:

2.1 Remark: Let

$$1 \longrightarrow K \longrightarrow G \longrightarrow H \longrightarrow 1$$

be an exact sequence of topological spaces and assume, that H is finite. Let X be a G-space. H acts on the fixed-point set X^K canonically. We have

$$(X^K)^H = X^G .$$

As H is finite, EG is an acyclic K-complex of finite type. We get

$$X^{hK} \simeq \mathrm{map}(EG,X)^K ,$$

where map(EG,X) is a G-space. Hence using the above equation and the exponential law for mapping spaces, we get the analogue:

$$X^{hG} \simeq (X^{hK})^{hH} .$$

2.2 Remark: Let $f:X_1 \longrightarrow X_2$ be a weak homotopy equivalence and a G-map between two G-spaces X_1, X_2 . The horizontal map in the diagram

$$\begin{array}{ccc} EG\times_G X_1 & \longrightarrow & EG\times_G X_2 \\ & \searrow \quad \swarrow & \\ & BG & \end{array}$$

is a weak homotopy equivalence. Because BG is a complex, the two spaces

$$\mathrm{map}(BG, EG\times_G X_i) \ , \quad i=1,2$$

are weak homotopy equivalent as well.

We denote with

$$\mathrm{map}(BG,BG)_{id}$$

the connected component of the identity and with

$$\mathrm{map}(BG, EG\times_G X_i)_s$$

the space of all maps, which are homotopic to a section. If we look at the two fibrations

$$X_i^{hG} \longrightarrow \mathrm{map}(BG, EG\times_G X_i)_s \longrightarrow \mathrm{map}(BG,BG)_{id} ,$$

it is easy to see, that the two homotopy fixed-point sets X_i^{hG} are weak homotopy equivalent.

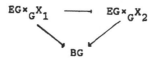

<u>**Proof of the theorem:**</u> i) reduction to the case of a torus. Let

$$1 \longrightarrow T \longrightarrow N \longrightarrow P \longrightarrow 1$$

be the exact sequence belonging to the p-toral group N. X^T is a finite P-complex. It is proved in [M] that

$$X^{N}{}^{\wedge}_{p} = (X^T)^{P}{}^{\wedge}_{p} \longrightarrow (X^{T}{}^{\wedge}_{p})^{hP}$$

is a weak homotopy equivalence. Setting

$$X^{\wedge hT}_{p} = map_T(EN, X^{\wedge}_{p}) \ ,$$

remark 2.1 implies a weak homotopy equivalence

$$X^{\wedge hN}_{p} \simeq_w (X^{\wedge hT}_{p})^{hN}$$

The map

$$X^{T}{}^{\wedge}_{p} \longrightarrow X^{\wedge hT}_{p}$$

is P-equivariant. Together with (2.2) we can reduce therefore the problem to the case of a torus.

ii) Let n be the dimension of T. We can think of $\mathbb{Z}/p^k \subset S^1$ as the group of the roots of unity with order p^k and define

$$\sigma_k := (\mathbb{Z}/p^k)^n \quad , \quad \sigma_\infty := \varinjlim \sigma_k$$

The homomorphism $\sigma_\infty \longrightarrow T$ induces a mod p-equivalence

$$B\sigma_\infty \longrightarrow T \ ,$$

which is the same as to say that the map

$$H_j(B\sigma_\infty;\mathbb{Z}/p) \longrightarrow H_j(BT;\mathbb{Z}/p)$$

is an isomorphism.

Now let X be a $\mathbb{Z}/p$-good connected finite T-complex. Then there are the following maps

$$X^{\wedge hT}_{p} \longrightarrow X^{\wedge h\sigma_\infty}_{p} \longrightarrow \varprojlim X^{\wedge h\sigma_k}_{p}$$

$$X^{T} \longrightarrow X^{\sigma_\infty} \longrightarrow \varprojlim X^{\sigma_k}$$

As T is a finite σ_k-complex for all k, X has also the structure of a finite σ_k-complex. Using Miller's Theorem [M] and the following three propositions, the proof will be finished in a straightforward way.

2.3 Proposition: Let X be a finite T complex. Then it is

$$X^T = \varprojlim X^{\sigma_k}$$

and the sequence of the fixed-point sets is a finite sequence.

2.4 Proposition: Let X be a $\mathbb{Z}/p$-good finite T-complex. Then the map

$$X_p^{\wedge hT} \longrightarrow X_p^{\wedge h\sigma_\infty}$$

is a weak homotopy equivalence.

2.5 Proposition: Let X be a finite T-complex. Then it is

$$\pi_n(X_p^{\wedge h\sigma_\infty}) \cong \varprojlim \pi_n(X_p^{\wedge h\sigma_k}) \quad .$$

3. Proofs of the Propositions 2.3 - 2.5

Proof of 2.3: If X is a finite T-complex, it consists of a finite number of cells of the form $T/A \times e_n$, where $A \subset T$ is a closed subgroup.

$T/A \times e_n$ belongs to X^T, if $A=T$, and it belongs to X^{σ_k} , if $\sigma_k \subset A$. Because σ_∞ is dense in T and because there is only a finite number of orbit types T/A, we get

$$X^{\sigma_k} = X^T$$

for k big enough.

3.1 Lemma: Let Y be a p-complete space. Let

$$Y \longrightarrow E \longrightarrow B$$

be a fibration, such that the action of $\pi_1 B$ on $H_n(Y;\mathbb{Z}/p)$ is nilpotent. Then the $\mathbb{Z}/p$-completion induces a homotopy equivalence

$$\Gamma(E \to B) \longrightarrow \Gamma(E_p^\wedge \to B_p^\wedge)$$

between the section spaces.

Proof: Under the above assumption the mod R fibre lemma [BK; II, 5] is applicable. We get a fibre square

(*)
$$\begin{array}{ccc} E & \longrightarrow & E^{\wedge}_p \\ \downarrow & & \downarrow \\ B & \longrightarrow & B^{\wedge}_p \end{array}$$

which induces a commutative diagram

(**)
$$\begin{array}{ccc} \mathrm{map}(B,E)_s & \longrightarrow & \mathrm{map}(B,B)_{id} \\ \downarrow & & \downarrow \\ \mathrm{map}(B^{\wedge}_p,E^{\wedge}_p)_s & \longrightarrow & \mathrm{map}(B^{\wedge}_p,B^{\wedge}_p)_{id} \end{array}$$

where the rows are fibrations and the columns are given by the completion. With the universal property of pullback diagrams, which fibre squares are, you can prove, that (**) is up to homotopy a fibre square too. The fibres of the rows in (**) are exactly the section spaces. This implies the Lemma.

<u>**3.2 Lemma:**</u> Let $E_i \longrightarrow B_i$, i=0,1 , be fibrations with p-complete fibre, in such a way that the diagram

$$\begin{array}{ccc} E_0 & \longrightarrow & E_1 \\ \downarrow & & \downarrow \\ B_0 & \longrightarrow & B_1 \end{array}$$

is a fibre square. Assume that the operation of $\pi_1(B_1)$ on $H_j(E_1;\mathbb{Z}/p)$ is nilpotent and that the map $B_0 \longrightarrow B_1$ is a mod p equivalence. Then the two section spaces

$$\Gamma(E_1 \to B_1) \longrightarrow \Gamma(E_0 \to B_0)$$

are weak homotopy equivalent.

<u>**Proof:**</u> The assumptions of the mod R fibre square lemma [BK; II, 5.3] are satified. We get up to homotopy a fibre square

$$\begin{array}{ccc} E_{0p}^{\wedge} & \longrightarrow & E_{1p}^{\wedge} \\ \downarrow & & \downarrow \\ B_{0p}^{\wedge} & \longrightarrow & B_{1p}^{\wedge} \end{array}$$

with homotopy equivalences in the rows. This implies that the associated section spaces of the fibre squares are weak homotopy equivalent. If you use 3.1, the proof will be finished.

<u>**Proof of 2.4:**</u> The diagram

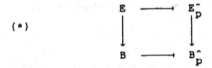

$$\begin{array}{ccc} E\sigma_\infty \times_{\sigma_\infty} X_p^\wedge & \longrightarrow & ET \times_T X_p^\wedge \\ \downarrow & & \downarrow \\ B\sigma_\infty & \longrightarrow & BT \end{array}$$

is a fibre square with a p-complete fibre in the columns. Moreover BT is 1-connected. Lemma 3.2 applies.

3.3 Lemma: Let $G_1 \subset G_2 \subset \ldots$ be a ascending sequence of groups and define $G_\infty := \bigcup_k G_k$. Let X be a G_∞-space. Then the map

$$X^{hG_\infty} \longrightarrow \underleftarrow{\mathrm{holim}}\, X^{hG_k}$$

is a weak homotopy equivalence.

Proof: For the definition of $\underleftarrow{\mathrm{holim}}$ see [BK; XI].

We choose the Milnor model for the spaces EG_∞ and EG_k. Then EG_∞ is exactly the union of the spaces EG_k or

$$EG_\infty = \underrightarrow{\lim}\, EG_k .$$

This implies that

$$X^{hG_\infty} = \mathrm{map}_{G_\infty}(EG\ ,X) = \underleftarrow{\lim}\, \mathrm{map}_{G_k}(EG_k,X) = X^{hG_k} .$$

On the other hand the maps $X^{hG_k} \longrightarrow X^{hG_{k-1}}$ are fibrations. According to [BK; XI] there is a weak homotopy equivalence

$$EG_\infty = \underleftarrow{\lim}\, X^{hG_k} \longrightarrow \underleftarrow{\mathrm{holim}}\, X^{hG_k} .$$

Proof of 2.5: Because of Lemma 3.3 there is an exact sequence

$$0 \longrightarrow \underleftarrow{\lim}^1\, \pi_{n+1}(X_p^{\wedge\, h\sigma_k}) \longrightarrow \pi_n(X_p^{\wedge\, h\sigma_\infty}) \longrightarrow \underleftarrow{\lim}\, \pi_n(X_p^{\wedge\, h\sigma_k}) \longrightarrow 0$$

for all base points [BK; XI, 7.4] . By [M] we get

$$\pi_n(X_p^{\wedge\, h\sigma_k}) \cong \pi_n(X^{\sigma_k}{}_p^\wedge) .$$

Proposition 2.3 implies that the $\underleftarrow{\lim}^1$-term must vanish.

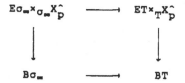

References:

[BK] A.K. Bousfield and D.M. Kan: Homotopy Limits, Completion, and Localisation; Lecture Notes in Math. 304, Springer 1972

[DZ] W.G. Dwyer and A. Zabrodsky: Maps between Classifying Spaces; preprint.

[M] H. Miller: The Fixed-Point Conjecture; to appear.

Dietrich Notbohm
Mathematisches Institut der
Georg-August-Universität
Bunsenstr. 3-5

D-3400 Göttingen
Bundesrepublik Deutschland

Simply connected manifolds without S^1-symmetry

V. Puppe

Several authors have studied the question of existence of manifolds with little or no symmetry (s.[1],[2],[6],[7],[16],[17]), e.g. E. Bloomberg has shown (s.[2]) that there exist closed manifolds which do not admit any effective topological (continuous) action of a compact Lie group. For his argument the presence of a rather complicated fundamental group is essential.

From a completely different point of view M. Atiyah and F.Hirzebruch had proved earlier (s.[1]) that a compact spin manifold M can not admit an effective differentiable S^1-action if the $\hat{A}$-genus $\hat{A}(M)$ is different from zero. It has been shown, though, that the differentiability assumption in their result is crucial, i.e. there exist examples of topological effective S^1-actions on spin manifolds with $\hat{A}(M) \neq 0$ (s.[3] VI. 9.6 and [4]).

Here we prove, using the connection between P.A. Smith-theory and deformation of algebras (s.[12],[13],[14]), that there exist simply connected, closed, oriented, differentiable manifolds M such that any closed, orientable manifold $\tilde{M}$ with $H^*(M;\mathbb{Q}) \cong H^*(\tilde{M};\mathbb{Q})$ (as algebras over $\mathbb{Q}$) has no topological S^1-symmetry, i.e. does not admit any non trivial topological S^1-action; in fact, there exist examples which admit non trivial topological $\mathbb{Z}/_p$-actions only for (at most) finitely many primes p (compare [11]).

S.Kwasik and R.Schultz have studied topological S^1-actions on 4-manifolds and - among other results - they show, by completely different methods, that there exist many closed simply connected 4-manifolds without topological S^1-symmetries (s.[19]).

I want to thank R. Buchweitz, J. Damon and A. Iarrobino for illuminating conversations on the deformation theory of Artin algebras. In fact, what is described in this note is more or less an interpretation of certain algebraic results about deformations of algebras (s.[9],[10]) in the context of S^1-action from the view point of [12],[13],[14].

If X is a paracompact, finitistic S^1-space which is totally non homologous to zero (TNHZ) in the Borel construction $X_G := EG \times_G X$ $(G=S^1)$ with respect to Čech cohomology with rational coefficients and if $\dim_{\mathbb{Q}} H^*(X;\mathbb{Q}) < \infty$, then the cohomology algebra $B := H^*(X^G;\mathbb{Q})$ of the fix point set X^G can be viewed as a deformation of the algebra $A := H^*(X;\mathbb{Q})$.

A one parameter family of deformations $A[t] = A \tilde{\otimes} \mathbb{Q}[t]$ (where "~" indicates the twisting of the multiplication) with $A_o = A$ and $A_1 \cong B$ (disregarding the grading is given by the cohomology $H^*(X_G;\mathbb{Q})$ of X_G considered as an algebra over $H^*(BG;\mathbb{Q}) = \mathbb{Q}[t]$, $\deg(t) = 2$ (s.[12] for details).

The property (TNHZ) is automatically fulfilled if $H^{odd}(X;\mathbb{Q}) = 0$ because then the Leray-Serre spectral sequence degenerates already for degree reasons.

This suggests the following program to exhibit manifolds without S^1-symmetry:

1. Find a rigid graded algebra A^* over $\mathbb{Q}$ with $A^{odd} = 0$ and $\dim_{\mathbb{Q}} A^* < \infty$, which fulfils Poincaré duality ("rigid" means roughly that all algebras obtained from A^* by deformation are isomorphic to A^*; s. [20] for a discussion of different notions of rigidity and how they relate to each other).
2. Realize A^* as the rational cohomology algebra of a manifold M.
3. Check that $H^*(M;\mathbb{Q}) \cong H^*(M^G;\mathbb{Q})$ implies $M = M^G$.

Yet there are several obstacles:

a) First of all it is not known (to me) and seems to be a very difficult question to decide whether there exist non trivial rigid algebras of the desired form (s. [20] for examples of non-commutative finite dimensional rigid algebras, in particular over $\mathbb{Z}/p$).

b) To realize a Poincaré algebra as the cohomology of a (simply connected) compact manifold one needs certain extra conditions to be satisfied if the dimension is divisible by 4 (s.[18]).

c) Even if one would find a non trivial rigid graded algebra A, the isomorphism between A and some deformation of A need not respect the grading (supposing the algebra obtained by deformation has an a priori grading).

But not every deformation of an algebra $A^* \cong H^*(X;\mathbb{Q})$ which is possible algebraically can be realized by an S^1-action on X. The one parameter families $A \tilde{\otimes} \mathbb{Q}[t]$ which correspond to S^1-actions on X have certain special properties (s.[12],[13]):

There exists a grading on A_1 (namely the one given by the isomorphism $A_1 \cong H^*(X^G;\mathbb{Q})$) such that $A \tilde{\otimes} \mathbb{Q}[t]$ embeds into the trivial family $A_1 \otimes \mathbb{Q}[t]$ as a graded algebra over $\mathbb{Q}[t]$ and the cokernel of this embedding is $\mathbb{Q}[t]$-torsion. In fact, this is just a reformulation of the localization theorem, which says that the morphism $H^*(X_G;\mathbb{Q}) \to H^*((X^G)_G;\mathbb{Q})$ induced by the inclusion $X^G \longrightarrow X$ becomes an isomorphism after localization at (0). The property (TNHZ) gives that $H^*(X_G;\mathbb{Q}) \to H^*((X^G)_G;\mathbb{Q})$ is injective. In particular $A \tilde{\otimes} \mathbb{Q}[t]$ is then a jump deformation (s.[8]) in the sense that all $A_\varepsilon := (A \tilde{\otimes} \mathbb{Q}[t] \otimes_{\mathbb{Q}[t]} \mathbb{Q}^\varepsilon$, where $\mathbb{Q}^\varepsilon$ is $\mathbb{Q}$ considered as a $\mathbb{Q}[t]$-module via $\mathbb{Q}[t] \to \mathbb{Q}$, $t \longmapsto \varepsilon$, are isomorphic for $\varepsilon \neq 0$; moreover

A_1 has a filtration such that the associated graded algebra is isomorphic to A_o (s.[13]).

Hence for the first part of our program we only need to know that A is rigid with respect to "g-deformations", i.e. deformations of the special kind described above. If A has that property we will say that A is g-rigid. It is shown in [5] and [8] that a non trivial g-deformation of A lowers the dimension of the second Hochschild cohomology of A with coefficients in A (the space of infinitesimal deformations), i.e. $\dim_{\mathbb{Q}} H^2(A_o,A_o) > \dim_{\mathbb{Q}} H^2(A_1,A_1)$. On the other hand there exist non smoothable graded Artin algebras (already defined over $\mathbb{Q}$) (s.[9],[10]), i.e. algebra which do not admit deformations to $\mathbb{Q}\times\ldots\times\mathbb{Q}$, in fact not even to $\mathbb{C}\times\ldots\times\mathbb{C}$ if one extends the ground field to $\mathbb{C}$. Therefore, if one starts with a non smoothable graded Artin algebra and considers all algebras which can be obtained from A by (iterated) g-deformations, there must exist non trivial g-rigid graded algebras (among the components, i.e. direct factors of the algebras obtained). By Quillen's results in rational homotopy (s.[15] and also [18])one can realize such an algebra as the rational cohomology of a simply connected finite CW-complex and hence obtains:

<u>Proposition 1</u>: There exist simply connected finite CW complex X such that the strictly commutative graded algebra $H^*(X;\mathbb{Q})$ is g-rigid ($H^{odd}(X;\mathbb{Q})=0$).

For every S^1-action on such a space X the rational cohomology of X^G is isomorphic as a filtered algebra to $H^*(X;\mathbb{Q})$ (s.[5]). Within the ("filtered") isomorphism type of $H^*(X;\mathbb{Q})$ one can choose a graded algebra A^* with "minimal degree", i.e. $\dim_{\mathbb{Q}}\left(\bigoplus_{i=o}^{q} A^i\right)$ should be maximal for each q. Let Y be a simply connected finite CW-complex with $H^*(Y;\mathbb{Q})=A^*$, then one gets the following:

<u>Corollary 1</u>: For any S^1-action on Y the inclusion $Y^G \to Y$ induces an isomorphism in rational cohomology.

<u>Proof</u>: By the choice of Y one has that $H^*(Y;\mathbb{Q})$ and $H^*(Y^G;\mathbb{Q})$ are isomorphic as graded vector spaces (and as filtered algebras). Therefore the morphism $H^*(Y;\mathbb{Q}) \tilde{\otimes} \mathbb{Q}[t] \to H^*(Y^G;\mathbb{Q}) \otimes \mathbb{Q}[t]$ induced by $Y^G \longrightarrow Y$, can only become an isomorphism after localization, if the evalution at t=0 (i.e. applying $- \otimes_{\mathbb{Q}[t]} \mathbb{Q}^o$), which gives $H^*(Y,\mathbb{Q}) \to H^*(Y^G,\mathbb{Q})$, is an isomorphism, too.

<u>Remark</u>: Of course Corollary 1 does not imply that any S^1-action on Y

must be trivial. Instead of Y one could as well take $Y \times D^2 \simeq Y$ ($D^2 := \{x \in \mathbb{R}^2; |X| \leq 1\}$) which clearly admits a non trivial S^1-action with $(Y \times D^2)^G = Y \times \{0\}$.

To exhibit simply connected manifolds without S^1-symmetry we are looking for connected Poincaré algebras over $\mathbb{Q}$ which are g-rigid. It is shown in [9] and [10] that there are non smoothable connected Poincaré algebras A (graded Gorenstein algebras) with $A^{odd} = 0$ and formal dimension of A equal to 6. This leads to the following

Proposition 2: If A is a non smoothable connected Poincaré algebra of formal dimension 6 with $A^{odd} = 0$, then A is g-rigid.

Proof: Let B be an algebra obtained from A by a g-deformation. Then $B = \prod_{i=1}^{k} B_i$ and B_i is a connected Poincaré algebra of even formal dimension ≤ 6 with $B_i^{odd} = 0$ for $i = 1,\dots,k$. It is easy to see that any connected Poincaré algebra C of formal dimension fd(C) = 2 or 4 and with $C^{odd} = 0$ is smoothable. (For fd(C) = 2 this is obvious since $C \cong \mathbb{Q}[x]/_{(x^2)}$; for fd(C) = 4 a somewhat round about but simple argument is to observe that $C \otimes \mathbb{C} \cong H^*(M;\mathbb{C})$ where M is a connected sum of a number of copies of $\mathbb{C}P^2$s and therefore admits an S^1-action with isolated fix points.) Hence there must be a component, say B_1, in B with $fd(B_1) = 6$. By the inequalities $\bigoplus_{i=q}^{\infty} A^i \geq \dim \bigoplus_{i=q}^{\infty} B^i$ there is precisely one component with formal dimension 6.

The top dimensional generator $a \in A^6$ is mapped to a non zero element in $B_1 \otimes 1 \subset B_1 \otimes \mathbb{Q}[t]$ by the morphism $A \tilde{\otimes} \mathbb{Q}[t] \to B \otimes \mathbb{Q}[t] \to B_1 \otimes \mathbb{Q}[t]$, where the last map is induced by the projection $B = \prod_{i=1}^{k} B_i \to B_1$. (Otherwise the first map in the composition would not become an isomorphism after localization.)

Evaluated at t = 0 one obtains a morphism $A \to B_1$ of connected Poincaré algebras which has non zero degree and hence is an isomorphism. This implies, of course, that $A \cong B \cong B_1$, since $\dim_{\mathbb{Q}} A = \dim_{\mathbb{Q}} B$.

Theorem 1: There exist simply connected, closed, oriented differentiable 6-dimensional manifolds M such that no closed, orientable manifold with the same rational cohomology algebra as M admits any non trivial S^1-action.

Proof: Choose a non smoothable connected Poincaré algebra A* of formal dimension 6 with $A^{odd} = 0$. By Sullivan's results (s.[18]) there exists a simply connected, closed, oriented 6-dimensional differentiable manifold M with $H^*(M,\mathbb{Q}) \cong A^*$. By Proposition 2 the algebra A^* is g-rigid; in fact the proof of Proposition 2 shows that for any S^1-action on a manifold $\tilde{M}$ with $H^*(\tilde{M};\mathbb{Q}) = A^*$ the inclusion of the fix point set $\tilde{M}^G \to \tilde{M}$ induces an isomorphism in rational cohomology. Since $\tilde{M}$ is closed and orientable this implies $\tilde{M}^G = \tilde{M}$, i.e. the action is trivial.

Remarks: a) For the above argument it is, of course, essential to assume $\tilde{M}$ to be closed. The manifold with boundary $M \times D^2$ and the open manifold $M \times \overset{\circ}{D}{}^2$ are homotopy equivalent to M and clearly admit non trivial S^1-actions.

b) The following example of a graded Gorenstein algebra is due to A. Iarrobino (s.[9] Ex.7,[10]). It was checked on a computer to admit only deformations to algebras of the same "type", in particular it is not smoothable.

$A := R/J$ with $R = \mathbb{Q}[a,b,c,d,e,f]$ and the ideal J is generated by $\{3ab-4ac-3bd, ad, ae, b^2-af, 12bc-9af-16bd-12ce, be, bf, 3c^2-4ac, 3cd-3ac-4ce-3df, cf, 3d^2-4bd, 12de-12bd-16df-9a^2, 3e^2-4ce, ef-ce-a^2, 3f^2-4df\}$.

If one assigns the degree 2 to all the generators a,b,c,d,e,f then J is a homogeneous ideal and A is a connected Poincaré algebra of formal dimension 6 with $A^{odd} = 0$. Iarrobino remarks that the above example should not be considered as a rare exception but one of many similarly constructed.

By [18] the algebra A can be realized as the rational cohomology of a closed oriented 6-dimensional differentiable manifold. Such a manifold does not admit any non trivial S^1-action.

We now discuss some implication of the above method to the non existence of non trivial cyclic group actions. If suffices to consider groups $G = \mathbb{Z}/_p$ of prime order p.

The following difficulties occur:

a) An analogue of Sullivan's result about realizing a rational Poincaré algebra as the cohomology of a manifold seems completely out of reach in the $\mathbb{Z}/_p$-case.
b) Even if one has a simply connected, closed, oriented manifold M with $H^{odd}(M;\mathbb{Z}/_p) = 0$ and such that $H^*(M;\mathbb{Z}/_p)$ is g-rigid in an appropriate sense (note that $H^*(B\,\mathbb{Z}/_p;\mathbb{Z}/_p)$ is not just the polynomial ring over $\mathbb{Z}/_p$ in one variable in case p is an odd prime), M need not be TNHZ in $EG \times_G M$ with respect to $H^*(-,\mathbb{Z}/_p)$ for all actions of $G = \mathbb{Z}/_p$ on M.

   There could be actions such that the Leray-Serre spectral sequence of the fibration $M \to EG \times_G M \to BG$ does not collapse from the E_2-term on and the action of G on $H^*(M;\mathbb{Z}/_p)$ may be non trivial either.

Hence to prove an analogous result to Theorem 1 above for a given fixed prime p ($G = \mathbb{Z}/_p$) by just immitating the proof does not seem to work. But it is possible to use similar arguments in order to show that there are simply connected, closed, oriented differentiable manifolds that do not admit any non trivial $\mathbb{Z}/_p$-action, for almost all (i.e. all but finitely many) primes p. This answers a question of P. Löffler and M. Raußen (s.[11]) to the negative.
I am grateful to T. Petrie for suggesting this possibility.

We use Example 7 of [9] (s. Remark b) above) to show the following:

Lemma: There exists a connected Poincaré algebra A^* of $fd(A^*) = 6$ with $A^{odd} = 0$, defined over $\mathbb{Z}$, such that $A_p^* := A^* \otimes \mathbb{Z}/_p$ is g-rigid for almost all primes.

By g-rigid we mean that any embedding (as graded algebras) of an one parameter family $A_p^* \tilde{\otimes} \mathbb{Z}/_p[t] \to B^* \otimes \mathbb{Z}/_p[t]$ ($deg(t) = 2$) into a trivial family with cokernel being $\mathbb{Z}/_p[t]$-torsion must actually be an isomorphism.

Proof: The algebra given as Example 7 in [9] can be written as $A^* \otimes \mathbb{Q}$, where A^* is a connected graded Poincaré algebra of $fd(A^*) = 6$ with $A^{odd} = 0$, defined over $\mathbb{Z}$ and free as a $\mathbb{Z}$-module. To prove that A_p^* is g-rigid we show that the part of the Hochschild cohomology $H_c^{2,-}(A_p^*;A_p^*)$ which classifies the infinitesimal deformations of negative weight (of commutative algebras) is zero for almost all primes p. An element in $H_c^{2,-}(A_p^*;A_p^*)$ is represented by a 2-cycle μ_p: $A_p^* \otimes A_p^* \to A_p^*$, i.e. a symmetric bilinear form of negativ degree (as a map of graded vector spaces) with $\delta\mu_p = 0$. Let μ: $A^* \otimes A^* \to A^*$ be a symmetric bilinear form such that $\mu \otimes id_{\mathbb{Z}/_p} = \mu_p$. Then $\delta\mu = p\cdot\xi$ for some $\xi \in C^3(A^*;A^*)$ and $\delta\xi = 0$ since A^* is torsion free, and $p[\xi] = 0$ in $H^3(A^*;A^*)$. If p is a prime which does not occur in the torsion of $H^3(A^*;A^*)$ then $[\xi]$ must be zero and there exists an $\eta \in C^2(A^*,A^*)$ such that $\delta\eta = \xi$ and therefore $\delta(\mu - p\eta) = 0$, i.e. $\tilde{\mu} := \mu - p\eta \in Z^2(A^*;A^*)$ and $\tilde{\mu} \otimes id_{\mathbb{Z}/_p} = \mu_p$. We can "symmetrize" $\tilde{\mu}$ to get $\tilde{\tilde{\mu}} = (\tilde{\mu} + \tilde{\mu}^t) \in Z_c^2(A^*,A^*)$, where $\tilde{\mu}^t(a_1,a_2) := \tilde{\mu}(a_2,a_1)$, with $\tilde{\tilde{\mu}} \otimes id_{\mathbb{Z}/_p} = 2\mu_p$. (More conceptually one might use Harrison or André-Quillen cohomology of commutative algebras instead of Hochschild cohomology. But since we have to exclude a finite number of primes anyway we may as well stick to odd primes p, making 2 a unit in $\mathbb{Z}/_p$.) By [9] and [10] $[\tilde{\tilde{\mu}}] = 0$ in $H_c^{2,-}(A^* \otimes \mathbb{Q}, A^* \otimes \mathbb{Q})$, i.e. there exists a morphism $\alpha_{\mathbb{Q}}$: $A^* \otimes \mathbb{Q} \to A^* \otimes \mathbb{Q}$ such that

$\delta\alpha = \tilde{\tilde{\mu}} \otimes id_{\mathbb{Q}}$. Let $\mathbb{Z}_{(p)}$ denote the localization of $\mathbb{Z}$ at p.

For almost all primes p the morphism $\alpha_{\mathbb{Q}}$ is already defined over $\mathbb{Z}_{(p)}$, i.e. there is a map $\alpha_{(p)}: A^* \otimes \mathbb{Z}_{(p)} \to A^* \otimes \mathbb{Z}_{(p)}$ such that $\delta\alpha_{(p)} = \tilde{\tilde{\mu}} \otimes id_{\mathbb{Z}_{(p)}}$. It follows that $\delta(\alpha_{(p)} \otimes id_{\mathbb{Z}/_p}) = 2\mu_p$, i.e. $[2\mu_p]$ and hence $[\mu_p] = 0$ in $H_c^{2,-}(A_p^*,A_p^*)$ for almost all primes. Since $H_c^{2,-}(A_p^*,A_p^*)$ is finitely generated on gets: $H_c^{2,-}(A_p^*,A_p^*) = 0$ for almost all primes p.

The argument now proceeds as in the rational case before to give that A_p^* is g-rigid for almost all primes.

As before we can realize $A^* \otimes \mathbb{Q}$ as the rational cohomology of a closed, oriented, differentiable 6-dimensional manifold M. For almost all primes p one has $H^*(M;\mathbb{Z}/_p) \cong H^*(M;\mathbb{Z}) \otimes \mathbb{Z}/_p$. It follows from the isomorphism $A^*\otimes\mathbb{Q} \cong H^*(M;\mathbb{Q}) \cong H^*(M;\mathbb{Z}) \otimes \mathbb{Q}$, that $A^* \otimes \mathbb{Z}_{(p)} \cong H^*(M;\mathbb{Z})\otimes\mathbb{Z}_{(p)}$ and hence $A_p^* \cong H^*(M;\mathbb{Z}) \otimes \mathbb{Z}/_p$ for almost all primes p, since the denominators of the rational coefficient matrix which gives the isomorphism between $A^* \otimes \mathbb{Q}$ and $H^*(M;\mathbb{Z}) \otimes \mathbb{Q}$, and the torsion of $H^*(M;\mathbb{Z})$ involve only finitely many primes. Let P denote the set of primes, such that A_p^* is g-rigid, $A_p^* = H^*(M;\mathbb{Z}/_p) \cong H^*(M;\mathbb{Z}) \otimes \mathbb{Z}/_p$ for $p \in P$ and $\mathbb{Z}/_p$ must act trivially on $H^*(M;\mathbb{Z}_{(p)})$ (which - for a given M - is the case if p is large enough).

Assume $\mathbb{Z}/_p$ acts on M for some $p \in P$.

The localization theorem for the equivariant cohomology given by the Borel construction works as well if we use coefficients $\mathbb{Z}_{(p)}$ instead of $\mathbb{Z}/_p$, i.e. the map $H^*(EG \times_G M; \mathbb{Z}_{(p)}) \to H^*(EG \times_G M^G;\mathbb{Z}_{(p)})$ induced by the inclusion $M^G \to M$ becomes an isomorphism after inverting the multiplication with the polynomial generator $t \in H^*(BG;\mathbb{Z}_{(p)}) \cong \mathbb{Z}_{(p)}[t]/p\cdot(t)$.

The group $\mathbb{Z}/_p$ acts trivially on $H^*(M;\mathbb{Z}_{(p)})$; hence the E_2-term of the Leray-Serre spectral sequence of the fibration $M \to EG \times_G M \to BG$ is given by $E_2 = H^*(BG; H^*(M;\mathbb{Z}_{(p)})) \cong H^*(BG;\mathbb{Z}_{(p)}) \otimes H^*(M;\mathbb{Z}_{(p)})$.

Since $H^{odd}(BG;\mathbb{Z}_{(p)}) = 0 = H^{odd}(M;\mathbb{Z}_{(p)})$ the spectral sequence collapses already for degree reasons (as in the S^1-case with rational coefficients). One therefore gets a morphism

$$H^*(M_G;\mathbb{Z}_{(p)}) \cong H^*(BG;\mathbb{Z}_{(p)}) \tilde{\otimes} H^*(M;\mathbb{Z}_{(p)}) \longrightarrow H^*((M^G)_G,\mathbb{Z}_{(p)}) \cong H^*(BG;\mathbb{Z}_{(p)}) \otimes H^*(M^G;\mathbb{Z}_{(p)})$$

which becomes an isomorphism after localization.

(Note that $H^*(M^G;\mathbb{Z}_{(p)})$ can not have p-torsion since $H^{odd}(M^G;\mathbb{Z}/_p) = 0$ s. e.g. [3], VII (2.2))

Tensoring this morphism with $\mathbb{Z}/_p$ gives an embedding $H^*(M;\mathbb{Z}/_p) \tilde{\otimes} \mathbb{Z}/_p[t] \longrightarrow H^*(M^G;\mathbb{Z}/_p) \otimes \mathbb{Z}/_p[t]$ such that the cokernel is $\mathbb{Z}/_p[t]$-torsion.

Since $A_p^* \cong H^*(M;\mathbb{Z}/_p)$ is g-rigid it now follows, that $M^G \to M$ induces an isomorphism $H^*(M;\mathbb{Z}/_p) \to H^*(M^G;\mathbb{Z}/_p)$ and hence we get:

Theorem 2: There exist simply-connected, orientable, closed 6-dimensional differentiable manifolds M such that for any closed orientable manifold $\tilde{M}$ with $H^*(\tilde{M};\mathbb{Q}) \cong M^*(M;\mathbb{Q})$ a non-trivial action of $\mathbb{Z}/_p$ on $\tilde{M}$ is only possible for at most a finite number of primes p.

References

[1] ATIYAH, M.F. and HIRZEBRUCH, F.: Spin manifolds and group actions, Essays on Topology and Related Topics (Mémoires dédiés à G. de Rham), 18-28. Berlin-Heidelberg-New York: Springer 1969

[2] BLOOMBERG, E.M.: Manifolds with no periodic homeomorphism. Trans.Amer.Math.Soc. 202, 67-78 (1975)

[3] BREDON, G.: Introduction to compact transformation groups. New York-London: Academic Press 1972

[4] BURGHELEA, D.: Free differentiable S^1 and S^3 actions on homotopy spheres, Ann.Sci.École Norm.Sup. (4) 5, 183-215 (1972)

[5] COFFEE, J.P.: Filtered and associated graded rings, Bull.Amer.Math.Soc. 78, 584-587 (1972)

[6] CONNER, P.E. and RAYMOND, F.: Manifolds with few periodic homeomorphisms, Proceedings of the Second Conference on Compact Transformation Groups (Univ. of Massachusetts, Amherst 1971) Part II. Springer Lecture Notes in Math. 299, 1-75 (1972)

[7] CONNER, P.E., RAYMOND, F. and WEINBERGER, P.: Manifolds with no periodic maps, Proceedings of the Second Conference on Compact Transformation Groups (Univ. of Massachusetts, Amherst 1971) Part II. Springer Lecture Notes in Math. 299, 81-108 (1972)

[8] GERSTENHABER, M.: On the deformation of rings and algebras IV. Ann. of Math. 99, 257-276 (1974)

[9] IARROBINO, A.: Compressed algebras and components of the punctual Hilbert scheme, Algebraic Geometry, Sitges 1983, Proceedings. Spinger Lecture Notes in Math. 1124, 146-166 (1985)

[10] IARROBINO, A. and EMSALEM, J.: Some zero-dimensional generic singularities; finite algebras having small tangent space, Compositio Math. 36, 145-188 (1978)

[11] LÖFFLER, P. und RAUSSEN, M.: Symmetrien von Mannigfaltigkeiten und rationale Homotopietheorie. Math.Ann. 271, 549-576 (1985)

[12] PUPPE, V.: Cohomology of fixed point sets and deformation of algebras, Manuscripta Math. 23, 343-354 (1978)

[13] PUPPE, V.: Deformation of algebras and cohomology of fixed point sets, Manuscripta Math. 30, 119-136 (1979)

[14] PUPPE, V.: P.A. Smith theory via deformations. Homotopie algébrique et algèbre locale, Luminy, 1982, Astérisque 113-114, Soc.Math. de France, 278-287 (1984)

[15] QUILLEN, D.: Rational homotopy theory, Ann. of Math. 90, 205-295 (1969)

[16] SCHULTZ, R.: Group actions on hypertoral manifolds. I. Topology Symposium Siegen 1979, Proceedings Springer Lecture Notes in Math. 788, 364-377 (1980)

[17] SCHULTZ, R.: Group actions on hypertoral manifolds.II. J. Reine Angew. Math. 325, 75-86 (1981)

[18] SULLIVAN, D.: Infinitesimal computations in topology. Publ. I.H.E.S. 47, 269-331 (1977)

[19] KWASIK,S. and SCHULTZ,R.: Topological circle actions on 4-manifolds. Preprint (1987)

[20] GERSTENHABER, M. and SCHACK, S.D.: Relative Hochschild cohomology, rigid algebras, and the Bockstein. J.Pure Appl.Algebra 43, 53-74 (1986)

2 × 2 - MATRICES AND APPLICATION TO LINK THEORY

by

Pierre VOGEL

In many subjects in topology , particularly in low dimensionnal topology , a great deal of the difficulty of the theory come from the presence of very big groups like : free groups , braid groups , mapping class groups, fundamental groups of surfaces or 3-dimensionnal manifolds, ... It is very difficult to make direct computations in such a group G . A possible way to study it is to consider homology groups $H_n(G)$. These functors H_n are derived functors of the abelianization functor H_1 and the morphism from $\mathbf{Z}[G]$ to $\mathbf{Z}[H_1(G)]$ is the universel representation of the algebra $\mathbf{Z}[G]$ to a commutative algebra . A possible way to construct other invariants is to consider representations in the algebra of 2×2-matrices with entries in a commutative ring . This method was already used in some particular cases . In [2] Culler and Shalen consider representations of the fundamental group of a surface or a 3-dimensionnal manifold in $SL_2(\mathbf{C})$ and obtains many interesting results about 3-dimensionnal manifolds . In [1] Casson considers representations of the fundamental group of a surface in SU_2 and constructs an invariant in $\mathbf{Z}$ for homology 3-spheres .

In this paper we consider representations from an algebra R in an algebra $M_2(A)$,where A is a commutative ring , and we construct functors $\mathcal{M}$ and C satisfying the following properties : if R is an algebra, C(R) is a commutative ring and $\mathcal{M}(R)$ is a C(R)-algebra . Moreover we have a natural representation from R to $\mathcal{M}(R)$ an this representation is in some sense the universel representation from R to the algebra of 2×2- matrices with entries in a commutative ring .The algebra $\mathcal{M}(R)$ is not exactly an algebra of 2×2- matrices but we have a trace map t and a determinant map δ from $\mathcal{M}(R)$ to $C(R)$ and if K is a $C(R)$-algebra which is an algebraicly closed field, $\mathcal{M}(R)\otimes K$ is, in almost all cases, isomorphic to $M_2(K)$.

It is well known that a braid with n components acts on the free group F_n on n letters. But this action on $F_2 = F(x,y)$ is not very interesting if the braid is pure and has only 2 components. On the other hand if we replace the braid by an embedding L of 2 intervals in $I\times\mathbb{R}^2$ which is standard on the boundary, L doesn't act on F_2 neither on the algebra $\mathbf{Z}[F_2]$ except if L is a braid.

In this paper we will prove that there exists a ring Λ, algebraic extension of a polynomial ring of 5 variables, and a morphism from the ring $C(\mathbf{Z}[F_2])$ to Λ, such that L acts on $\mathcal{M}(\mathbf{Z}[F_2])\otimes\Lambda$ by conjugation by an element on the form $u+vxy$, where u and v belongs to Λ. The pair (u,v) in Λ^2 is well defined up to a scalar and depends only on the concordance class of L. This invariant is explicitly computable, as it is shown in an example, and it is absolutely not trivial.

§1 - Functors $\mathcal{M}$ and C

Let R be the algebra of 2×2 matrices with entries in a commutative ring A. The trace tr and the determinant det are maps from R to A satisfying the following:

i) tr is A-linear and det is A-quadratic.

ii) det is multiplicative.

iii) for every x,y in R:

$$\mathrm{tr}(xy) - \mathrm{tr}(x)\,\mathrm{tr}(y) + \det(x+y) - \det(x) - \det(y) = 0$$

Moreover, for every matrix in R, we have the Cayley-Hamilton formula:

iv) $\quad x^2 - \mathrm{tr}(x)\,x + \det(x) = 0$

On the other hand, we have a map: $x \to \bar{x}$ from R to R defined by:

$$\bar{x} = \mathrm{tr}(x) - x$$

The map: $x \to \bar{x}$ is an (anti-) involution of R and satisfies the following:

$\forall x \in R \qquad x + \bar{x} = \mathrm{tr}(x)$

$\qquad\qquad\quad x\bar{x} = \bar{x}x = \det(x)$

Definition 1.1 A quasi 2×2 matrix algebra is an algebra R over a commutative ring A equipped with an involution $\bar{\ }$ and maps t and δ from R to A satisfying the following:

P0 $\forall x \in R \quad t(x) = x + \bar{x}$

$\delta(x) = x\bar{x} = \bar{x}x$

P1 t is A-linear and δ is A-quadratic

P2 δ is multiplicative

P3 $\forall x,y \in R \quad t(xy) - t(x)\,t(y) = \delta(x) + \delta(y) - \delta(x+y)$

P4 $\forall x \in R \quad x^2 - t(x)\,x + \delta(x) = 0$

P5 $\forall x,y \in R \quad xy + yx - x\,t(x) - t(x)\,y - t(xy) + t(x)\,t(y) = 0$

Remark 1.2 Properties P4 and P5 are obvious consequences of property P0 . Moreover , if A is included in R , P1 , P2 , P3 are also consequences of P0 .

Remark 1.3 If A is a commutative ring, then $M_2(A)$ is a quasi 2×2 matrix algebra over R. But a quasi 2×2 matrix algebra over A is generally not isomorphic to $M_2(A)$. For instance, if A is the field $\mathbb{R}$ and R is the quaternionic skew field $\mathbb{H}$ endowed with the standard involution, it is easy to check that R is a quasi 2×2 matrix algebra not isomorphic to $M_2(\mathbb{R})$.

Let us denote by $\mathcal{A}$ (resp. $\mathcal{A}_2$) the category of algebras (resp. quasi 2×2 matrix algebras). If R and S are algebras over commutative rings A and B, a morphism into $\mathcal{A}$ from (R,A) to (S,B) is a couple of compatible homomorphisms from R to S and A to B. A morphism is a morphism in $\mathcal{A}_2$ if it respects traces, determinants and involutions.

Theorem 1.4 There exists a functor $(\mathcal{M},C)$ from $\mathcal{A}$ to $\mathcal{A}_2$ and a morphism η from the identity functor of $\mathcal{A}$ to $(\mathcal{M},C)$ satisfying the following:

For each A-algebra R , each morphism φ from R to a quasi 2×2 matrix algebra M over B factorizes uniquely through the C(R)-algebra $\mathcal{M}(R)$.

<u>Proof</u> Suppose that R is an algebra over a commutative ring A. Denote by A' the ring $A[R \quad R]$. If x belongs to R, the two corresponding elements in $R \quad R \subset A[R \quad R]$ will be denoted by $t(x)$ and $\delta(x)$ respectively.

So, we get maps t and δ from R to A'. If we force t and δ to satisfy properties P1, P2 and P3, we get a quotient A'' of A'.

Now we set:

$$R' = R \underset{A}{\otimes} A''$$

We have a A''-linear map $\overline{}$ from R' to itself defined by:

$$\forall x \in R \quad \overline{x \otimes 1} = 1 \otimes t(x) - x \otimes 1$$

Let R'' be the quotient of R' by the two-sided ideal generated by the following elements:

$$\overline{xy} - \overline{y}\,\overline{x} \qquad , x \in R' \ , \ y \in R'$$

$$x\,\overline{x} - \delta(x) \qquad , x \in R'$$

The A''-algebra R'' is clearly a quasi 2×2 matrix algebra. Moreover, it is the universal one. Now we set:

$$C(R) = A''$$

$$\mathcal{M}(R) = R''$$

and $(\mathcal{M}, C)$ is a functor from $\mathcal{A}$ to $\mathcal{A}_2$ satisfying the desired property.

Let us consider the following example:

R is the group ring $\mathbf{Z}[F(x,y)]$, where $F(x,y)$ is the free group generated by x and y; R is an algebra over $\mathbf{Z}$.

<u>Theorem 1.5</u> In this case, we have:

$$C(R) = \mathbf{Z}[\, t(x), t(y), t(xy), \delta(x), \delta(y), \delta(x)^{-1}, \delta(y)^{-1} \,]$$

and $\mathcal{M}(R)$ is a free $C(R)$-module with basis $1, x, y$ and xy.

<u>Proof</u> Denote by $C_1(R)$ the subring of $C(R)$ generated by $t(x)$, $t(y)$, $t(xy)$, $\delta(x)$, $\delta(y)$, $\delta(x)^{-1}$ and $\delta(y)^{-1}$.

<u>Claim 1</u> for every $u \in F(x,y)$, $t(u)$ lies in $C_1(R)$.

The proof is by induction on the length $l(u)$ of the word u in $F(x,y)$.

Suppose that $t(u)$ lies in $C_1(R)$ for every u in $F(x,y)$ of length less that n, and let u be a word in $F(x,y)$ of length n.

If u contains a power x^p, with $p \neq 0,1$:

$$u = v x^p w$$

by the Cayley-Hamilton formula (property P4), x^p belongs to:

$$C_1(R) \oplus x\, C_1(R)$$

If p is less than -1, vw and vxw have length less than n and $t(u)$ lies in $C_1(R)$ by induction. If p is -1, vw has length less than n, so:

$$t(u) \in C_1(R) \iff t(vxw) \in C_1(R)$$

The same holds if u contains a non trivial power of y. Thus it is enough to consider the case where u does not contain x^p or y^p ($p \neq 0$ ou 1). Hence, the word u has the following form:

$$u = xyxy\ldots$$

$$u = yxyx\ldots$$

If n is bigger than 3, u contains $(xy)^2$ or $(yx)^2$ and $t(u)$ belongs to $C_1(R)$. In the other case, we have:

$$n \leq 1 \Rightarrow t(u) \in C_1(R)$$

$$t(xy) = t(yx) \in C_1(R)$$

$$t(xyx) = t(x^2 y) \in C_1(R)$$

$$t(yxy) = t(y^2 x) \in C_1(R)$$

and the claim is proved.

Claim 2 For every u in R, $t(u)$ and $\delta(u)$ belong to $C_1(R)$.

Let u be an element of R . Since t is linear, $t(u)$ lies in $C_1(R)$. Since δ is quadratic, $\delta(u)$ belongs to $C_1(R)$ for every u in R if and only if:

$$\forall u \in F(x,y) \quad \delta(u) \in C_1(R)$$

$$\forall u,v \in F(x,y) \quad \delta(u+v) - \delta(u) - \delta(v) \in C_1(R)$$

But that is easy to check because of property P3.

Let $\mathcal{M}_1(R)$ be the subring of $\mathcal{M}(R)$ generated by R and $C_1(R)$. An easy consequence of claims 1 and 2 is:

Claim 3 $\mathcal{M}_1(R)$ is a quasi 2×2 matrix algebra over $C_1(R)$.

By the universal property of $\mathcal{M}(R)$, we have:

$$C(R) = C_1(R)$$

$$\mathcal{M}(R) = \mathcal{M}_1(R)$$

On the other hand, it is easy to check that:

$$C_1(R) + x\,C(R) + y\,C(R) + xy\,C(R)$$

is an algebra. Then:

$$\mathcal{M}(R) = C(R) + x\,C(R) + y\,C(R) + xy\,C(R)$$

Now, consider the representation ρ from R to $M_2(\mathbb{C})$ defined by:

$$\rho(x) = \begin{pmatrix} u & 0 \\ 0 & v \end{pmatrix} \qquad \rho(y) = \begin{pmatrix} a & b \\ 1 & c \end{pmatrix}$$

where u, v, a, b, c are complex numbers, and:

$$u \neq 0 \ , \ v \neq 0 \ , \ ac - b \neq 0$$

By universal property, we have maps:

$$\rho_* : \mathcal{M}(R) \to M_2(\mathbb{C})$$

$$\rho_* : C(R) \to \mathbb{C}$$

and we check:

$$\rho_*(t(x)) = u + v \qquad \rho_*(\delta(x)) = uv$$

$$\rho_*(t(y)) = a + c \qquad \rho_*(\delta(y)) = ac - b$$

$$\rho_*(t(xy)) = au + cv$$

If a,b, c, u, v are chosen to be algebraically independant , $\rho_*(t(x))$, $\rho_*(t(y))$, $\rho_*(t(xy))$, $\rho_*(\delta(x))$, $\rho_*(\delta(y))$ are algebraically independant too and C(R) is the polynomial ring $\mathbb{Z}[\, t(x), t(y), t(xy), \delta(x), \delta(y), \delta(x)^{-1}, \delta(y)^{-1}\,]$. Moreover, 1, $\rho_*(x)$, $\rho_*(y)$, $\rho_*(xy)$ are linearly independant, so:

$$\mathcal{M}(R) = C(R) \oplus x\,C(R) \oplus y\,C(R) \oplus xy\,C(R)$$

§2 - Relation with representations.

Definition 2.1 let K be a field and R be a ring. Two representations ρ and ρ' from R to $M_2(K)$ are called almost conjugate if either ρ and ρ' are conjugate or ρ (resp ρ') is extension of 1 - dimensional representations α and β (resp α' and β') and:

$$\alpha = \alpha' \text{ and } \beta = \beta'$$

or: $$\alpha = \beta' \text{ and } \beta = \alpha'$$

The set of representations from R to $M_2(K)$ modulo almost conjugation will be denoted by $R_2(R)$.

Proposition 2 - 2 Let K be a field. Two almost conjugate representations from a ring R to $M_2(K)$ induce the same morphism from C(R) to K.

Proof Let ρ and ρ' be almost conjugate representations from R to $M_2(K)$. If ρ and ρ' are conjugate, we have a commutative diagram:

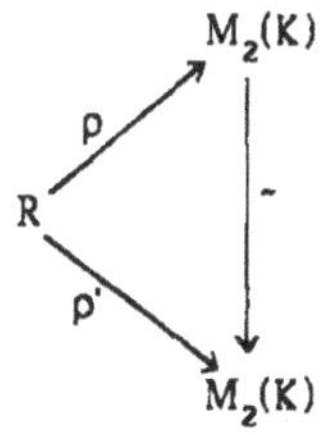

and by the universal property, we have diagrams:

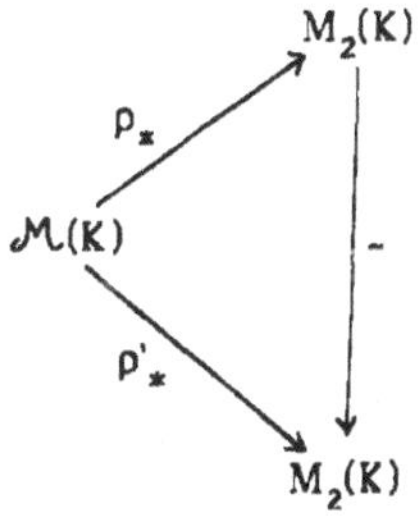

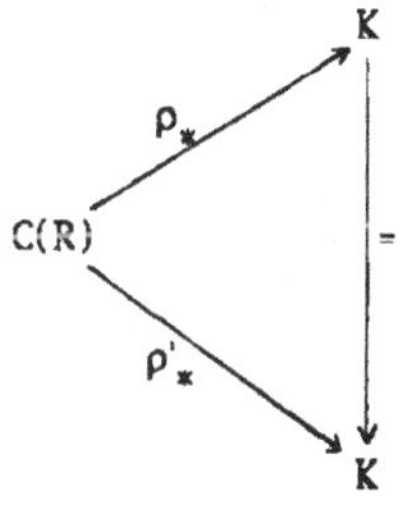

Then morphisms ρ_* and ρ'_* from C(R) to K are the same.

If ρ and ρ' are not conjugate, ρ and ρ' are conjugate to representations ρ_1 and ρ'_1 from R to the subring $M'_2(K)$ of upper triangular matrices in $M_2(K)$. Moreover, the diagonal evaluation gives a map from $M'_2(K)$ to K^2 and we get a commutative diagram:

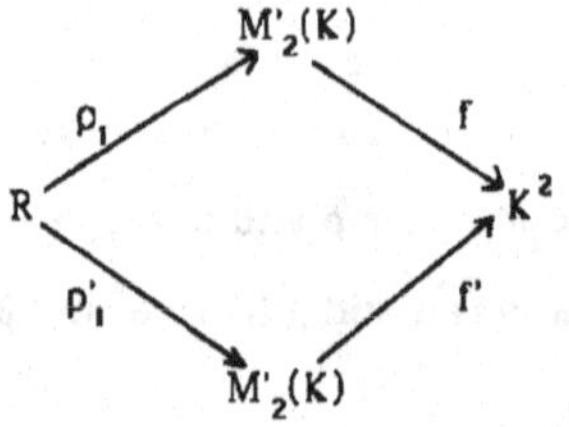

On the other hand, $M'_2(K)$ and K^2 are quasi 2×2 matrix algebras over K and f, f' and the inclusion $M'_2(K)$ in $M_2(K)$ are morphisms in $\mathcal{A}_2$. Then, if we apply the functor C, we get : $\rho_* = \rho_{1*} = \rho'_{1*} = \rho'_*$

<u>Theorem 2-3</u> let R be a ring and K be a field. Let f be a morphism from C(R) to K. Then there exists an extension L of K and a representation from R to $M_2(L)$ inducing f. Moreover, L can be chosen to be K or a quadratic extension of K or, if K has characteristic 2, a subfield of $\sqrt{K}$

<u>Theorem 2-4</u> Let R be a ring and K a field. Let ρ and ρ' be representations from R to $M_2(K)$. Then, ρ and ρ' are quasi conjugate if and only if ρ and ρ' induce the same morphism from C(R) to K.

<u>Proof</u> If f is a character from C(R) to K, let us denote by $R_2(R,f)$ the almost conjugacy classes of representations from R to $M_2(K)$ inducing f from C(R) to K.

Let Λ be the following K-algebra :

$$\Lambda = \mathcal{M}(R) \underset{C(R)}{\otimes} K$$

where the C(R)-algebra structure of K is given by f.

The algebra Λ is a quasi 2×2 matrix algebra over K. by the universal property of $\mathcal{M}(R)$ we have:

$$R_2(R,f) = R_2(\Lambda,Id)$$

Then, if we want to prove theorem 2-3, it is enough to show that $R_2(\Lambda\otimes L\ ,\ Id)$ is not empty for some algebraic extension L of K. Theorem 2-4 is equivalent to the fact that $R_2(\Lambda,\ Id)$ has at most one element.

<u>Case 1</u> Suppose that the characteristic of K is different from 2 and:

$\forall x \in \Lambda \;\; t(x)^2 = 4\,\delta(x)$

In this case, denote by f the map $(1/2)t$ from Λ to K. Since t is linear, f is linear too.

On the other hand, f^2 is equal to δ and f^2 is multiplicative. Then, for every $x, y \in K$ there exists $\varepsilon = \pm 1$ such that:

$f(xy) = \varepsilon\, f(x)\, f(y)$

Since $f(xy)$ and $f(x)\,f(y)$ are bilinear, it is easy to see that ε doesn't depend on x and y, and f is multiplicative. The morphism:

$$x \rightsquigarrow \begin{pmatrix} f(x) & 0 \\ 0 & f(x) \end{pmatrix}$$

belongs to $R_2(\Lambda, \mathrm{Id})$ and theorem 2-3 is proved in this case (with L = K).

Let ρ be an element of $R_2(\Lambda, \mathrm{Id})$. Since $t(x)^2$ is equal to $4\,\delta(x)$ for every x in Λ, $\rho(x)$ is either the scalar matrix $f(x)$ or this matrix plus some nilpotent matrix. Then, if ρ is not the scalar representation f there exists some element x_0 in Λ such that $\rho(x_0)$ is the matrix:

$$\rho(x_0) = \begin{pmatrix} f(x_0) & 1 \\ 0 & f(x_0) \end{pmatrix}$$

in some basis in K^2.

Let $x \in \Lambda$ and $\begin{pmatrix} a & b \\ c & d \end{pmatrix}$ be the matrix $\rho(x)$. We have:

$a + d = t(x) = 2\,f(x)$

$f(x_0)\,a + c + f(x_0)\,d = t(x_0\,x) = 2\,f(x_0\,x) = 2\,f(x_0)\,f(x)$

Then c is zero, and ρ is the following morphism:

$$\rho = \begin{pmatrix} f & g \\ 0 & f \end{pmatrix}$$

for some map g from Λ to K, and ρ is almost conjugate to:

$$\begin{pmatrix} f & 0 \\ 0 & f \end{pmatrix}$$

Therefore, theorem 2-4 is proved in this case.

<u>Case 2</u> Suppose that K is of characteristic 2 and the trace t is null on Λ.

In this case, denote by f the map $\sqrt{\delta}$ from Λ to $\sqrt{K}$ and by L the image of f. It is easy to see that f is an algebraic homomorphim and the scalar representation f is an element of $R_2(\Lambda \otimes L, \mathrm{Id})$. Hence, theorem 2-3 is proved in this case.

Let ρ be an element of $R_2(\Lambda, Id)$. If $\rho(x)$ is a scalar matrix, for every x in Λ, L is equal to K and ρ is the scalar representation f.

If L is equal to K and $\rho(x_0)$ is not a scalar matrix for some $x_0 \in \Lambda$, we can show, as in the first case, that there exists a map g from Λ to K such that ρ is conjugate to the representation $\begin{pmatrix} f & g \\ 0 & f \end{pmatrix}$ and then ρ is almost conjugate to $\begin{pmatrix} f & 0 \\ 0 & f \end{pmatrix}$.

Suppose that L is different from K. Let $x_0 \in \Lambda$ such that $f(x_0)$ is not in K. Then $\rho(x_0)$ is not a scalar matrix and, as above, $\rho \otimes L$ is conjugate to a representation ρ':

$$\rho' = \begin{pmatrix} f & g \\ 0 & f \end{pmatrix}$$

such that $g(x_0)$ is not zero.

If x is an element of Λ, $\rho'(x)$ is a linear combination of 1 and $\rho'(x_0)$, and there exist $a, b \in K$ such that:

$$\rho(x) = a + b\,\rho(x_0)$$

$$\Rightarrow \quad f(x) = a + b\,f(x_0)$$

Therefore L is the extension $K[f(x_0)]$ of K and there exist unique functions α and β from L to K such that:

$$\forall u \in L \quad u = \alpha(u) + \beta(u)\,f(x_0)$$

and we have:

$$\forall x \in \Lambda \quad \rho(x) = \alpha(f(x)) + \beta(f(x))\,\rho(x_0)$$

The conjugacy class of ρ is the conjugacy class of $\rho(x_0)$ which is the conjugacy class of:

$$\begin{pmatrix} 0 & 1 \\ \delta(x_0) & 0 \end{pmatrix}$$

So $R_2(\Lambda, Id)$ has at most 1 element.

<u>Case 3</u> We suppose that we are not in cases 1 or 2 and that:

$$\forall x,y \in \Lambda \quad t(xy)^2 - t(xy)\,t(x)\,t(y) + \delta(x)\,t(y)^2 + \delta(y)\,t(x)^2 - 4\,\delta(x)\,\delta(y) = 0$$

Since we are not in cases 1 or 2, there exists $x_0 \in \Lambda$ such that:

$$t(x_0)^2 - 4\,\delta(x_0) \neq 0$$

Let λ and μ be two elements of a quadratic extension L of K such that:

$$\lambda + \mu = t(x_0) \qquad \lambda\mu = \delta(x_0)$$

Since $t(x_0)^2 \neq 4\,\delta(x_0)$, λ is different from μ. Then, for every $x \in \Lambda$, the following equations:

$$a + b = t(x)$$

$$\lambda a + \mu b = t(x_0 x)$$

have a unique solution. Define maps f and g from Λ to L by:

$$f(x) = a$$

$$g(x) = b$$

Since we have:

$$t(x_0 x)^2 - t(x_0 x)\, t(x_0)\, t(x) + \delta(x_0)\, t(x)^2 + \delta(x)\, [t(x_0)^2 - 4\, \delta(x_0)] = 0$$

it is not difficult to compute $\delta(x)$. After computation, we get:

$$\delta(x) = f(x)\, g(x)$$

Clearly, f and g are K-linear.

Let x and y be two elements in Λ. We set:

$$a = f(x) \quad b = g(x)$$

If $a \neq b$ there exist unique elements α and β in L such that:

$$\alpha + \beta = t(y) \quad \text{and} \quad a\alpha + b\beta = t(xy)$$

and we have as above: $\delta(y) = \alpha\beta$

Consequently, we have:

$$\alpha = f(y) \quad \text{and } \beta = g(y)$$

or: $$\alpha = g(y) \quad \text{and } \beta = f(y)$$

Suppose that: $\alpha = g(y)$ and $\beta = f(y)$

Let $u \in L$ such that: $u\lambda + a \neq u\mu + b$

Then $t(u x_0 + x)^2$ is different from $4\,\delta(u x_0 + u)$ and , as above, there exist unique elements α' and β' in L such that:

$$\alpha' + \beta' = t(y)$$

$$(u\lambda + a)\,\alpha' + (u\mu + b)\,\beta' = t(\,(ux_0 + x)\,y\,)$$

$$\alpha'\beta' = \delta(y)$$

And we have:

$$t(\,(ux_0 + x)\,y\,) = (u\lambda + a)\,\alpha + (u\mu + b)\,\beta$$

or: $$t(\,(ux_0 + x)\,y\,) = (u\lambda + a)\,\beta + (u\mu + b)\,\alpha$$

In other words:

$$u(\,\lambda\beta + \mu\alpha\,) + a\alpha + b\beta = u(\,\lambda\alpha + \mu\beta\,) + a\alpha + b\beta$$

or: $$u(\,\lambda\beta + \mu\alpha\,) + a\alpha + b\beta = u(\,\lambda\beta + \mu\alpha\,) + a\beta + b\alpha$$

i.e. $\quad u(\alpha-\beta)(\lambda-\mu) = 0 \quad$ or $\quad (\alpha-\beta)(a-b) = 0$

and this is impossible.

hence, if $f(x) \neq g(x)$, we have:

$$t(xy) = f(xy) + g(xy) = f(x)\,f(y) + g(x)\,g(y)$$

Of course, the same holds if $f(y) \neq g(y)$.

If $f(x) = g(x)$ and $f(y) = g(y)$, we have:

$$t(xy)^2 - 4\,t(xy)\,f(x)\,f(y) + 4\,f(x)^2 f(y)^2 = 0$$

$$\Rightarrow [\,t(xy) - 2\,f(x)\,f(y)\,]^2 = 0$$

$$\Rightarrow t(xy) = 2\,f(x)\,f(y) = f(x)\,f(y) + g(x)\,g(y)$$

Therefore for every x and y in Λ, we have :

$$f(xy) + g(xy) = f(x)\,f(y) + g(x)\,g(y)$$

and : $\quad f(xy)\,g(xy) = f(x)\,g(x)\,f(y)\,g(y)$

Hence we have two possibilities :

$$f(xy) = f(x)\,f(y) \qquad \text{and} \qquad g(xy) = g(x)\,g(y)$$

or : $\quad f(xy) = g(x)\,g(y) \qquad$ and $\qquad g(xy) = f(x)\,f(y)$

Suppose that $f(x)$ and $g(x)$ are different . Let $y \in \Lambda$. If $f(xy)$ is equal to $g(x)$ $g(y)$, we have :

$$f(x(y+1)) = g(x)\,g(y) + f(x)$$

and : $\quad f(x(y+1)) = f(x)\,(f(y)+1) \qquad$ or $\qquad f(x(y+1)) = g(x)\,(g(y)+1)$

Since $f(x)$ and $g(x)$ are different , we get :

$$g(x)\,g(y) = f(x)\,f(y)$$

The same holds if $f(y) \neq g(y)$ and then in any case .

Finally we have :

$$\forall\, x, y \in \Lambda \quad f(xy) = f(x)\,f(y) \qquad g(xy) = g(x)\,g(y)$$

and $\begin{pmatrix} f & 0 \\ 0 & g \end{pmatrix}$ is a representation in $R_2(\Lambda\otimes L\,,\, Id)$.

Now suppose that ρ is a representation in $R_2(\Lambda\otimes L\,,\, Id)$. The representation $\rho\otimes L$ is conjugate to a representation ρ' such that :

$$\rho'(x_0) = \begin{pmatrix} \lambda & 0 \\ 0 & \mu \end{pmatrix}$$

Suppose that x and y are two elements in Λ and that:

$$\rho'(x) = \begin{pmatrix} a & b \\ c & d \end{pmatrix} \qquad \rho'(y) = \begin{pmatrix} \alpha & \beta \\ \gamma & \delta \end{pmatrix}$$

We have:

$$a + d = t(x) = f(x) + g(x)$$

$$\lambda a + \mu d = t(x_0 x) = f(x_0 x) + g(x_0 x)$$

$$\Rightarrow a = f(x) \quad d = g(x)$$

and this implies:

$$\delta(x) = ad - bc = ad \Rightarrow bc = 0$$

Then p'x) and $\rho'(y)$ are triangular.

Suppose that: $c = 0 \quad b \neq 0 \quad \beta = 0 \quad \gamma \neq 0$

then we have:

$$f(xy) = a\alpha + b\gamma = f(x)f(y) = a\alpha$$

which is impossible. Hence, ρ' has the following form:

$$\begin{pmatrix} f & \varphi \\ 0 & g \end{pmatrix} \quad \text{or} \quad \begin{pmatrix} f & 0 \\ \varphi & g \end{pmatrix}$$

for some map φ from Λ to L.

if we change f and g, we may as well suppose that:

$$\rho' = \begin{pmatrix} f & \varphi \\ 0 & g \end{pmatrix}$$

Suppose that φ is zero. Then, for every $x \in \Lambda$, $\rho'(x)$ is a linear combination of 1 and $\rho'(x_0)$

and there exist two functions α and β from Λ to K such that:

$$\rho(x) = \alpha(x) + \rho(x_0)\,\beta(x)$$

Moreover, we have:

$$t(x) = 2\,\alpha(x) + \beta(x)\,t(x_0)$$

$$t(x_0 x) = \alpha(x)\,t(x_0) + \beta(x)\,t(x_0^2)$$

and $\alpha(x)$ and $\beta(x)$ depend only on $t(x)$, $t(x_0)$, $\delta(x)$, $\delta(x_0)$ and $t(x_0 x)$.

Since $\rho(x_0)$ is conjugate to:

$$\begin{pmatrix} t(x_0) & -1 \\ \delta(x_0) & 0 \end{pmatrix}$$

ρ is conjugate to:

$$\alpha \begin{pmatrix} 1 & 0 \\ 0 & 1 \end{pmatrix} + \beta \begin{pmatrix} t(x_0) & -1 \\ \delta(x_0) & 0 \end{pmatrix}$$

If φ is non zero, there exists $x_1 \in \Lambda$ such that $\varphi(x_1) \neq 0$. We have:

$$\rho'(x_0) = \begin{pmatrix} \lambda & 0 \\ 0 & \mu \end{pmatrix} \qquad \rho'(x_1) = \begin{pmatrix} \alpha & u \\ 0 & \beta \end{pmatrix}.$$

Since $\rho(x_0)$ is conjugate to:

$$\begin{pmatrix} t(x_0) & -1 \\ \delta(x_0) & 0 \end{pmatrix}$$

there exists a matrix:

$$\begin{pmatrix} a & b \\ c & d \end{pmatrix}$$

in $GL_2(L)$ such that:

$$\begin{pmatrix} a & b \\ c & d \end{pmatrix}\begin{pmatrix} \lambda & 0 \\ 0 & \mu \end{pmatrix} = \begin{pmatrix} t(x_0) & -1 \\ \delta(x_0) & 0 \end{pmatrix}\begin{pmatrix} a & b \\ c & d \end{pmatrix}$$

and:

$$\begin{pmatrix} a & b \\ c & d \end{pmatrix}\begin{pmatrix} \alpha & u \\ 0 & \beta \end{pmatrix}\begin{pmatrix} a & b \\ c & d \end{pmatrix}^{-1} \in M_2(K)$$

After computation we get:

$$a \neq 0 \qquad b \neq 0$$

$$\frac{\alpha\lambda - \beta\mu}{\lambda-\mu} - \frac{au\mu}{b(\lambda-\mu)} \in K \qquad \frac{\alpha-\beta}{\lambda-\mu} - \frac{au}{b(\lambda-\mu)} \in K$$

$$\lambda\mu\frac{\alpha-\beta}{\lambda-\mu} - \frac{au\mu^2}{b(\lambda-\mu)} \in K \qquad \frac{\lambda\beta - \alpha\mu}{\lambda-\mu} + \frac{au\mu}{b(\lambda-\mu)} \in K$$

Suppose that L is different from K. Then L is a quadratic extension of K and we have a Galois action $\bar{\ }$ on L:

$$\bar{\mu} = \lambda \qquad \bar{\alpha} = \beta$$

So we get:

$$\frac{au\mu}{b(\lambda-\mu)} \in K \quad \text{and} \quad \frac{au}{b(\lambda-\mu)} \in K \Rightarrow \mu \in K$$

and then L is equal to K.

So L is equal to K and ρ (= ρ') is quasi conjugate to:

$$\begin{pmatrix} f & \varphi \\ 0 & g \end{pmatrix}$$

and theorems 2-3 and 2-4 are proved in this case.

<u>Case 4</u> We suppose that we are not in case 1 or 2 or 3.

For any x and y in a quasi 2×2 matrix algebra, set:

$$\Delta(x,y) = t(xy)^2 - t(xy)\,t(x)\,t(y) + t(x)^2\,\delta(y) + t(y)^2\,\delta(x) - 4\,\delta(x)\,\delta(y)$$

In this case there exist x and y in Λ such that $\Delta(x,y)$ is not zero. Let Λ_1 be the subalgebra of Λ generated by x and y. Clearly Λ_1 is generated as a K-vector space by $1, x, y, xy$. Suppose we have a relation:

$$a + bx + cy + d\,xy = 0 \qquad a, b, c, d \in K$$

Then we get:

$$2a + b\,t(x) + c\,t(y) + d\,t(xy) = 0$$

$$a\,t(x) + b\,t(x^2) + c\,t(xy) + d\,t(x^2y) = 0$$

$$a\,t(y) + b\,t(xy) + c\,t(y^2) + d\,t(xy^2) = 0$$

$$a\,t(xy) + b\,t(x^2y) + c\,t(xy^2) + d\,t(x^2y^2) = 0$$

It is not difficult to check the following:

$$t(x^2) = t(x)^2 - 2\,\delta(x)$$

$$t(y^2) = t(y)^2 - 2\,\delta(y)$$

$$t(x^2y) = t(x)\,t(xy) - \delta(x)\,t(y)$$

$$t(xy^2) = t(y)\,t(xy) - \delta(y)\,t(x)$$

$$t(x^2y^2) = t(xy)\,t(x)\,t(y) - t(x^2)\,\delta(y) - t(y^2)\,\delta(x) + 2\,\delta(x)\,\delta(y)$$

and the determinant of this system is $\Delta(x,y)^2$ which is not zero. Therefore, $(1, x, y, xy)$ is a basis of Λ_1.

Let $a + bx + cy + dxy$ be an element of the center of Λ_1. We have:

$$x\,(a + bx + cy + dxy) = (a + bx + cy + dxy)\,x$$

$$\Rightarrow (c+dx)\,(xy - yx) = 0$$

But we have the following formula:

$$\delta(xy - yx) = (xy - yx)\,(\overline{xy - yx}) = \delta(xy) + \delta(yx) + t(xy^2x) - t(xy)\,t(yx)$$

$$= 2\,\delta(x)\,\delta(y) + t(x^2y^2) - t(xy)^2 = -\,\Delta(x,y)$$

Then we get:

$$(c+dx)\,\Delta(x,y) = 0 \;\Rightarrow\; c = d = 0$$

and $a + bx$ which commutes with y is a multiple of 1. Therefore, the center of Λ_1 is K.

On the other hand, it is not difficult to see there is no character from Λ_1 to K. Then Λ_1 is simple and $\Lambda_1 \otimes L$ is isomorphic to $M_2(L)$ for some quadratic extension L of K. Consequently there exist elements e_{ij} in $\Lambda_1 \otimes L$, $i = 1,2$ $j = 1,2$ such that:

$$e_{ij}\,e_{i'j'} = 0 \text{ if } j \neq i'$$

$$= e_{ij'} \text{ if } j = i'$$

Let us define the following maps f_{ij} from Λ to L:

$$\forall i,j \quad \forall x \in \Lambda \quad f_{ij}(x) = t(x\, e_{ij})$$

Claim For every x,y in Λ and every i,j in {1,2} we have:

$$f_{ij}(xy) = f_{i1}(x)\, f_{1j}(y) + f_{i2}(x)\, f_{2j}(y)$$

Proof of the claim: For every u,v in Λ we have:

$$\delta(u\, e_{11} + v\, e_{11}) = \delta(u\, e_{11}) + \delta(v\, e_{11}) + t(u\, e_{11})\, t(v\, e_{11}) - t(u\, e_{11}\, v\, e_{11})$$

but δ is multiplicative and $\delta(e_{11})$ is zero. Thus we have:

$$t(u\, e_{11})\, t(v\, e_{11}) = t(u\, e_{11}\, v\, e_{11})$$

and this implies:

$$\begin{aligned} t(xy\, e_{11}) &= t(x\,(e_{11}+e_{22})\, y\, e_{11}) = t(x\, e_{11}\, y\, e_{11}) + t(x\, e_{22}\, y\, e_{11}) \\ &= t(x\, e_{11})\, t(y\, e_{11}) + t(x\, e_{21}\, e_{11}\, e_{12}\, y\, e_{11}) \\ &= t(x\, e_{11})\, t(y\, e_{11}) + t(x\, e_{21}\, e_{11})\, t(e_{12}\, y\, e_{11}) \\ &= t(x\, e_{11})\, t(y\, e_{11}) + t(x\, e_{21})\, t(y\, e_{12}) \end{aligned}$$

So we have

$$\begin{aligned} f_{ij}(xy) &= t(xy\, e_{ji}) = t(e_{1i}\, x\, y\, e_{j1}\, e_{11}) \\ &= t(e_{1i}\, x\, e_{11})\, t(y\, e_{j1}\, e_{11}) + t(e_{1i}\, x\, e_{21})\, t(y\, e_{j1}\, e_{12}) \\ &= f_{i1}(x)\, f_{1j}(y) + f_{i2}(x)\, f_{2j}(y) \end{aligned}$$

and the claim is proved.

As a consequence of the claim, we have a morphism f from Λ to $M_2(L)$:

$$f = \begin{pmatrix} f_{11} & f_{12} \\ f_{21} & f_{22} \end{pmatrix}$$

and it is not difficult to see that f is a morphism in the category $\mathcal{A}_2$ (i.e. it preserves trace and determinant).

Now let u and v be two elements of Λ such that $f(u) = 0$. We have:

$$t(u) = t(f(u)) = 0$$

$$t(uv) = t(f(uv)) = 0$$

$$\Rightarrow u + \overline{u} = 0$$

$$uv + \overline{vu} = 0$$

and this implies: $\quad uv = \overline{v}\, u$

Let a,b in Λ. We have:

$$u(ab-ba) = \overline{a}\,u\,b - \overline{ba}\,u = \overline{a}\,\overline{b}\,u\,a - \overline{ba}\,u = 0$$

In particular:

$$u(e_{12}\,e_{21} - e_{21}\,e_{12}) = u\,(e_{11} - e_{22}) = 0$$

But $e_{11}-e_{22}$ is invertible . Then u is trivial and f induces a monomorphism from Λ to $M_2(L)$.

That proves theorem 2-3.

If we have a representation from Λ to $M_2(K)$, Λ is not a skew field and Λ is isomorphic to $M_2(K)$. Hence two representations from Λ to $M_2(K)$ are conjugate and theorem 2-4 is proved.

§3 - An invariant for links.

Definition 3-1

A link of n intervals is an embedding of $I\times\{1, 2, \dots, n\}$ to $I\times\mathbb{R}^2$ which is standard on the boundary.

Two links are concordant if there is an embedding F from $I^2\times\{1, \dots, n\}$ to $I\times I\times\mathbb{R}^2$ standard on $\partial I\times I\times\{1, \dots, n\}$ and inducing f_i on $I\times\{i\}\times\{1, \dots, n\}$ for $i = 0,1$.

The set of concordance of links of n intervals is a set C_n, which is actually a group for the juxtaposition law [3].

Let L be a link of n intervals. Denote by X the complement of L and by X_0 and $X_1$the top part and the bottom part of ∂X. Let x_i (resp x'_i) be the element of $\pi_1(X_0)$ (resp $\pi_1(X_1)$) which turns around the ith component of L in X_0 (resp X_1). The fundamental group $\pi_1(X_0)$ is a free group with basis $x_1, \dots, x_n$. The same holds for $\pi_1(X_1)$. But $\pi_1(X)$ is generally not free. We only know the following [3]:

There exists a universal group G_n depending only on n and a morphism ε from $\pi_1(X_0)$ to G_n such that for any link L, ε extends uniquely on $\pi_1(X)$. Moreover, there exists a unique automorphism τ_L depending on the concordance class of a link L on G_n such that:

$$\forall i= 1, \dots, n \quad \tau_L(x_i) = x'_i$$

This automorphism satisfies the following:

for every i, $\tau_L(x_i)$ is conjugate to x_i and $\tau_L(x_1x_2 \dots x_n) = x_1x_2 \dots x_n$

In fact, G_n is the algebraic closure of $\pi_1(X_0)$ in the sense of Levine [4].

The problem is that G_n is completely unknown and it is therefore difficult to give a description of some automorphism of G_n.

From now on, we will suppose that L is a link with 2 components. We set:

$x_1 = x \quad x_2 = y$

Then $\pi_1(X_0)$ is the free group $F(x,y)$.

Notation 3-2 We set the following in the ring $C(\mathbf{Z}[F(x,y)])$:

$a = t(x) \quad b = t(y) \quad c = t(xy)$

$\alpha = \delta(x) \quad \beta = \delta(y)$

A denotes the ring $C(\mathbf{Z}[F(x,y)]) = \mathbf{Z}[a, b, c, \alpha, \alpha^{-1}, \beta, \beta^{-1}]$ and $\mathcal{M}$ is the algebra $\mathcal{M}(\mathbf{Z}[F(x,y)])$. Δ is the element of A defined by:

$$\Delta = c^2 - abc + a^2\beta + b^2\alpha - 4\alpha\beta$$

S is the multiplicative subset of A which consists of polynomials $P(a^2\alpha^{-1}, b^2\beta^{-1})$ of $\mathbf{Z}[a^2\alpha^{-1}, b^2\beta^{-1}] \subset A$ such that $P(4,4) = 1$.

$\hat{A}$ is the completion of $S^{-1}A$ with respect to the ideal generated by Δ:

$$\hat{A} = \lim S^{-1} A_{/\Delta^n}$$

Λ is the subring of $\hat{A}$ which consists of all elements of $\hat{A}$ algebraic over A.

Theorem 3-3 let L be a link of 2 intervals. Then the morphism from $\mathbf{Z}[\pi_1(X_0)]$ to $\mathcal{M}$ extends uniquely to a morphism from $\mathbf{Z}[\pi_1(X)]$ to $\mathcal{M}\otimes\Lambda$. Moreover there exists a unique automorphism φ_L from $\mathcal{M}\otimes\Lambda$ to itself such that:

$\varphi_L(x) = x' \quad \varphi_L(y) = y'$

Furthermore there exist elements u,v in Λ, unique up to multiplication by a scalar in Λ such that:

$$\forall z \in \mathcal{M} \quad \varphi_L(z) = (u+vxy)\, z\, (u+vxy)^{-1}$$

The automorphism φ_L depends only on the concordance class of L and the correspondance $L \to \varphi_L$ is a representation of the group C_2 to $\mathrm{Aut}(\mathcal{M}\otimes\Lambda)$.

Remark 3-4 In fact the morphism from $\mathbf{Z}[\pi_1(X_0)]$ to $\mathcal{M}$ extends uniquely to a morphism from $\mathbf{Z}[G_2]$ to $\mathcal{M}\otimes\Lambda$, and we have a canonical representation from G_2 to $(\mathcal{M}\otimes\Lambda)^*$

The proof of theorem 3-3 is quite long and will be divided in several lemmas.

Lemma 3-5 Let (K, K_0) be a pair of finite complexes. We suppose that K_0 is homotopy equivalent to a bouquet of two cercles and that $K/_{K_0}$ is contractible. Let x and y be the generators of $\pi_1(K_0)$. Let ε be the augmentation map:

$$\varepsilon: \mathbf{Z}[x, x^{-1}, y, y^{-1}] \to \mathbf{Z} \quad \varepsilon(x) = \varepsilon(y) = 1$$

Then we have:

$$\forall i \leq 2 \quad H_i(\pi_1(K), \pi_1(K_0); \varepsilon^{-1}(1)^{-1}\,\mathbf{Z}[x, x^{-1}, y, y^{-1}]) = 0$$

Proof: Let B be the ring: $B = \varepsilon^{-1}(1)^{-1}\,\mathbf{Z}[x, x^{-1}, y, y^{-1}]$

We have an augmentation map: $B \to \mathbf{Z}$

Since B is noetherian, and (K, K_0) is finite, $H_*(K, K_0 ; B)$ is finitely generated.

Let $H_p(K, K_0 ; B)$ be the first non trivial homology group of (K, K_0), if it exists. Since $H_*(K, K_0 ; \mathbf{Z})$ vanishes, $H_p(K, K_0 ; B)$ is killed by some element of B going to 1 in $\mathbf{Z}$

Therefore $H_p(K, K ; B)$ vanishes too and (K, K_0) is B- acyclic. But, for $i \leq 2$ $H_i(\pi_1(K), \pi_1(K_0); B)$ is a quotient of $H_i(K, K_0 ; B)$. That proves the lemma.

Lemma 3-6 Let X be the complement of a link of 2 intervals. Let M be a $\mathbf{Z}[x, x^{-1}, y, y^{-1}]$-module such that every element of $\varepsilon^{-1}(1)$ acts bijectively on M. Then:

$$\forall i \leq 2 \quad H^i(\pi_1(X), \pi_1(X_0); M) = H^i(G_2, F(x,y); M) = 0$$

Proof: The module M is a module over the ring:

$$B = \varepsilon^{-1}(1)^{-1}\,\mathbf{Z}[x, x^{-1}, y, y^{-1}]$$

Then, by the universal coefficient spectral sequence, it is enough to prove:

$$\forall i \leq 2 \quad H_i(\pi_1(X), \pi_1(X_0); B) = H_i(G_2, F(x,y); M) = 0$$

The first part of that is proved in lemma 3-5.

There exists a sequence of finite complexes [3]:

$$X_0 \subset K_1 \subset K_2 \subset \dots$$

such that:

$$\forall i \quad K_i/_{X_0} \text{ is contractible}$$

$$G_2 = \pi_1(\cup K_n)$$

So we have: $\forall i \;\; H_i(\cup K_n, X_0; B) = \lim H_i(K_n, X_0; B) = 0$

and the lemma can be easily deduced.

<u>Lemma 3-7</u> Let $n \geq 1$ be an integer, and Γ_n be the group of units of the algebra $S^{-1}\mathcal{M}/_{I^n}$, where I is the two-sided ideal of $S^{-1}\mathcal{M}$ generated by $xy-yx$. Let X be the complement of a link of two intervals. Then the morphism from $F(x,y)$ to Γ_n factorizes uniquely trough G_2 and $\pi_1(X)$.

<u>Proof:</u> This lemma will be proved by induction on n.

For $n=1$, Γ_1 is commutative and the lemma is obvious since:

$$H_1(F(x,y); \mathbf{Z}) = H_1(\pi_1(X); \mathbf{Z}) = H_1(G_2; \mathbf{Z})$$

On the other hand, we have an exact sequence (for $n \geq 1$):

$$1 \to 1 + I^n/_{I^{n+1}} \to \Gamma_{n+1} \to \Gamma_n \to 1$$

Let G be the group $\pi_1(X)$ or G_2. By induction we have a commutative diagram:

$$\begin{array}{ccccccc} 1 \to 1 + I^n/_{I^{n+1}} \to & \Gamma_{n+1} & \to & \Gamma_n & \to 1 \\ & & & \uparrow & \\ & F(x,y) & \to & G & \end{array}$$

and we want to prove that there exists a unique morphism from G to Γ_{n+1} which makes the diagram commute:

$$\begin{array}{ccc} \Gamma_{n+1} & \to & \Gamma_n \\ \uparrow & & \uparrow \\ F(x,y) & \to & G \end{array}$$

Since the multiplicative group $1 + I^n/_{I^{n+1}}$ is commutative, it is enough to prove that $1 + I^n/_{I^{n+1}}$ is by the conjugation action a B-module.

Let ω be the element $xy-yx$ of $S^{-1}\mathcal{M}$. It is easy to check the following:

$$x\omega = \omega\bar{x} \qquad \bar{x}\omega = \omega x$$

$$y\omega = \omega\bar{y} \qquad \bar{y}\omega = \omega y$$

Then I^n is generated by ω^n and $I^n/_{I^{n+1}}$ is additively isomorphic to

$$S^{-1}\mathcal{M}/_{\omega} \approx S^{-1}\mathbf{Z}[x, x^{-1}, y, y^{-1}, \bar{x}, \bar{x}^{-1}, \bar{y}, \bar{y}^{-1}]$$

On the other hand we have:

$$x(1+\omega^n u)x^{-1} = 1+\omega^n x' u x^{-1}$$

$$y(1+\omega^n u)x^{y1} = 1+\omega^n y' u y^{-1}$$

where x' and y' are x and y if n is even, and $\bar{x}$ and $\bar{y}$ if n is odd.

Then $1+I^n/_{I^{n+1}}$ is isomorphic to $S^{-1}\mathbf{Z}[x, x^{-1}, y, y^{-1}, \bar{x}, \bar{x}^{-1}, \bar{y}, \bar{y}^{-1}]$ and $F(x,y)$ and G acts on it (via $H_1(F(x,y)) = H_1(G)$) trivially if n is even and in the following way if n is odd:

$$x(u) = \bar{x}x^{-1}u \qquad y(u) = \bar{y}y^{-1}u$$

If n is even, $1+I^n/_{I^{n+1}}$ is a B-module.

Suppose now that n is odd. Let $P(x,y) \in \mathbf{Z}[x, x^{-1}, y, y^{-1}]$ such that $P(1,1)$ is 1. Then $P(x,y)$ acts on $S^{-1}\mathbf{Z}[x, x^{-1}, y, y^{-1}, \bar{x}, \bar{x}^{-1}, \bar{y}, \bar{y}^{-1}]$ by multiplication by $P(\bar{x}x^{-1}, \bar{y}y^{-1})$. It is not difficult to prove the following:

$$(\bar{x}x^{-1})^2 = \frac{a^2-2\alpha}{\alpha}\bar{x}x^{-1} - 1$$

$$(\bar{y}y^{-1})^2 = \frac{b^2-2\beta}{\beta}\bar{y}y^{-1} - 1$$

Therefore P has the following form:

$$P = U + V\bar{x}x^{-1} + W\bar{y}y^{-1} + T\,\bar{x}x^{-1}\bar{y}y^{-1}$$

$$U, V, W, T \in \mathbf{Z}[a^2\alpha^{-1}, b^2\beta^{-1}]$$

and: $\quad U(4,4) + V(4,4) + W(4,4) + T(4,4) = 1$

Set: $\quad P' = U + V\bar{x}x^{-1} + W y\bar{y}^{-1} + T\,\bar{x}x^{-1}y\bar{y}^{-1}$

We have:

$PP' = (U+V\bar{x}x^{-1})^2 + (W+T\bar{x}x^{-1})^2 + (U+V\bar{x}x^{-1})(W+T\bar{x}x^{-1})(y\bar{y}^{-1}+\bar{y}y^{-1})$ Thus we have: $\quad PP' = U' + V'\bar{x}x^{-1} \quad U', V' \in \mathbf{Z}[a^2\alpha^{-1}, b^2\beta^{-1}]$

$$U'(4,4) + V'(4,4) = 1$$

In the same way, $PP'(U'+V'x\bar{x}^{-1})$ is a polynomial U'' in $\mathbf{Z}[a^2\alpha^{-1}, b^2\beta^{-1}]$ such that:

$U''(4,4) = 1$

Therefore, $PP'(U'+V'x\bar{x}^{-1})$ belongs to S and P is invertible in $S^{-1}\mathbf{Z}[x,x^{-1},y,y^{-1},\bar{x},\bar{x}^{-1},\bar{y},\bar{y}^{-1}]$. Thus $1 + I^n/_{I^{n+1}}$ is a B-module.

Lemma 3-8 Let X be the complement of a link of two intervals. Then the morphism from $\mathbf{Z}[F(x,y)]$ to $\mathcal{M}\otimes\Lambda$ factorizes uniquely through $\mathbf{Z}[G_2]$ and $\mathbf{Z}[\pi_1(X)]$.

Proof: We have:

$$\omega^2 = (xy-yx)(xy-yx) = -\delta(xy-yx)$$
$$= -\delta(xy) - \delta(yx) - t(xyyx) + t(xy)\,t(yx) = \Delta$$

Therefore we have:

$$\lim (S^{-1}\mathcal{M}/_{I^n}) = \lim (S^{-1}\mathcal{M}/_{\Delta^n}) = \mathcal{M}\otimes\hat{\Lambda}$$

and the map from $\mathbf{Z}[F(x,y)]$ to $\mathcal{M}\otimes\hat{\Lambda}$ factorizes uniquely through $\mathbf{Z}[G_2]$ and $\mathbf{Z}[\pi_1(X)]$.

Let u be an element of G_2 or $\pi_1(X)$. If u lies in G_2, u is contained in a finitely generated subgroup G of G_2 such that $F(x,y) \subset G$ is normally surjective. If u is in $\pi_1(X)$, set $G = \pi_1(X)$.

In all cases $F(x,y) \to G$ is normally surjective, and G is generated by x,y and elements $z_1, \dots, z_n$ in $[G,G]$, and we have:

$$\forall i = 1, \dots, n \quad z_i \in [F(x,y), G]$$

So there exist words $W_i(z_1, \dots, z_n)$ in the subgroup $[F(x,y), F(x,y,z_1, \dots, z_n)]$ of the free group $F(x,y,z_1, \dots, z_n)$ and we have:

$$\forall i = 1, \dots, n \quad z_i = W_i(z_1, \dots, z_n)$$

But we have a canonical map from G to $\mathcal{M}\otimes\hat{\Lambda}$. Then $z_1, \dots, z_n$ can be considered as elements in $\mathcal{M}\otimes\hat{\Lambda}$, and W_i is a word in $\mathcal{M}\otimes\hat{\Lambda}$ which involves $z_1, \dots, z_n, z_1^{-1}, \dots z_n^{-1}$. We can replace z_i^{-1} by $\bar{z}_i\,\delta(z_i)^{-1}$, and by multiplying the relation above by a product of $\delta(z_i)$, we get:

$$\forall i = 1, \dots, n \quad z_i \prod_j (z_j \bar{z}_j)^{\alpha_{ij}} = W'_i(z_1, \dots, z_n, \bar{z}_1, \dots \bar{z}_n)$$

But z_i is congruent to 1 mod $\omega = xy - yx$:

$$z_i = 1 + \omega u_i$$

So we get equations:

$$(E_i) \quad \Phi_{0i}(u) + \omega\,\Phi_{1i}(u) + \dots + \omega^q\,\Phi_{qi}(u) = 0$$

where Φ_{pi} is a polynomial function of degree p depending on $u = (u_1, \dots, u_n)$ with values in $\mathcal{M}\otimes\hat{A}$ and coefficients in A.

On the other hand, there is a unique morphism from the group of presentation $< x , y , z_1, \dots , z_n : z_i = W_i(z_1, \dots, z_n) >$ to $\mathcal{M}\otimes\hat{A}$ which is standard on F(x,y). Then equations (E_i) have a unique solution in $(\mathcal{M}\otimes\hat{A})^n$ and this solution has algebraic coordinates (over A). Therefore z_i belongs to $\mathcal{M}\otimes\Lambda$ and the image of G in $\mathcal{M}\otimes\hat{A}$ is included in $\mathcal{M}\otimes\Lambda$. That proves the lemma.

We are now able to prove the first part of theorem 3-3. Let us consider the following diagram:

$$\begin{array}{lll} \mathbf{Z}[F(x,y)] & & \\ & \searrow & \\ & & \mathbf{Z}[\pi_1(X)] \rightarrow \mathcal{M}\otimes\Lambda \\ & \nearrow & \\ \mathbf{Z}[F(x',y')] & & \end{array}$$

The composition map φ: $\mathbf{Z}[F(x',y')] \rightarrow \mathcal{M}\otimes\Lambda$ goes to a quasi 2×2 matrix algebra. Then φ induces morphisms $\tilde{\varphi}$, $\tilde{\varphi}'$:

$$\tilde{\varphi} : \mathcal{M}(\mathbf{Z}[F(x',y')]) \rightarrow \mathcal{M}\otimes\Lambda$$

$$\tilde{\varphi}' : \ C(\mathbf{Z}[F(x',y')]) \rightarrow \Lambda$$

On the other hand, we have:

$\varphi(x')$ is conjugate to x in $\mathcal{M}\otimes\Lambda$

$\varphi(y')$ is conjugate to y in $\mathcal{M}\otimes\Lambda$

$\varphi(x'y') = xy$

Therefore we have:

$$\tilde{\varphi}' (t(x')) = t(\varphi(x')) = t(x)$$

$$\tilde{\varphi}' (t(y')) = t(\varphi(y')) = t(y)$$

$$\tilde{\varphi}' (\delta(x')) = \delta(\varphi(x')) = \delta(x)$$

$$\tilde{\varphi}' (\delta(y')) = \delta(\varphi(y')) = \delta(y)$$

$$\tilde{\varphi}' (t(x'y')) = t(\varphi(x'y')) = \delta(xy)$$

If we identify x' and x, y' and y, $\tilde{\varphi}$ is a map from $\mathcal{M}$ to $\mathcal{M}\otimes\Lambda$ inducing the inclusion $A \subset \Lambda$ in the coefficient ring. Thus $\tilde{\varphi}$ extends to an endomorphism φ_L of the Λ-algebra $\mathcal{M}\otimes\Lambda$.

Suppose that L and L' are two links of 2 intervals such that L and L' are concordant. Let X and X' be the complements of L and L' in $I\times\mathbb{R}^2$ and Y be the complement of the cobordism in $I \times I \times \mathbb{R}^2$. We have a commutative diagram:

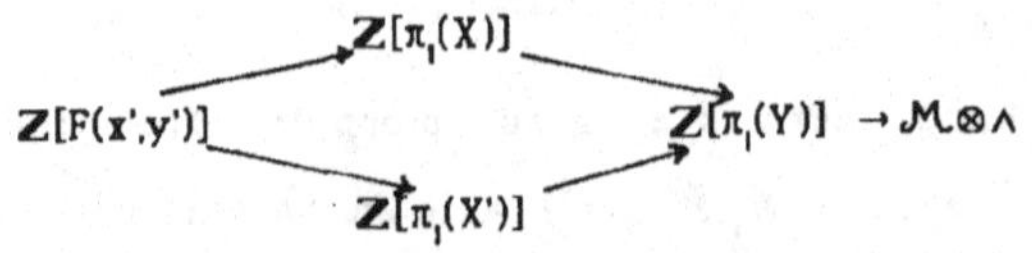

Therefore links L and L' induce the same morphism from $\mathbb{Z}[F(x',y')]$ to $\mathcal{M}\otimes\Lambda$ and φ_L and $\varphi_{L'}$ are the same. Hence, φ_L depends only on the concordance class of L.

Suppose that L and L' are two links of 2 intervals. Let L" be the juxtaposition of L and L'. Let X, X', X" be the complements of L, L', L" in $I \times \mathbb{R}^2$. We have the diagram:

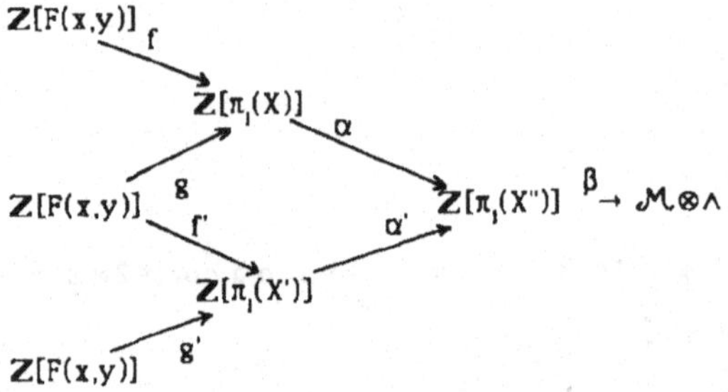

Then we have:

$$\varphi_L \circ (\beta \circ \alpha \circ f) = \beta\, \alpha\, g$$

$$\varphi_{L'} \circ (\beta \circ \alpha' \circ f') = \beta\, \alpha'\, g'$$

$$\varphi_{L''} \circ (\beta \circ \alpha \circ f) = \beta\, \alpha'\, g'$$

and this implies:

$$\varphi_{L''} = \varphi_{L'} \circ \varphi_L$$

Thus φ_L is an automorphism of $\mathcal{M}\otimes\Lambda$and φ is a representation of the group C_2 to $\mathrm{Aut}(\mathcal{M}\otimes\Lambda)$

Now, the last thing to do is to prove that φ_L is the conjugation by some element in $\Lambda\oplus\Lambda xy$ and that will be a consequence of the following lemma:

<u>Lemma 3-9</u> Let φ be an automorphism of the algebra $\mathcal{M}\otimes\Lambda$ which keeps xy fixed. Then there exists an element ε in $\Lambda\oplus\Lambda xy$, unique up to a scalar, such that φ is the conjugation by ε.

<u>Proof:</u> Let us denote by Λ' the following algebra:

$$\Lambda' = \Lambda[\lambda]/_{\lambda^2 - a\lambda + \alpha}$$

We have a Galois action of this extension:

$$\lambda \rightsquigarrow \overline{\lambda} = a - \lambda$$

Let K and K' be the quotient fields of Λ and Λ'. Let μ and θ be the elements of K' defined by:

$$\mu + \overline{\mu} = b$$

$$\lambda\mu + \overline{\lambda}\ \overline{\mu} = c$$

$$\mu\overline{\mu} - \theta = \beta$$

Actually θ lies in K. Thus we have a representation ρ from $\mathcal{M}$ to $M_2(K')$:

$$x \rightsquigarrow \rho(x) = \begin{pmatrix} \lambda & 0 \\ 0 & \overline{\lambda} \end{pmatrix} \qquad y \rightsquigarrow \rho(y) = \begin{pmatrix} \mu & \theta \\ 1 & \overline{\mu} \end{pmatrix}$$

Let Ω be the matrix:

$$\begin{pmatrix} 0 & \theta \\ 1 & 0 \end{pmatrix}$$

in $M_2(K) \subset M_2(K')$

It is easy to see the following: $\forall z \in \mathcal{M} \quad \Omega\, \rho(z) = \overline{\rho(z)}\, \Omega$

where $\overline{}$ denotes the Galois action on K' extended to $M_2(K')$. And then, $\mathcal{M} \otimes K$ is isomorphic to the subring R of matrices $A \in M_2(K')$ such that:

$$\Omega\, A = \overline{A}\, \Omega$$

The automorphism φ induces an automorphism φ_0 on R and on $R \otimes K' \approx M_2(K')$ and this automorphism keeps the center fixed. Therefore there exists a matrix ε_0 in $GL_2(K')$ such that:

$$\forall A \in R \quad \varphi_0(A) = \varepsilon_0\, A\, \varepsilon_0^{-1}$$

That means:

$$\forall A \in M_2(K') \quad \Omega\, A = \overline{A}\, \Omega \Rightarrow \Omega\, \varepsilon_0\, A\, \varepsilon_0^{-1} = \overline{\varepsilon_0}\, \overline{A}\, \overline{\varepsilon_0}^{-1}\, \Omega$$

$$\Longleftrightarrow \forall A \in M_2(K') \quad \Omega\, A = \overline{A}\, \Omega \Rightarrow \Omega\, \varepsilon_0\, A\, \varepsilon_0^{-1} = \overline{\varepsilon_0}\, \Omega\, A\, \Omega^{-1}\, \overline{\varepsilon_0}^{-1}\, \Omega$$

$$\Rightarrow \Omega^{-1}\, \overline{\varepsilon_0}^{-1}\, \Omega\, \varepsilon_0\, A = A\, \Omega^{-1}\, \overline{\varepsilon_0}^{-1}\, \Omega\, \varepsilon_0$$

Then, for any A in R, A commutes with $\Omega^{-1}\, \overline{\varepsilon_0}^{-1}\, \Omega\, \varepsilon_0$. But $R \otimes K'$ is isomorphic to $M_2(K')$. Hence $\Omega^{-1}\, \overline{\varepsilon_0}^{-1}\, \Omega\, \varepsilon_0$ is central and there exists $k \in K'$ such that:

$$\Omega\, \varepsilon_0 = k\, \overline{\varepsilon_0}\, \Omega$$

On the other hand, $\varphi_0(xy)$ is equal to xy and ε_0 commutes with xy. Then there exist u,v $\in K'$ such that: $\varepsilon_0 = u + v\, \rho(xy)$ and this implies:

$$\Omega\, \varepsilon_0 = \Omega\, (u + v\, \rho(xy)) = (u + v\, \overline{\rho(xy)})\, \Omega$$

$$= (k\,\overline{u} + k\,\overline{v}\;\overline{\rho(xy)})\,\Omega$$

$$\Rightarrow u = k\,\overline{u} \qquad v = k\,\overline{v}$$

Since ε_0 is invertible, u (or v) is not zero. But ε_0 is defined up to multiplication by a scalar. Therefore we may as well suppose that u (or v) is equal to 1 and u and v belong to K. After multiplication by some element in Λ we will get:

$$\varepsilon_0 = u + v\,\rho(xy) \qquad u,v \in \Lambda$$

Let us set: $\varepsilon = u + v\,x\,y \in \mathcal{M}\otimes\Lambda$

We have the following:

$$\forall z \in \mathcal{M}\otimes\Lambda \quad \varphi(z)\,\varepsilon = \varepsilon\,z$$

and theorem 3-3 is proved.

Remark 3-10 It is not clear that $u + v\,x\,y$ can be chosen to be a unit in $\mathcal{M}\otimes\Lambda$. But we have the following:

Proposition 3-11 Let L be a link of 2 intervals. Then the automorphism φ_L is the conjugation by an element $u + v\,x\,y \in \mathcal{M}\otimes\Lambda$ such that $u+v$ goes to 1 by the augmentation map from Λ to $\mathbf{Z}$ sending a, b, c to 2 and α, β to 1.

Proof: The automorphism φ_L is the conjugation by an element

$$\varepsilon = u + v\,x\,y \in \Lambda\oplus\Lambda xy$$

Since $\overline{\varepsilon}\,\delta(\varepsilon)^{-1}$ is the inverse of ε in $\mathcal{M}\otimes K$, where K is the fraction field of Λ, we have the following:

$$\varepsilon\,y\,\overline{\varepsilon} = 0 \quad \mathrm{mod}\ \delta(\varepsilon) \qquad \overline{\varepsilon}\;\overline{x}\;\varepsilon = 0 \quad \mathrm{mod}\ \delta(\varepsilon)$$

but we have:

$$\varepsilon\,y\,\overline{\varepsilon} = \varepsilon\,t(y\,\overline{\varepsilon}) - \varepsilon^2\,\overline{y} = t(\varepsilon\,\overline{y}\,)\,\varepsilon - t(\varepsilon)\,\varepsilon\,\overline{y} + \delta(\varepsilon)\,\overline{y}$$

$$\overline{\varepsilon}\;\overline{x}\,\varepsilon = \overline{\varepsilon}\;t(\,\overline{x}\,\varepsilon) - \overline{\varepsilon}^{2}\,x = t(\,\overline{x}\,\varepsilon)\,\overline{\varepsilon} - t(\varepsilon)\,\overline{\varepsilon}\,x + \delta(\varepsilon)\,x$$

This implies:

$$t(\varepsilon\,\overline{y}\,)\,(u + vxy) - t(\varepsilon)\,(\,u\,\overline{y} + \beta v x) = 0 \quad \mathrm{mod}\ \delta(\varepsilon)$$

$$t(\,\overline{x}\,\varepsilon)\,(u + vxy) - t(\varepsilon)\,(ux + \alpha v\,\overline{y}\,) = 0 \quad \mathrm{mod}\ \delta(\varepsilon)$$

Since $(1, x, \overline{y}, xy)$ and $(1, x, \overline{y}, \overline{xy}\,)$ are Λ-basis of $\mathcal{M}\otimes\Lambda$, we get:

$$u\,t(\varepsilon) = 0 \quad \mathrm{mod}\ \delta(\varepsilon) \qquad v\,t(\varepsilon) = 0 \ \mathrm{mod}\ \delta(\varepsilon)$$

$$u\,t(\varepsilon\,\overline{y}\,) = 0\ \mathrm{mod}\ \delta(\varepsilon) \qquad v\,t(\varepsilon\,\overline{y}\,) = 0\ \mathrm{mod}\ \delta(\varepsilon)$$

$$u\, t(\overline{x}\,\varepsilon) \equiv 0 \bmod \delta(\varepsilon) \qquad v\, t(\overline{x}\,\varepsilon) \equiv 0 \bmod \delta(\varepsilon)$$

For $x_1, \dots, x_p$ in Λ, denote by $< x_1, \dots, x_p >$ the ideal generated by $x_1, \dots, x_p$. We have:

$$< u,v > . < t(x), t(\varepsilon\, \overline{y}), t(\overline{x}\,\varepsilon) > \subset < \delta(\varepsilon) >$$

$$\Rightarrow \qquad < u,v > . < 2u + cv, bu + a\beta v, au + b\alpha v > \subset < \delta(\varepsilon) >$$

It is easy to check the following:

$$< 2u + cv, bu + a\beta v, au + b\alpha v > \supset < b^2\alpha - a^2\beta, 2a\beta - bc, 2b\alpha - ac > . < u,v >$$

Then we have:

$$< u,v >^2 < b^2\alpha - a^2\beta, 2a\beta - bc, 2b\alpha - ac > \subset < \delta(\varepsilon) >$$

let w be an element of $< u,v >^2$. There exist X and Y in Λ such that:

$$w(b^2\alpha - a^2\beta) = X\, \delta(\varepsilon) \qquad w(2a\beta - bc) = Y\, \delta(\varepsilon)$$

Therefore $X(2a\beta - bc)$ is divisible by $b^2\alpha - a^2\beta$.

Let B be the subring of $\hat{A}$ defined by:

$$B = S^{-1}\mathbf{Z}[a, b, \alpha, \alpha^{-1}, \beta, \beta^{-1}]\, [[\Delta]] \quad \text{(see notation 3-2)}$$

We have: $\hat{A} = B \oplus cB$. Then there exist $X_0, X_1 \in B$ such that: $X = X_0 + X_1 c$

and we deduce:

$$2\alpha\beta X_0 - b\, X_1 (\Delta - a^2\beta - b^2\alpha + 4\alpha\beta) \equiv 0 \quad \bmod\ b^2\alpha - \alpha^2\beta$$

$$-b\, X_0 + 2a\beta\, X_1 - ab^2 X_1 \equiv 0 \qquad \bmod\ b^2\alpha - \alpha^2\beta$$

$$\Rightarrow \quad X_1\, b^2\Delta \equiv 0 \qquad \bmod\ b^2\alpha - \alpha^2\beta$$

Then X_1 is divisible by $b^2\alpha - \alpha^2\beta$ (in B) and X_0 also. Therefore X is divisible by $b^2\alpha - \alpha^2\beta$ in $\hat{A}$ and then in Λ. This implies that w itself is divisible by $\delta(\varepsilon)$, and we have:

$$< u,v >^2 \subset < \delta(\varepsilon) >$$

Thus there exist three elements $r, s, t \in \Lambda$ such that:

$$u^2 = r\, \delta(\varepsilon) = r\, (u^2 + cuv + \alpha\beta\, v^2)$$

$$uv = s\, \delta(\varepsilon) \qquad v^2 = t\, \delta(\varepsilon)$$

It is easy to check that:

$$s^2 = rt \qquad\qquad r + cs + \alpha\beta\, t = 1$$

Moreover, r, s, t depend only on the homothety class of ε and on the automorphism φ.

Let us denote by r_0, s_0, t_0 the images of r, s, t by the augmentation morphism from Λ to $\mathbf{Z}$ which sends α, β to 1 and a, b, c to 2. We have:

$$s_0^{\,2} = s_0\, t_0$$

$$r_0 + 2\, s_0 + t_0 = 1$$

Therefore it is easy to see that there exists a unique integer θ satisfying the following:

$$r_0 = (1-\theta)^2 \qquad s_0 = \theta(1-\theta) \qquad t_0^2 = \theta^2$$

Now, if we consider another automorphism φ', we get another integer θ' and it is easy to check that $\theta+\theta'$ is the integer corresponding to $\varphi\,\varphi'$. On the other hand, if φ is the conjugation by xy, we have $u = 0$, $v = 1$ and the corresponding integer is $\theta = 1$

Therefore, there exists an integer n such that the corresponding integer of $\varepsilon(xy)^n$ is zero. Denote by $\varepsilon' = u' + v'\,xy$ this new element of $\mathcal{M}\otimes\Lambda$ and by r', s', t' the corresponding elements in Λ constructed as above. We have:

$$r'_0 = 1 \qquad s'_0 = 0$$

and $r' + s'\,xy$ goes to a unit in $\mathcal{M}\otimes\mathbf{Z}$. Hence ε is a multiple of:

$$(r' + s'\,xy)(xy)^{-n}$$

which is invertible in $\mathcal{M}\otimes\mathbf{Z}$

§4-An example.

Consider the link L given by the following picture:

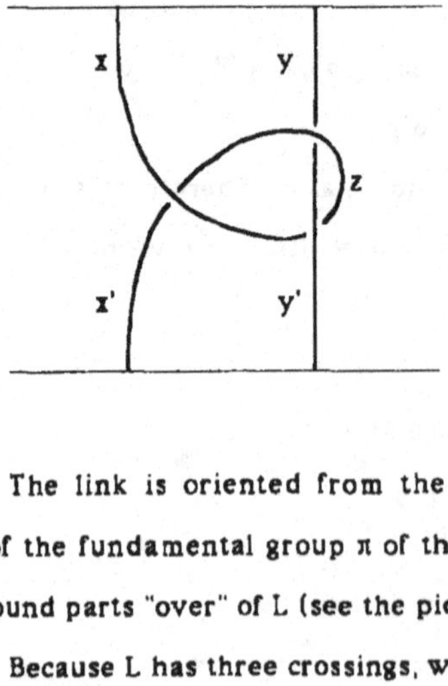

The link is oriented from the top to the bottom and x, y, x', y', z are elements of the fundamental group π of the complement of L corresponding to paths turning around parts "over" of L (see the picture)

Because L has three crossings, we have the following relations:

$$x\,z = x'x \qquad x\,y' = y'\,z \qquad y\,z = z\,y'$$

and we deduce:

$$x\,y' = y'\,x^{-1}x'\,x$$

Thanks to theorem 3-4, there exists an element ε in $\Lambda\oplus\Lambda xy$ such that:

$$x' = \varepsilon\,x\,\varepsilon^{-1} \qquad y' = \varepsilon\,y\,\varepsilon^{-1} \quad \text{in } \mathcal{M}\otimes\Lambda$$

Therefore we have in $\mathcal{M}\otimes\Lambda$:

$$x\,\varepsilon\,y\,\varepsilon^{-1} = \varepsilon\,y\,\varepsilon^{-1}\,x^{-1}\,\varepsilon\,x\,\varepsilon^{-1}\,x$$

If we multiply on the left by $\overline{\varepsilon}\,\delta(\varepsilon)$, we get:

$$\overline{\varepsilon}\,x\,\varepsilon\,y\,\overline{\varepsilon} = y\,\overline{\varepsilon}\,x^{-1}\,\varepsilon\,x\,\overline{\varepsilon}\,x$$

and this implies:

$$\overline{\varepsilon}\,x\,\varepsilon\,y\,\overline{\varepsilon}\,\overline{x} = y\,\overline{\varepsilon}\,\overline{x}\,\varepsilon\,x\,\overline{\varepsilon}$$

Let f be the antiinvolution of $\mathcal{M}\otimes\Lambda$ sending

x to x and y to y (and xy to yx) and let $\varepsilon' = f(\varepsilon)$. We have:

$$\varepsilon\,x = x\,\varepsilon' \qquad y\,\varepsilon = \varepsilon'\,y$$

$$\overline{\varepsilon}\,x = x\,\overline{\varepsilon'} \qquad y\,\overline{\varepsilon} = \overline{\varepsilon'}\,y$$

Then we have:

$$\overline{\varepsilon}\,x\,\varepsilon\,y\,\overline{\varepsilon}\,\overline{x} = x\,\overline{\varepsilon'}\,\varepsilon\,y\,\overline{\varepsilon}\,\overline{x} = x\,\overline{\varepsilon'}\,\varepsilon\,\overline{\varepsilon'}\,y\,\overline{x}$$

$$y\,\overline{\varepsilon}\,\overline{x}\,\varepsilon\,x\,\overline{\varepsilon} = y\,\overline{\varepsilon}\,\overline{x}\,x\,\varepsilon'\,\overline{\varepsilon} = x\,\overline{x}\,y\,\overline{\varepsilon}\,\varepsilon'\,\overline{\varepsilon}$$

and we get:

$$\overline{\varepsilon'}\,\varepsilon\,\overline{\varepsilon'}\,y\,\overline{x} = \overline{x}\,y\,\overline{\varepsilon}\,\varepsilon'\,\overline{\varepsilon} = f(\,\overline{\varepsilon'}\,\varepsilon\,\overline{\varepsilon'}\,y\,\overline{\varepsilon}\,)$$

We have: $\overline{\varepsilon'}\,\varepsilon\,\overline{\varepsilon'} = t(\varepsilon\,\overline{\varepsilon'})\,\overline{\varepsilon'} - \overline{\varepsilon'}\,\varepsilon'\,\overline{\varepsilon} = t(\varepsilon\,\overline{\varepsilon'})\,\overline{\varepsilon'} - \delta(\varepsilon')\,\overline{\varepsilon}$

let us set:

$$U = t(\varepsilon\,\overline{\varepsilon'}) \qquad V = \delta(\varepsilon') \qquad \varepsilon = u + v\,xy$$

So we have:

$$\begin{aligned}
\overline{\varepsilon'}\,\varepsilon\,\overline{\varepsilon'}\,y\,\overline{x} &= U\,\overline{\varepsilon'}\,y\,\overline{x} - V\,\overline{\varepsilon}\,y\,\overline{x} \\
&= U(u\,y\,\overline{x} + v\,\overline{x}\,\overline{y}\,y\,\overline{x}) - V(u\,y\,\overline{x} + v\,\overline{y}\,\overline{x}\,y\,\overline{x}) \\
&= U(u\,y\,\overline{x} + \beta\,v\,\overline{x}^{2}) - V(u\,y\,\overline{x} + t(y\,\overline{x})v\,\overline{y}\,\overline{x} - v\overline{y}\,\overline{x}\,x\,y) \\
&= U(u\,y\,\overline{x} + \beta\,v\,\overline{x}^{2}) - V(u\,y\,\overline{x} + t(y\,\overline{x})v\,\overline{y}\,\overline{x} - v\,\alpha\,\overline{y}^{2})
\end{aligned}$$

and the equation:

$$\varepsilon'\,\varepsilon\,\varepsilon'\,y\,\overline{x} = f(\varepsilon'\,\varepsilon\,\varepsilon'\,y\,\overline{x})$$

gives rise to the following:

$$U\,u(y\,\overline{x} - \overline{x}\,y) - V\,u(y\,\overline{x} - \overline{x}\,y) - V\,v\,t(y\,\overline{x})(\overline{y}\,\overline{x} - \overline{x}\,\overline{y}) = 0$$

$$\Rightarrow (U\,u - V\,u + V\,v\,t(y\,\overline{x}))(y\,\overline{x} - \overline{x}\,y) = 0$$

and we get:

$$U\,u - V\,u + V\,v(ab - c) = 0$$

On the other hand, we have:

$$U = t(\varepsilon\,\overline{\varepsilon'}) = t((u + v\,xy)(u + v\,\overline{x}\,\overline{y}))$$

$$= 2u^2 + 2uvc + v^2 t(xy\overline{x}\,\overline{y})$$

$$= 2u^2 + 2uvc + v^2(\Delta + 2\alpha\beta)$$

(with: $\Delta = c^2 - abc + a^2\alpha + b^2\beta - 4\alpha\beta$)

$$V = \delta(\varepsilon') = u^2 + uvc + \alpha\beta v^2$$

Now it is easy to obtain the following equation:

$$(u + (ab-c)v)(u^2 + cuv + \alpha\beta v^2) + \Delta u v^2 = 0$$

Modulo the augmentation ideal of Λ, we get:

$$(u + 2v)(u^2 + 2uv + v^2) \equiv 0$$

But we know that u and v can be chosen such that $u+v$ is not congruent to zero. Then $u^2 + cuv + \alpha\beta v^2$ is not zero modulo Δ, and we have:

$$u \equiv (c-ab)\, v \mod \Delta$$

In this example, we can choose v to be 1, and u is the unique element in Λ, congruent to $c-ab$ modulo Δ and satisfying the following equation:

$$(u + ab - c)(u^2 + cu + \alpha\beta) + \Delta u = 0$$

Actually this equation doesn't have any solution in A. The element u belongs to a cubic extension of A included in Λ, and it seems to be very difficult to find a subring of Λ, smaller than Λ where we can do all this construction for all links.

Rererences

[1] A . J . CASSON , oral communication . See also :

A . MARIN , L'invariant de Casson , preprint

[2] M . CULLER and P . B . SHALEN , Varieties of group representations and splittings of 3-manifolds . Ann. of Math. 117, n°1 (1983), pp. 109-146

[3] J . Y . LE DIMET , Cobordisme d'enlacements de disques. To appear

[4] J . P . LEVINE , Link concordance and algebraic closure of groups . Preprint

Université de Nantes

Département de Mathématiques

2 rue de la Houssinière

F-44072 NANTES Cedex 03

LECTURE NOTES IN MATHEMATICS

Edited by A. Dold and B. Eckmann

Some general remarks on the publication of proceedings of congresses and symposia

Lecture Notes aim to report new developments - quickly, informally and at a high level. The following describes criteria and procedures which apply to proceedings volumes. The editors of a volume are strongly advised to inform contributors about these points at an early stage.

§1. One (or more) expert participant(s) of the meeting should act as the responsible editor(s) of the proceedings. They select the papers which are suitable (cf. §§ 2, 3) for inclusion in the proceedings, and have them individually refereed (as for a journal). It should not be assumed that the published proceedings must reflect conference events faithfully and in their entirety. Contributions to the meeting which are not included in the proceedings can be listed by title. The series editors will normally not interfere with the editing of a particular proceedings volume - except in fairly obvious cases, or on technical matters, such as described in §§ 2, 3. The names of the responsible editors appear on the title page of the volume.

§2. The proceedings should be reasonably homogeneous (concerned with a limited area). For instance, the proceedings of a congress on "Analysis" or "Mathematics in Wonderland" would normally not be sufficiently homogeneous.

One or two longer survey articles on recent developments in the field are often very useful additions to such proceedings - even if they do not correspond to actual lectures at the congress. An extensive introduction on the subject of the congress would be desirable.

§3. The contributions should be of a high mathematical standard and of current interest. Research articles should present new material and not duplicate other papers already published or due to be published. They should contain sufficient information and motivation and they should present proofs, or at least outlines of such, in sufficient detail to enable an expert to complete them. Thus resumes and mere announcements of papers appearing elsewhere cannot be included, although more detailed versions of a contribution may well be published in other places later.

Surveys, if included, should cover a sufficiently broad topic, and should in general not simply review the author's own recent research. In the case of surveys, exceptionally, proofs of results may not be necessary.

"Mathematical Reviews" and "Zentralblatt für Mathematik" require that papers in proceedings volumes carry an explicit statement that they are in final form and that no similar paper has been or is being submitted elsewhere, if these papers are to be considered for a review. Normally, papers that satisfy the criteria of the Lecture Notes in Mathematics series also satisfy this

.../...

requirement, but we would strongly recommend that the contributing authors be asked to give this guarantee explicitly at the beginning or end of their paper. There will occasionally be cases where this does not apply but where, for special reasons, the paper is still acceptable for LNM.

§4. Proceedings should appear soon after the meeeting. The publisher should, therefore, receive the complete manuscript within nine months of the date of the meeting at the latest.

§5. Plans or proposals for proceedings volumes should be sent to one of the editors of the series or to Springer-Verlag Heidelberg. They should give sufficient information on the conference or symposium, and on the proposed proceedings. In particular, they should contain a list of the expected contributions with their prospective length. Abstracts or early versions (drafts) of some of the contributions are very helpful.

§6. Lecture Notes are printed by photo-offset from camera-ready typed copy provided by the editors. For this purpose Springer-Verlag provides editors with technical instructions for the preparation of manuscripts and these should be distributed to all contributing authors. Springer-Verlag can also, on request, supply stationery on which the prescribed typing area is outlined. Some homogeneity in the presentation of the contributions is desirable.

Careful preparation of manuscripts will help keep production time short and ensure a satisfactory appearance of the finished book. The actual production of a Lecture Notes volume normally takes 6 -8 weeks.

Manuscripts should be at least 100 pages long. The final version should include a table of contents and as far as applicable a subject index.

§7. Editors receive a total of 50 free copies of their volume for distribution to the contributing authors, but no royalties. (Unfortunately, no reprints of individual contributions can be supplied.) They are entitled to purchase further copies of their book for their personal use at a discount of 33.3 %, other Springer mathematics books at a discount of 20 % directly from Springer-Verlag. Contributing authors may purchase the volume in which their article appears at a discount of 33.3 %.

Commitment to publish is made by letter of intent rather than by signing a formal contract. Springer-Verlag secures the copyright for each volume.

Vol. 1201: Curvature and Topology of Riemannian Manifolds. Proceedings, 1985. Edited by K. Shiohama, T. Sakai and T. Sunada. VII, 336 pages. 1986.

Vol. 1202: A. Dür, Möbius Functions, Incidence Algebras and Power Series Representations. XI, 134 pages. 1986.

Vol. 1203: Stochastic Processes and Their Applications. Proceedings, 1985. Edited by K. Itô and T. Hida. VI, 222 pages. 1986.

Vol. 1204: Séminaire de Probabilités XX, 1984/85. Proceedings. Edité par J. Azéma et M. Yor. V, 639 pages. 1986.

Vol. 1205: B.Z. Moroz, Analytic Arithmetic in Algebraic Number Fields. VII, 177 pages. 1986.

Vol. 1206: Probability and Analysis, Varenna (Como) 1985. Seminar. Edited by G. Letta and M. Pratelli. VIII, 280 pages. 1986.

Vol. 1207: P.H. Bérard, Spectral Geometry: Direct and Inverse Problems. With an Appendix by G. Besson. XIII, 272 pages. 1986.

Vol. 1208: S. Kaijser, J.W. Pelletier, Interpolation Functors and Duality. IV, 167 pages. 1986.

Vol. 1209: Differential Geometry, Peñíscola 1985. Proceedings. Edited by A.M. Naveira, A. Ferrández and F. Mascaró. VIII, 306 pages. 1986.

Vol. 1210: Probability Measures on Groups VIII. Proceedings, 1985. Edited by H. Heyer. X, 386 pages. 1986.

Vol. 1211: M.B. Sevryuk, Reversible Systems. V, 319 pages. 1986.

Vol. 1212: Stochastic Spatial Processes. Proceedings, 1984. Edited by P. Tautu. VIII, 311 pages. 1986.

Vol. 1213: L.G. Lewis, Jr., J.P. May, M. Steinberger, Equivariant Stable Homotopy Theory. IX, 538 pages. 1986.

Vol. 1214: Global Analysis – Studies and Applications II. Edited by Yu.G. Borisovich and Yu.E. Gliklikh. V, 275 pages. 1986.

Vol. 1215: Lectures in Probability and Statistics. Edited by G. del Pino and R. Rebolledo. V, 491 pages. 1986.

Vol. 1216: J. Kogan, Bifurcation of Extremals in Optimal Control. VIII, 106 pages. 1986.

Vol. 1217: Transformation Groups. Proceedings, 1985. Edited by S. Jackowski and K. Pawalowski. X, 396 pages. 1986.

Vol. 1218: Schrödinger Operators, Aarhus 1985. Seminar. Edited by E. Balslev. V, 222 pages. 1986.

Vol. 1219: R. Weissauer, Stabile Modulformen und Eisensteinreihen. III, 147 Seiten. 1986.

Vol. 1220: Séminaire d'Algèbre Paul Dubreil et Marie-Paule Malliavin. Proceedings, 1985. Edité par M.-P. Malliavin. IV, 200 pages. 1986.

Vol. 1221: Probability and Banach Spaces. Proceedings, 1985. Edited by J. Bastero and M. San Miguel. XI, 222 pages. 1986.

Vol. 1222: A. Katok, J.-M. Strelcyn, with the collaboration of F. Ledrappier and F. Przytycki, Invariant Manifolds, Entropy and Billiards; Smooth Maps with Singularities. VIII, 283 pages. 1986.

Vol. 1223: Differential Equations in Banach Spaces. Proceedings, 1985. Edited by A. Favini and E. Obrecht. VIII, 299 pages. 1986.

Vol. 1224: Nonlinear Diffusion Problems, Montecatini Terme 1985. Seminar. Edited by A. Fasano and M. Primicerio. VIII, 188 pages. 1986.

Vol. 1225: Inverse Problems, Montecatini Terme 1986. Seminar. Edited by G. Talenti. VIII, 204 pages. 1986.

Vol. 1226: A. Buium, Differential Function Fields and Moduli of Algebraic Varieties. IX, 146 pages. 1986.

Vol. 1227: H. Helson, The Spectral Theorem. VI, 104 pages. 1986.

Vol. 1228: Multigrid Methods II. Proceedings, 1985. Edited by W. Hackbusch and U. Trottenberg. VI, 336 pages. 1986.

Vol. 1229: O. Bratteli, Derivations, Dissipations and Group Actions on C*-algebras. IV, 277 pages. 1986.

Vol. 1230: Numerical Analysis. Proceedings, 1984. Edited by J.-P. Hennart. X, 234 pages. 1986.

Vol. 1231: E.-U. Gekeler, Drinfeld Modular Curves. XIV, 107 pages. 1986.

Vol. 1232: P.C. Schuur, Asymptotic Analysis of Soliton Problems. VIII, 180 pages. 1986.

Vol. 1233: Stability Problems for Stochastic Models. Proceedings, 1985. Edited by V.V. Kalashnikov, B. Penkov and V.M. Zolotarev. VI, 223 pages. 1986.

Vol. 1234: Combinatoire énumérative. Proceedings, 1985. Edité par G. Labelle et P. Leroux. XIV, 387 pages. 1986.

Vol. 1235: Séminaire de Théorie du Potentiel, Paris, No. 8. Directeurs: M. Brelot, G. Choquet et J. Deny. Rédacteurs: F. Hirsch et G. Mokobodzki. III, 209 pages. 1987.

Vol. 1236: Stochastic Partial Differential Equations and Applications. Proceedings, 1985. Edited by G. Da Prato and L. Tubaro. V, 257 pages. 1987.

Vol. 1237: Rational Approximation and its Applications in Mathematics and Physics. Proceedings, 1985. Edited by J. Gilewicz, M. Pindor and W. Siemaszko. XII, 350 pages. 1987.

Vol. 1238: M. Holz, K.-P. Podewski and K. Steffens, Injective Choice Functions. VI, 183 pages. 1987.

Vol. 1239: P. Vojta, Diophantine Approximations and Value Distribution Theory. X, 132 pages. 1987.

Vol. 1240: Number Theory, New York 1984–85. Seminar. Edited by D.V. Chudnovsky, G.V. Chudnovsky, H. Cohn and M.B. Nathanson. V, 324 pages. 1987.

Vol. 1241: L. Gårding, Singularities in Linear Wave Propagation. III, 125 pages. 1987.

Vol. 1242: Functional Analysis II, with Contributions by J. Hoffmann-Jørgensen et al. Edited by S. Kurepa, H. Kraljević and D. Butković. VII, 432 pages. 1987.

Vol. 1243: Non Commutative Harmonic Analysis and Lie Groups. Proceedings, 1985. Edited by J. Carmona, P. Delorme and M. Vergne. V, 309 pages. 1987.

Vol. 1244: W. Müller, Manifolds with Cusps of Rank One. XI, 158 pages. 1987.

Vol. 1245: S. Rallis, L-Functions and the Oscillator Representation. XVI, 239 pages. 1987.

Vol. 1246: Hodge Theory. Proceedings, 1985. Edited by E. Cattani, F. Guillén, A. Kaplan and F. Puerta. VII, 175 pages. 1987.

Vol. 1247: Séminaire de Probabilités XXI. Proceedings. Edité par J. Azéma, P.A. Meyer et M. Yor. IV, 579 pages. 1987.

Vol. 1248: Nonlinear Semigroups, Partial Differential Equations and Attractors. Proceedings, 1985. Edited by T.L. Gill and W.W. Zachary. IX, 185 pages. 1987.

Vol. 1249: I. van den Berg, Nonstandard Asymptotic Analysis. IX, 187 pages. 1987.

Vol. 1250: Stochastic Processes – Mathematics and Physics II. Proceedings 1985. Edited by S. Albeverio, Ph. Blanchard and L. Streit. VI, 359 pages. 1987.

Vol. 1251: Differential Geometric Methods in Mathematical Physics. Proceedings, 1985. Edited by P.L. García and A. Pérez-Rendón. VII, 300 pages. 1987.

Vol. 1252: T. Kaise, Représentations de Weil et GL_2 Algèbres de division et GL_n. VII, 203 pages. 1987.

Vol. 1253: J. Fischer, An Approach to the Selberg Trace Formula via the Selberg Zeta-Function. III, 184 pages. 1987.

Vol. 1254: S. Gelbart, I. Piatetski-Shapiro, S. Rallis. Explicit Constructions of Automorphic L-Functions. VI, 152 pages. 1987.

Vol. 1255: Differential Geometry and Differential Equations. Proceedings, 1985. Edited by C. Gu, M. Berger and R.L. Bryant. XII, 243 pages. 1987.

Vol. 1256: Pseudo-Differential Operators. Proceedings, 1986. Edited by H.O. Cordes, B. Gramsch and H. Widom. X, 479 pages. 1987.

Vol. 1257: X. Wang, On the C*-Algebras of Foliations in the Plane. V, 165 pages. 1987.

Vol. 1258: J. Weidmann, Spectral Theory of Ordinary Differential Operators. VI, 303 pages. 1987.

Vol. 1259: F. Cano Torres, Desingularization Strategies for Three-Dimensional Vector Fields. IX, 189 pages. 1987.

Vol. 1260: N.H. Pavel, Nonlinear Evolution Operators and Semigroups. VI, 285 pages. 1987.

Vol. 1261: H. Abels, Finite Presentability of S-Arithmetic Groups. Compact Presentability of Solvable Groups. VI, 178 pages. 1987.

Vol. 1262: E. Hlawka (Hrsg.), Zahlentheoretische Analysis II. Seminar, 1984–86. V, 158 Seiten. 1987.

Vol. 1263: V.L. Hansen (Ed.), Differential Geometry. Proceedings, 1985. XI, 288 pages. 1987.

Vol. 1264: Wu Wen-tsün, Rational Homotopy Type. VIII, 219 pages. 1987.

Vol. 1265: W. Van Assche, Asymptotics for Orthogonal Polynomials. VI, 201 pages. 1987.

Vol. 1266: F. Ghione, C. Peskine, E. Sernesi (Eds.), Space Curves. Proceedings, 1985. VI, 272 pages. 1987.

Vol. 1267: J. Lindenstrauss, V.D. Milman (Eds.), Geometrical Aspects of Functional Analysis. Seminar. VII, 212 pages. 1987.

Vol. 1268: S.G. Krantz (Ed.), Complex Analysis. Seminar, 1986. VII, 195 pages. 1987.

Vol. 1269: M. Shiota, Nash Manifolds. VI, 223 pages. 1987.

Vol. 1270: C. Carasso, P.-A. Raviart, D. Serre (Eds.), Nonlinear Hyperbolic Problems. Proceedings, 1986. XV, 341 pages. 1987.

Vol. 1271: A.M. Cohen, W.H. Hesselink, W.L.J. van der Kallen, J.R. Strooker (Eds.), Algebraic Groups Utrecht 1986. Proceedings. XII, 284 pages. 1987.

Vol. 1272: M.S. Livšic, L.L. Waksman, Commuting Nonselfadjoint Operators in Hilbert Space. III, 115 pages. 1987.

Vol. 1273: G.-M. Greuel, G. Trautmann (Eds.), Singularities, Representation of Algebras, and Vector Bundles. Proceedings, 1985. XIV, 383 pages. 1987.

Vol. 1274: N. C. Phillips, Equivariant K-Theory and Freeness of Group Actions on C*-Algebras. VIII, 371 pages. 1987.

Vol. 1275: C.A. Berenstein (Ed.), Complex Analysis I. Proceedings, 1985–86. XV, 331 pages. 1987.

Vol. 1276: C.A. Berenstein (Ed.), Complex Analysis II. Proceedings, 1985–86. IX, 320 pages. 1987.

Vol. 1277: C.A. Berenstein (Ed.), Complex Analysis III. Proceedings, 1985–86. X, 350 pages. 1987.

Vol. 1278: S.S. Koh (Ed.), Invariant Theory. Proceedings, 1985. V, 102 pages. 1987.

Vol. 1279: D. Ieşan, Saint-Venant's Problem. VIII, 162 Seiten. 1987.

Vol. 1280: E. Neher, Jordan Triple Systems by the Grid Approach. XII, 193 pages. 1987.

Vol. 1281: O.H. Kegel, F. Menegazzo, G. Zacher (Eds.), Group Theory. Proceedings, 1986. VII, 179 pages. 1987.

Vol. 1282: D.E. Handelman, Positive Polynomials, Convex Integral Polytopes, and a Random Walk Problem. XI, 136 pages. 1987.

Vol. 1283: S. Mardešić, J. Segal (Eds.), Geometric Topology and Shape Theory. Proceedings, 1986. V, 261 pages. 1987.

Vol. 1284: B.H. Matzat, Konstruktive Galoistheorie. X, 286 pages. 1987.

Vol. 1285: I.W. Knowles, Y. Saitō (Eds.), Differential Equations and Mathematical Physics. Proceedings, 1986. XVI, 499 pages. 1987.

Vol. 1286: H.R. Miller, D.C. Ravenel (Eds.), Algebraic Topology. Proceedings, 1986. VII, 341 pages. 1987.

Vol. 1287: E.B. Saff (Ed.), Approximation Theory, Tampa. Proceedings, 1985–1986. V, 228 pages. 1987.

Vol. 1288: Yu. L. Rodin, Generalized Analytic Functions on Riemann Surfaces. V, 128 pages, 1987.

Vol. 1289: Yu. I. Manin (Ed.), K-Theory, Arithmetic and Geometry. Seminar, 1984–1986. V, 399 pages. 1987.

Vol. 1290: G. Wüstholz (Ed.), Diophantine Approximation and Transcendence Theory. Seminar, 1985. V, 243 pages. 1987.

Vol. 1291: C. Mœglin, M.-F. Vignéras, J.-L. Waldspurger, Correspondances de Howe sur un Corps p-adique. VII, 163 pages. 1987.

Vol. 1292: J.T. Baldwin (Ed.), Classification Theory. Proceedings, 1985. VI, 500 pages. 1987.

Vol. 1293: W. Ebeling, The Monodromy Groups of Isolated Singularities of Complete Intersections. XIV, 153 pages. 1987.

Vol. 1294: M. Queffélec, Substitution Dynamical Systems – Spectral Analysis. XIII, 240 pages. 1987.

Vol. 1295: P. Lelong, P. Dolbeault, H. Skoda (Réd.), Séminaire d'Analyse P. Lelong – P. Dolbeault – H. Skoda. Seminar, 1985/1986. VII, 283 pages. 1987.

Vol. 1296: M.-P. Malliavin (Ed.), Séminaire d'Algèbre Paul Dubreil et Marie-Paule Malliavin. Proceedings, 1986. IV, 324 pages. 1987.

Vol. 1297: Zhu Y.-l., Guo B.-y. (Eds.), Numerical Methods for Partial Differential Equations. Proceedings. XI, 244 pages. 1987.

Vol. 1298: J. Aguadé, R. Kane (Eds.), Algebraic Topology, Barcelona 1986. Proceedings. X, 255 pages. 1987.

Vol. 1299: S. Watanabe, Yu.V. Prokhorov (Eds.), Probability Theory and Mathematical Statistics. Proceedings, 1986. VIII, 589 pages. 1988.

Vol. 1300: G.B. Seligman, Constructions of Lie Algebras and their Modules. VI, 190 pages. 1988.

Vol. 1301: N. Schappacher, Periods of Hecke Characters. XV, 160 pages. 1988.

Vol. 1302: M. Cwikel, J. Peetre, Y. Sagher, H. Wallin (Eds.), Function Spaces and Applications. Proceedings, 1986. VI, 445 pages. 1988.

Vol. 1303: L. Accardi, W. von Waldenfels (Eds.), Quantum Probability and Applications III. Proceedings, 1987. VI, 373 pages. 1988.

Vol. 1304: F.Q. Gouvêa, Arithmetic of p-adic Modular Forms. VIII, 121 pages. 1988.

Vol. 1305: D.S. Lubinsky, E.B. Saff, Strong Asymptotics for Extremal Polynomials Associated with Weights on $\mathbb{R}$. VII, 153 pages. 1988.

Vol. 1306: S.S. Chern (Ed.), Partial Differential Equations. Proceedings, 1986. VI, 294 pages. 1988.

Vol. 1307: T. Murai, A Real Variable Method for the Cauchy Transform, and Analytic Capacity. VIII, 133 pages. 1988.

Vol. 1308: P. Imkeller, Two-Parameter Martingales and Their Quadratic Variation. IV, 177 pages. 1988.

Vol. 1309: B. Fiedler, Global Bifurcation of Periodic Solutions with Symmetry. VIII, 144 pages. 1988.

Vol. 1310: O.A. Laudal, G. Pfister, Local Moduli and Singularities. V, 117 pages. 1988.

Vol. 1311: A. Holme, R. Speiser (Eds.), Algebraic Geometry, Sundance 1986. Proceedings. VI, 320 pages. 1988.

Vol. 1312: N.A. Shirokov, Analytic Functions Smooth up to the Boundary. III, 213 pages. 1988.

Vol. 1313: F. Colonius, Optimal Periodic Control. VI, 177 pages. 1988.

Vol. 1314: A. Futaki, Kähler-Einstein Metrics and Integral Invariants. IV, 140 pages. 1988.

Vol. 1315: R.A. McCoy, I. Ntantu, Topological Properties of Spaces of Continuous Functions. IV, 124 pages. 1988.

Vol. 1316: H. Korezlioglu, A.S. Ustunel (Eds.), Stochastic Analysis and Related Topics. Proceedings, 1986. V, 371 pages. 1988.

Vol. 1317: J. Lindenstrauss, V.D. Milman (Eds.), Geometric Aspects of Functional Analysis. Seminar, 1986–87. VII, 289 pages. 1988.

Vol. 1318: Y. Felix (Ed.), Algebraic Topology – Rational Homotopy. Proceedings, 1986. VIII, 245 pages. 1988.

Vol. 1319: M. Vuorinen, Conformal Geometry and Quasiregular Mappings. XIX, 209 pages. 1988.